PROCESS PLANT DESIGN

HEINEMANN CHEMICAL ENGINEERING SERIES

Heat Transfer
Process Control
Process Plant Design

HEINEMANN CHEMICAL ENGINEERING SERIES

Process
Plant
Design

J. R. BACKHURST
B.SC., PH.D., C.ENG., M.I.CHEM.E., A.R.I.C.

J. H. HARKER
B.SC., PH.D., C.ENG., M.I.CHEM.E., A.M.INST.F., A.R.I.C.

Department of Chemical Engineering,
The University of Newcastle upon Tyne

AMERICAN ELSEVIER
PUBLISHING COMPANY, INC.
NEW YORK

American Edition published by
American Elsevier Publishing Company, Inc.
52 Vanderbilt Avenue
New York, New York 10017

Library of Congress Catalog Card No: 72–12561

ISBN 0 444 19566 1

© J. R. Backhurst and J. H. Harker 1973

First published 1973

Published by Heinemann Educational Books Ltd
48 Charles Street, London w1x 8AH

Printed in Great Britain

Preface

The graduate chemical engineer, leaving university or polytechnic at the end of his period of academic endeavour, finds that the harsh industrial world, which is to provide him with excitement, stimulation, frustration and a means of livelihood for the rest of his working life, produces a whole host of problems totally unrelated to his past experience and, in some cases, quite beyond his capabilities. No longer is he dealing with frictionless pulleys and water flowing in clean smooth pipes; instead real problems, such as metering boiling phenol or the pumping of vegetable soup, are the order of the day. No longer is there a unique answer to each problem— indeed there may be several solutions or perhaps even no answer at all. In his new environment, the graduate finds that the intricacies of boundary layer theory or unsteady state thermal diffusion are of little help in sizing the supports for a distillation column or in selecting the best pump for dealing with 98 per cent sulphuric acid. This gulf between chemical engineering as taught from an academic viewpoint and chemical engineering as practised in the industrial sphere is bridged in the final months of the undergraduate course by several topics, not least of which is plant design. The preparation of a plant design is a major feature of most of the courses in chemical engineering in the UK and, hitherto, an important part of the examining function of the Institution of Chemical Engineers in London. This topic plays a vital role in translating the theories and principles of chemical engineering into pieces of hardware and process plant in general and it embodies the wide range of qualities and skills required of the modern chemical engineer.

It is the aim of this book to provide an introduction to the basic principles of plant design and to show how the fundamentals of design can be blended with commercial aspects to produce a final specification; how textbook parameters can be applied to the solution of real problems and how a training in chemical engineering can best be utilized in the industrial sphere. It is a long step from the study of bubble formation in boiling to the optimization of evaporator operation, and yet it is hoped that this book will at least point the reader in the right direction. In recent years, chemical engineering has lost much of its applied flavour and become, to some extent, a convenient and stimulating stamping ground for the applied mathematician; it is hoped that this book will at least go some way towards stemming the flood, if not to partially redressing the balance.

At the outset, it has been assumed that the reader knows how to calculate a heat transfer coefficient and the height of an absorber, for example, and the bulk of the book is concerned with the translation of such parameters into plant items which are ultimately linked into the production unit. We have attempted to follow a fairly logical sequence in which flowsheets, heat and mass balances, for example, are considered before attention is paid to the design of plant items, exchangers, columns and so on. Because of the vital role of economics in any design function, costing is dealt with early in the book and the principles further developed as appropriate. Rarely is the plant designer concerned with the design of smaller and standard items of equipment, and hence considerable emphasis is placed on the selection of such items. It is thought that this section may prove of particular value to the engineer in industry, especially if he has not the backing of comprehensive technical manuals produced by the larger companies. Finally an attempt is made to draw together the many facets of equipment design into one specification for the complete plant, and the many aspects relating to the completed unit are introduced in a final section.

A particularly important feature of the book, as indeed with all the texts in this series, is the inclusion of relationships and data in the SI system of units, although British Units are retained in order to enhance the usefulness of the data. The selection of material for inclusion has been largely based on personal preference, though the needs of the final year student and the young engineer in a wide range of industries have been borne in mind at all times.

No book of this type can possibly be written in isolation and we have drawn extensively on standard texts and important articles in the literature. Due reference is given in the appropriate section of the book and we are particularly grateful to the following organizations, who have kindly supplied material and have allowed us to reproduce freely from their literature and data in this text: Apex Construction Ltd, London; A.P.V.-Mitchell Ltd, Kestner-Metal Propeller Division, Croydon, Surrey; Hydronyl Ltd, Stoke-on-Trent, Staffordshire; Lightnin' Mixers Ltd, Poynton, Stockport, Cheshire; Pennwalt Ltd, Camberley, Surrey; Q.V.F. Ltd, Stoke-on-Trent, Staffordshire; Standard-Messo Duisburg, Germany, and Messrs van den Bergh and Partners Ltd, Holland.

In conclusion, we wish to pay tribute to Professor J. M. Coulson, who for many years has guided our thoughts and studies and has greatly encouraged our efforts in many directions. Our thanks are also due to Professor F. A. Holland, who has contributed in no small way to the production of this volume, to Miss K. Heads and the secretarial staff of the Department of Chemical Engineering at Newcastle University, who have patiently produced the final typescript, and to our colleagues and especially our students, who have provided stimulation and inspiration and channelled our efforts into what we trust is a worthwhile direction.

1973   J. R. B.
                                                             J. H. H.

Units

Throughout this text, where a value or an equation is quoted in square brackets thus [ ], this signifies that the value or equation is in SI units, the principles of which are discussed in Chapter 1. Factors enabling conversion from British units (or Anglo-American, A.A.) are included as an Appendix.

Contents

Part 4: Overall Considerations

Chapter 11: The Complete Plant 373

Part 1: Process Evaluation

Chapter 1: Preliminary Concepts

1.1 Process Plant Design

The Institution of Chemical Engineers defines chemical engineering as 'that branch of engineering, which is concerned with processes, which change the chemical composition or physical properties of material in bulk'. This definition embraces a very broad range of processes and products; these are the concern of the present text, which deals with the design of plant in which such processes are carried out, whether they be for peas or penicillin, paint or phosphoric acid. At the outset it is important to differentiate between process design and plant design, and this may best be done by considering the various stages in preparing the specification for a complete process plant.

Given that there is a demand for a product at a certain price, and that it can be produced in the laboratory from easily available starting materials at the bench scale, the steps in preparing the plant design may be summarized as follows:

(i) *Process Development*

It may be that the desired product can be produced by several routes, for example by direct reaction of two starting materials, or via one or more intermediates, some of which have a significant marketable value, thus rendering that particular route commercially attractive. In any event, it is necessary to evaluate or develop a process for each route. Essentially, process development is the translation of the bench scale chemistry into a means whereby the material can be produced on a large scale. In this operation, a great many problems arise because of the need to convert from batch processing to continuous production, with its inherent commerical advantages. Traditionally, a processing unit has been considered as a train of equipment in which unit operations such as distillation, filtration, drying and so on are carried out. In many such cases, the evaluation of a process for carrying out these operations is merely the selection of the best type of equipment for the job. Where Man's ingenuity and inventiveness are best applied, however, is in devising new means for carrying out unconventional operations. If, for example, a reaction is to be carried out in the vapour phase using a solid catalyst, the most efficient means must be sought of replenishing the catalyst and removing spent material. It may be that some elaborate screw arrangement can be developed for transporting the solid material, or it may prove cheaper to use manpower to cope with the operation. Such data are obtained and basic ideas tested at the technical scale, if not pilot scale of operation, in the laboratory, at which time, physical properties, reaction rates and so on are evaluated, together with plant selection for a particular processing stage. It is in this type of work that new materials of construction are developed, new types of tower packings produced, for example, and advances made generally in the technological field.

When a process has been developed, the whole unit is tested on the pilot plant, especially where the investment in the full-scale plant is very large, and the detailed operating conditions evaluated. For example, questions such as 'Can the yield be increased by reaction at higher pressure?', 'Will this add to the cost of the reactor?', 'Is the fullest use made of heat recovery and energy conservation in general?' must be answered at this stage. When the whole process has been evaluated and optimized in this way, the process development is complete and the operation passes to the next stage.

Process development is a costly business, especially when one realizes that failures must be paid for by successes, and the time is coming when it is just not worthwhile developing a process for certain products. Although development costs are very high, major returns on the investment must be assured, and it is a useful guide to appreciate that the larger organizations will not consider a new process unless the product can be made for less than 50 per cent of the current market price. For this reason, many contractors and design organizations carry out no process development, but buy in a process, usually from a large oil company or chemical manufacturer, for which they pay licence fees. For the large sums involved the contractor does not get a single nut and bolt, but a manual of data on the process, together with operating conditions and details of the basic design.

(ii) *Heat and Mass Balances*

Assuming that the data on how the process is to be carried out is available, either from previous development work or from another organization, heat and mass balances must be prepared for the particular plant in question. Normally the calculation is made backwards, that is from the desired production rate of the finished material. Although the basic chemistry provides the means of assessing the amounts of feed material and various intermediates, a great deal of know-how is involved in the preparation of the balances and it is here that experience can play a vital role in the design. For example, how much solvent will be carried away in the filter cake? What are the evaporation losses from the reaction vessel? How much heat will be lost from the drier? Much of this sort of information will become apparent during the development of the process, though a great deal will depend on the skill and experience of the designer. In many ways, design is more of an art than a precise science and the best designer is often the best guesser. Having said that, it is most important that the heat and mass balances should be carried with the utmost accuracy, as all further design calculations are based on these data.

(iii) *Plant Design*

At this stage, the design of the major items of equipment on the plant may be carried out—reactors, columns, heat exchangers and so on. Usually this is done in sufficient detail to enable accurate costing to be carried out and a reasonable amount of mechanical design may be involved. It is this aspect of the procedure with which this book is mainly concerned that is, the translation of basic chemical engineering principles into pieces of hardware. During this stage, as with all design, economy of finance is the overriding principle, and optimization techniques must be employed wherever possible to utilize the capital invested in plant and materials most effectively. It may be that the conditions required for the process to be carried out necessitate the development of a particular piece of equipment before the design can be completed, and considerable laboratory work may be involved. Indeed, one of the main functions of the plant contractor is the continual improvement of plant items and the development of new ideas with a view to increasing the efficiency or reducing the costs of a particular operation. The development of more efficient tower packings with lower pressure drop, new filter cloths, fluidized bed combustors or more compact heat exchangers are fairly obvious examples.

Many plant items of a standardized nature—pumps, instruments, valves and possibly heat exchangers, boilers and so on—will be bought in from suppliers and these can be costed at this stage. The design of an item such as a pump is a job for the specialist and rarely the function of the plant designer. What is important, however, is the knowledge and experience which will enable the correct selection of a piece of equipment, and some guidelines to this end are given in other sections of this book.

At this stage the design of the complete plant unit can be completed, taking into account factors such as structures, control and instrumentation, safety, plant layout, services, maintenance and general amenities, and the process flowsheets and layout diagrams finalized.

(iv) Costing

The final stage in the specification of a plant design, and the most significant, is costing, which is dealt with in Chapter 2 of this book. The aim is to obtain the total cost per unit mass of product as accurately as is feasible in a reasonable time. This incorporates the depreciation and interest on the capital invested, the costs of materials, services and labour, and overheads including sales, marketing, administration and so on. Many other factors, some not so obvious, must be included in this cost and special mention is made of the quality and quantity of manpower—what are current rates of pay in competing industries?—research and development costs—Are process development costs involved? Will any major modifications be required during the operating life of the plant?—the political situation—Will there be any changes in policy as far as development grants are concerned? The availability of finance is most important—Will interest rates fluctuate widely? Are power costs stable? From this it will be seen that not only is the immediate cost of product vital, but also the cost during the lifetime of the plant.

Having arrived at a suitable cost, possibly averaged over the plant lifetime, it must now be considered whether or not the product can be sold at this price and still show a worthwhile profit to the company. If the product already exists, the price may be compared with the market value (or preferably the manufacturing cost of another company), though where a new product is involved or an improved existing product market research techniques must be employed to assess the likely profit. An important consideration in this respect is the likely selling price during the plant life—will the price fall or will the demand decrease due to the efforts of competitors? If there is an anticipated increase in demand then overload factors must be incorporated into the design at the outset. It is usually more economic to run at below full output for two to three years rather than have to build a new plant at the end of this time—though this is a very complex question indeed. Many of these considerations are outside the activities of the plant designer, though he must be aware of their implications and how they affect his work.

If, at the end of the day, the cost per unit weight of product is too high and the whole project appears unprofitable, then the plant design must be re-examined to see if any improvements can be made—recycling of streams, greater conservation of energy and, in general, complete optimization of the entire plant. Should the situation still be impracticable the process itself must be questioned. Has the right route been chosen? Would a route involving high pressure and temperature or corrosive fluids be more profitable, because the by-products have a higher value? This type of consideration is discussed in the next section, though the important point is that not only must the plant be as economic and efficient as possible, but the whole operation from mining of raw materials to the sale of the finished product must be completely optimized.

1.2 Process Evaluation

It would be ideal if, for each possible reaction by which the end product could be manufactured in the laboratory, a process could be developed involving some technical scale practical work, the most efficient plant designed and costed, and the selling price for the product computed in each case. This is obviously not feasible, as, by the time this operation was completed there would probably be no demand for the product and certainly the prospective customer for the process plant would have gone elsewhere. Even the computer is not the complete answer to the problem. There are, however, many fairly obvious factors which indicate at the outset the probable choice of the most likely route and hence the most economic process. In the main, these are specific to a particular product, though the following considerations are of general significance in selecting a route and evaluating the process: Are the operating conditions hazardous? If this is the case, fairly elaborate control equipment will be involved, together with vital safety features, all of which add to the investment. Are the fluids involved corrosive or erosive? This, together with extremes of temperature and pressure, also adds considerably to the capital cost of the plant. Mention has already been made of quality of manpower and this introduces the cost of training. Are large quantities of cooling water, electric power or steam required? Is the disposal of effluent (including atmospheric dispersion) likely to be a problem? A significant consideration is the availability of physical and chemical data—this has an important effect on research and development costs. Similarly, it is usually fairly obvious from the basic chemistry whether or not the plant items are of standard design. If a special filter or reactor has to be developed, for example, this will add significantly to the product cost. Other related factors include the provision of special buildings, maintenance costs and the capital investment in spares, location and, a most important consideration, the company familiarity with process of this type.

On the basis of this type of reasoning, impossible and unlikely routes can be ruled out and processes developed accordingly. This may prove most difficult with new products and new techniques, electrochemical synthesis on a large scale for example, and yet even with well-established processes which are well documented, there are many variables to be considered. To take two simple examples:

(i) Steam Reforming

In the production of synthesis gas or towns gas, light distillate fuels are reacted with steam in the presence of a solid catalyst contained in tubes, heated in a furnace. Although the principles are well understood, and indeed the process is widely used, one must ask questions such as 'Is the catalyst the most effective?' 'Would the increased yield produced by a more effective catalyst offset any increase in cost?' 'Would increase in temperature be advantageous and is this possible with newer materials of construction?' The question of possible recycle must be resolved and also whether it is better to purify the feed or to remove impurities from the product gas. These type of questions were vital during the development of processes for carrying out this reaction, and the choice of a particular process depends on the type and range of feed which can be handled, variation in output and the time before the plant is on stream.

(ii) Wet Process Phosphoric Acid

The chemistry of the production of phosphoric acid from phosphate rock (calcium triphosphate) is well known. The calcium triphosphate is converted to soluble monophosphate by reaction with weak phosphoric acid:

$$Ca_3(PO_4)_2 + 4H_3PO_4 \rightarrow 3CaH_4(PO_4)_2$$

This is then digested with concentrated sulphuric acid to form the phosphoric acid, together with a precipitate of calcium sulphate:

$$3CaH_4(PO_4)_2 + 3H_2SO_4 + 6H_2O \rightarrow 3CaSO_4.2H_2O + 6H_3PO_4$$

There are several processes for carrying out these reactions and basically they differ according to the form in which the calcium sulphate is precipitated. This depends on the temperature of the reaction mixture and the concentration of the phosphoric acid produced—high values give semi-hydrate or even anhydrite rather than gypsum. The more important processes are summarized in Table 1–1. Processes involving gypsum are more widespread because they are easier to control and the conditions are less severe.

Thus even for this relatively straightforward reaction, a number of processes have been developed and many factors must be considered in evaluation of a process. All of these factors radically affect the final product and also the operation of the process. For example, the reaction

TABLE 1–1. PROCESS FOR THE PRODUCTION OF WET PROCESS PHOSPHORIC
ACID

Process	Form of calcium sulphate	Phosphoric acid strength ($\%P_2O_5$)	Type of filter used
Chemico	Gypsum	32	Horizontal rotary
Nordengren	Anhydrite	42–45	Tray-belt
KPV	Semi-hydrate	40	Travelling pan
S. Gobain	Gypsum	28	Vacuum pan
Prayon	Gypsum	30	Travelling pan

vessels take many forms and, although they are basically a number of stirred tanks, each process incorporates certain unique design features, all aimed at promoting an efficient yield. Questions as to ease of fabrication, power costs and special foundations must be answered, as well as queries relating to the control of the reaction. The more complex this is, the greater the manpower requirements and hence the higher the operating costs, which may easily offset any increase in yield. The other main feature differentiating the various processes is the design of the filter, and in evaluating the process one must query the efficiency of each type, the blinding tendency of the cloth and hence shutdown frequency, floor space, power and maintenance requirements. A further question is that of the final acid strength. A stronger acid, whilst saving on evaporation costs, is more corrosive and hence greater capital investment is required. Is it not more economic to produce a weak acid and to confine the materials problems to the evaporation section?

It will be seen that in process evaluation, even when a well-established reaction is involved, the situation is complex and there is no substitute for practical experience, whether it be the performance of full-sized plants or pilot-scale experiments. The latter of course are essential, where a new process is being developed.

1.3 Mass and Heat Balances

Once a process for carrying out a reaction has been selected, the first step in preparing a plant design is the compilation of mass and heat balances and the drawing up of a process flowsheet. Before this can be completed, the chemical and physical data must be available, either from the literature or by laboratory investigation if it is not obtained during the process development stage. One difficulty is that whereas the literature may provide idealized data on various reactions and the kinetics involved, in preparing mass and energy balances a knowledge of the properties of 'real' fluids is essential and in many cases laboratory tests are the only source of such information. For example, 'How much water is left when a given filter cake is washed?' 'What is the specific heat of a suspension of frozen peas in water?' In the absence of such data, the designer may have to resort to an inspired guess and it is here that experience can prove so valuable.

It has already been pointed out, that accuracy in drawing up heat and mass balances is essential as all subsequent calculations are based on these flowsheets. It is suggested that all calculated information should be checked wherever possible. One way in which this may be done is to make a balance for each component separately and to ascertain whether or not this agrees with the total mass balance. This approach is illustrated in the following example.

Example 1–1

Several lime kilns are heated by producer gas of the following volumetric composition:

$$CO_2 \ 9.5\%, \quad CO \ 20.1\%, \quad H_2 \ 11.3\%, \quad CH_4 \ 2.8\%, \quad N_2 \ 56.3\%$$

The flue gas leaving the kilns has the following composition by volume:

$$CO_2 \ 27.75\%, \quad O_2 \ 2.25\%, \quad N_2 \ 70.0\%$$

Calculate the weight of lime produced per unit weight of carbon burnt in the producer assuming that the limestone is 90 per cent calcium carbonate.

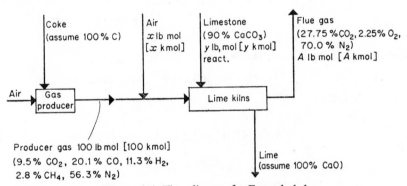

FIGURE 1–1. Flow diagram for Example 1–1

The general layout of the plant is shown in Figure 1–1. Working on the basis of 100 lb mol [100 kmol] producer gas fed to the kilns, we can assume that x lb mol [x kmol] air are used in the combustion of the producer gas and that y lb mol [y kmol] of calcium carbonate are reacted in the kiln (1 lb mol = molecular weight in lb [1 kmol = molecular weight in kg]).

The molecular weight of $CaCO_3$ is 100 and that of CaO, 56. Thus in y lb [y kg] of $CaCO_3$, there are $12y/100$ lb C [$12y/100$ kg C] which appear in the flue gas as CO_2. This is equivalent to $y/100$ lb atom [$y/100$ kmol]C.

Finally, if we assume that the flue gas leaving the kilns is A lb mol/100 lb mol producer gas [A kmol/100 kmol producer gas], then we can now make a mass balance for each component,

C balance

$$\text{(C in producer gas)} + \text{(C in limestone)} = \text{(C in flue gas)}$$
$$(9.5 + 20.1 + 2.8) + y/100 = 27.75A/100$$

from which
$$y = 27.75A - 3240 \qquad (1\text{–}1)$$

N_2 balance

$$(N_2 \text{ in producer gas}) + (N_2 \text{ in combustion air}) = (N_2 \text{ in flue gas})$$
$$56.3 + 79x/100 = 70.0A/100$$
$$\therefore x = 0.885A - 71.2$$
$$(1\text{--}2)$$

H_2 balance

$$(H_2 \text{ in producer gas}) = (H_2 \text{ in flue gas})$$

Hence $(11.3 + 5.6) = 16.9$ lb mol [kmol] H_2 produces 16.9 lb mol [kmol] water, which appears in the flue gas; this contains 8.45 lb mol [kmol] O_2.

O_2 balance

$$(O_2 \text{ in producer gas}) + (O_2 \text{ in air}) + (O_2 \text{ in limestone})$$
$$= (O_2 \text{ in lime}) + (O_2 \text{ in dry flue gas}) + (O_2 \text{ in water vapour}).$$
$$(9.5 + 20.1/2) + 21x/100 + 48y/(100 \times 32)$$
$$= 16/(56 \times 32) + 2.25A/100 + 8.45$$
$$\therefore 21x + 0.628y = 2.25A + 6495 \qquad (1\text{--}3)$$

Substituting for y from (1–1) and for x from (1–2) in (1–3)

$A = 365$ lb mol/100 lb mol [$= 365$ kmol/100 kmol] producer gas

Substituting for A in (1–1),

$y = 6810$ lb $CaCO_3$/100 lb mol

$[= 6810$ kg $CaCO_3$/100 kmol] producer gas

Thus

weight of limestone fed $= 6810 \times 100/90$

$= 7566$ lb/100 lb mol [$= 7566$ kg/100 kmol] producer gas

100 lb [100 kg] $CaCO_3$ react to form 56 lb [56 kg] lime

and hence lime produced $= 56 \times 6810/100$

$= 3814$ lb/100 lb mol [$= 3814$ kg/100 kmol] producer gas

It now remains to convert this value to a basis of unit weight of C fed to the producer.

In 100 lb [100 kg] C, there are

$$100/12 = 8.33 \text{ lb atom [kmol]C.}$$

In 100 lb mol [100 kmol] producer gas, there are

$$(9.5 + 20.1 + 2.8) = 32.4 \text{ lb mol [32.4 kmol] C}$$

Thus gas formed $= 8.33 \times 100/32.4$

$= 25.7$ lb mol/100 lb C [$= 25.7$ kmol/100 kg C] burned

lime produced $= 3814 \times 25.7/100$

$=$ 980 lb/100 lb C [$=$ 980 kg/100 kg C] burned

$= 9.80$ lb/lb C [kg/kg C].

This result could now be checked by means of a calcium balance as suggested previously.

This example is rather specialized in that the working has been in lb mol and kmol rather than mass, though this approach does avoid conversion problems between units. The conditions have been idealized in that several factors have been omitted, such as carbon lost in the ash from the producer, air leakage, especially with the lime, and water vapour in the incoming air. It is important to realize that the construction of flowsheets showing heat and mass balances is not a mechanical interpretation of the chemist's data and all factors must be included. For example, a report on the preparation of an organic chemical in the laboratory might include a wealth of information on equilibrium constants, reaction rates and so on, but omit to mention that large quantities of sulphur dioxide are evolved during the reaction and that the product is to be thoroughly washed with water. In the laboratory these problems are solved with the fume cupboard and the tap respectively, and yet it is vital to include both the gas and the water requirements in the mass balance. It may sound trivial to say that a plant is not elastic and what goes in must come out, though it is important to realize that both matter and energy leave the plant by routes other than the main product stream. The following example, which is worked more conventionally in terms of mass, illustrates the need to assume certain factors such as the carry-over of filtrate in the cake. The data is based on an example given by Ross.[1]

Example 1-2

Aqueous sodium carbonate containing 9.55 per cent w/w solids, is to be converted to caustic soda by the addition of crushed quicklime in a heated, agitated vessel. Batches of 195 200 lb [88 800 kg] of sodium carbonate are to be processed in each 8 hour [28.8 ks] shift and the feed will enter at 110°F [316 K]. After the stoichiometric amount of quicklime is added, the mixture will be heated to 180°F [355 K] and laboratory tests show that the reaction will be 90 per cent complete in 2 hour [7.2 ks]. The slurry will then be decanted and filtered. The following assumptions may be made:

(i) The insoluble matter in the sludge leaving the decanter is wetted with twice its own weight of water.

(ii) The filter cake contains 50 per cent w/w of insoluble matter and only 0.5 per cent w/w of each sodium salt, when the wash water is 5 lb/lb [5 kg/kg] of dry insoluble matter.

Prepare a mass balance.

Considering one batch, 195 200 lb [88 800 kg] sodium carbonate solution contains

$$195\ 200 \times 9.55/100 = 18\ 650 \text{ lb sodium carbonate}$$

$$[88\ 800 \times 9.55/100 = 8480 \text{ kg sodium carbonate}]$$

and $\qquad (195\ 200 - 18\ 650) = 176\ 550 \text{ lb water}$

$$[88\ 800 - 8480 = 80\ 320 \text{ kg water}]$$

The reaction taking place is

$$Na_2CO_3 + Ca(OH)_2 \rightarrow CaCO_3 + 2NaOH$$
$$\quad 106 \qquad\quad 74 \qquad\qquad 100 \qquad 80$$

Thus for complete reaction, the weight of calcium hydroxide required is

$$74 \times 18\,650/106 = 13\,010\,lb$$
$$[74 \times 8480/106 = 5920\,kg]$$

This will be produced from $(13\,010 \times 56/74) = 9850\,lb\ [(5920 \times 56/74)$ $= 4480\,kg]$ quicklime with the consumption of $(13\,010 - 9850) = 3160\,lb$ $[(5920 - 4480) = 1440\,kg]$ water according to the reaction

$$CaO + H_2O \rightarrow Ca(OH)_2$$
$$\quad 56 \qquad 18 \qquad\quad 74$$

As the reaction is only 90 per cent complete, however, the actual weights of calcium carbonate and caustic soda formed are

$$CaCO_3:\quad 18\,650 \times (100/106) \times 90/100 = 1580\,lb$$
$$[8480 \times (100/106) \times 90/100 = 7200\,kg]$$
$$NaOH:\quad 18\,650 \times (80/106) \times 90/100 = 12\,780\,lb$$
$$[8480 \times (80/106) \times 90/100 = 5760\,kg]$$

and the amounts of unused reactants are

$$Ca(OH)_2:\quad 13\,010 \times 10/100 = 1301\,lb$$
$$[5920 \times 10/100 = 592\,kg]$$
$$Na_2CO_3:\quad 1860 \times 10/100 = 1865\,lb$$
$$[8480 \times 10/100 = 848\,kg]$$

Thus at the end of the reaction, the causticizer will contain

$CaCO_3$	15 850 lb	[7200 kg]
NaOH	12 780 lb	[5760 kg]
$Ca(OH)_2$	1301 lb	[592 kg]
Na_2CO_3	1865 lb	[848 kg]
H_2O	(176 550 − 3160)	[(80 320 − 1440)
	= 173 390 lb	= 78 880 kg]

A total of 205 216 lb [93 280 kg] which is equal to the combined inputs of sodium carbonate solution and quicklime. Assuming that the solubilities of calcium carbonate and calcium hydroxide can be ignored in the presence of sodium carbonate and caustic soda, then the total insoluble matter is

$$(1301 + 15\,850) = 17\,151\,lb\ [(592 + 7200) = 7792\,kg]$$

Hence after decantation, the sludge will contain

$$(2 \times 17\,151) = 34\,302\,lb\ water$$
$$[(2 \times 7792) = 15\,584\,kg\ water]$$

that is

$$(34\,302 \times 100/173\,390) = 19.8\%$$
$$[15\,584 \times 100/78\,880 = 19.8\%]$$

of the total water leaving the causticizer. Therefore, the sludge will also contain

$$12\ 780 \times 19.8/100 = 2500\ \text{lb NaOH}$$
$$[5760 \times 19.8/100 = 1138\ \text{kg NaOH}]$$
and
$$1865 \times 19.8/100 = 370\ \text{lb Na}_2\text{CO}_3$$
$$[848 \times 19.8/100 = 168\ \text{kg Na}_2\text{CO}_3]$$

Thus at this stage, the sludge passing to the filter will contain

H_2O	34 302 lb	[15 584 kg]
$CaCO_3$	15 850 lb	[7200 kg]
$Ca(OH)_2$	1301 lb	[592 kg]
Na_2CO_3	370	[168 kg]
NaOH	2500 lb	[1138 kg]
	54 323 lb	[24 682 kg]

The dry insoluble matter is $(15\ 850 + 1301) = 17\ 151$ lb $[(7200 + 592) = 7792$ kg] and hence the wash water will be $(5 \times 17\ 151) = 85\ 755$ lb $[(5 \times 7792) = 38960$ kg]. It is assumed that the cake will contain 50 per cent of dry soluble matter and hence the total weight of the cake will be $(17\ 151 \times 100/50) = 34\ 302$ lb $[7792 \times 100/50 = 15\ 584$ kg]. Finally the sodium salts in the cake will be

$$\text{NaOH:} \quad 34\ 302 \times 0.5/100 = 172\ \text{lb}$$
$$[15\ 584 \times 0.5/100 = 78\ \text{kg}]$$
$$\text{Na}_2\text{CO}_3\text{:} \quad 34\ 302 \times 0.5/100 = 172\ \text{lb}$$
$$[15\ 584 \times 0.5/100 = 78\ \text{kg}]$$

and the weight of water in the cake is

$$343 - (172 + 172 + 15\ 850 + 1301) = 16\ 800\ \text{lb}$$
$$[15\ 584 - (78 + 78 + 7200 + 592) = 7636\ \text{kg}]$$

All other flows may now be calculated by difference and the mass balance drawn up as shown in Figure 1–2, which is based on the mass flow for one shift. In this diagram, intermediate flows between plant units are included. Although not as important as the quantities entering and leaving the plant, these will be useful in sizing pipework at a later stage in the design. There are, of course, many other ways of laying out the mass flowsheet, and in many cases the data may be included on the process flowsheet, together with operating conditions and thermal data. The layout in Figure 1–2 is particularly useful where complex arrangements of streams with recycles, for example, are concerned.

The compilation of a heat balance is a relatively simple operation providing the data relating to specific heats and enthalpies of the various streams are available. In many ways, the provision of such data, especially for dubious slurries and emulsions is rather more difficult than estimation of densities for the mass balance. Progress can be made, however, with

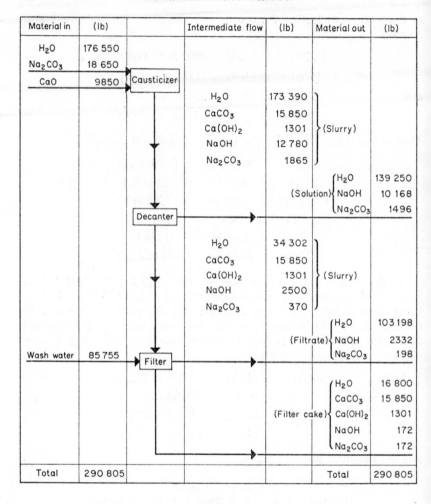

Material in	(lb)		Intermediate flow	(lb)	Material out	(lb)
H_2O	176 550					
Na_2CO_3	18 650					
CaO	9850	Causticizer				
			H_2O	173 390		
			$CaCO_3$	15 850		
			$Ca(OH)_2$	1301	(Slurry)	
			NaOH	12 780		
			Na_2CO_3	1865		
					(Solution) H_2O	139 250
					NaOH	10 168
					Na_2CO_3	1496
		Decanter				
			H_2O	34 302		
			$CaCO_3$	15 850		
			$Ca(OH)_2$	1301	(Slurry)	
			NaOH	2500		
			Na_2CO_3	370		
					(Filtrate) H_2O	103 198
					NaOH	2332
					Na_2CO_3	198
Wash water	85 755	Filter				
					(Filter cake) H_2O	16 800
					$CaCO_3$	15 850
					$Ca(OH)_2$	1301
					NaOH	172
					Na_2CO_3	172
Total	290 805				Total	290 805

(a)

FIGURE 1–2(a). Mass balance for Example 1–2 (British Units). Basis: 1 shift
of 8 hours

sensible assumptions such as taking the specific heat of dilute aqueous
solutions as 1.0 BTU/lb °F [4.18 kJ/kg K] and making due allowance for
such assumptions in sizing the equipment at a later stage. In the heat
balance one is concerned with the relative heat contents of streams of
material flowing in and out of the system, and the crux of the matter is to
calculate the heat content above a sensible datum temperature. It is conventional to use 32°F [273 K] as a datum in such calculations, though in
certain cases this can pose problems, especially where low temperatures
are involved as negative enthalpies may be introduced. In refrigeration
units, −40°F [233 K] is a useful datum, especially as this is equivalent to
−40°C [233 K]. Care must be taken in using enthalpy charts to ensure

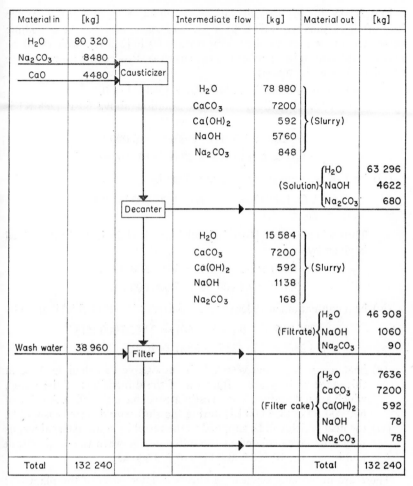

Material in	[kg]		Intermediate flow	[kg]	Material out	[kg]
H_2O	80 320					
Na_2CO_3	8480					
CaO	4480	Causticizer				
			H_2O	78 880		
			$CaCO_3$	7200		
			$Ca(OH)_2$	592	(Slurry)	
			NaOH	5760		
			Na_2CO_3	848		
					(Solution) H_2O	63 296
					NaOH	4622
					Na_2CO_3	680
		Decanter				
			H_2O	15 584		
			$CaCO_3$	7200		
			$Ca(OH)_2$	592	(Slurry)	
			NaOH	1138		
			Na_2CO_3	168		
					(Filtrate) H_2O	46 908
					NaOH	1060
Wash water	38 960	Filter			Na_2CO_3	90
					(Filter cake) H_2O	7636
					$CaCO_3$	7200
					$Ca(OH)_2$	592
					NaOH	78
					Na_2CO_3	78
Total	132 240				Total	132 240

(b)

FIGURE 1-2(b) Mass balance for Example 1-2 (SI Units). Basis: 1 shift of 2.88 ks

that all values calculated for the heat balance are based on the same datum and it is usually safer to compute the enthalpy of a stream from:

(mass flow) (mean specific heat) (temperature of stream
— chosen datum temperature)

A useful approach is to select as a datum the minimum temperature at which streams enter or leave the process—in this way the enthalpy of such streams is zero and the amount of calculation (and chances of error) is reduced. This procedure is illustrated in the following example.

Example 1-3

Prepare a heat balance for the process outlined in Example 1-2 assuming that ambient temperature is 60°F [288 K]. The heat evolved in slaking quicklime is 27 900 BTU/lb mol [64 900 kJ/kmol].

Basis: 1 shift of 8 h [2.88 ks]

In order to simplify the calculations, it will be assumed that the specific heat of all streams is 1.0 BTU/lb °F [4.18 kJ/kg K] and that the quicklime enters the process at ambient temperature 60°F [288 K], which will be taken as the datum in calculating heat contents. Proceeding now with the calculation of each stream:

(i) heat input of quicklime above 60°F [288 K] = 0.

(ii) heat input of sodium carbonate solution above 60°F [288 K] is given by

$$195\,200 \times 1.0\,(110 - 60) = 976\,000 \text{ BTU}$$
$$[88\,800 \times 4.187\,(316 - 288) = 10\,407\,400 \text{ kJ}]$$

(iii) heat of reaction in slaking quicklime is

$$\underset{56}{CaO} + \underset{18}{H_2O} = \underset{74}{Ca(OH)_2} + 27\,900\,[64\,900]$$

Thus when 8850 lb [4480 kg] quicklime is slaked, the heat evolved is given by

$$9850 \times 27\,900/56 = 4\,907\,300 \text{ BTU}$$
$$[4480 \times 64\,900/56 = 5\,600\,000 \text{ kJ}]$$

(iv) heat content of slurry leaving causticizer at 180°F [355 K] is given by

$$205\,216 \times 1.0\,(180 - 60) = 24\,626\,000 \text{ BTU}$$
$$[93\,280 \times 4.18\,(335 - 288) = 26\,165\,000 \text{ kJ}]$$

(v) heat losses from the vessel. It is impossible to calculate these at this stage as they are a function of the dimensions of the vessel, which are not known, and an estimate will have to suffice. A value of 600 000 BTU [633 000 kJ] during the shift would seem reasonable.

(vi) the heat which will be supplied to the vessel by some external means may now be calculated as the difference between the total output and input, i.e. 10 558 700 BTU [10 790 600 kJ].

There are no cooling or heating effects in later parts of the plant and hence the heat balance ends at this stage. The calculated data are laid out in a similar way to those for the mass balance, as shown in Figure 1–3. In Examples 1–2 and 1–3 certain simplifying assumptions were made, and before passing to the next stage of the design it is most important that the validity of these should be questioned. For example, one must ask whether or not the solubilities of calcium hydroxide and calcium carbonate are in fact negligible and to ascertain, by laboratory tests if necessary, the specific heat of the material in the reactor, on which the entire heat balance depends. The working out of a mass balance often highlights several disadvantages of the process, and any modifications should be considered and incorporated at this stage. In the present case, one should consider adding the solution from the decanter to the filtrate—will this be suitable for process purposes or should one or the other be concentrated? Are there any advantages to be gained in starting with a more concentrated solution of sodium carbonate—could solid soda ash be added as an alternative? This type of query must be resolved before equipment design can be started.

The heat balance in Example 1–3 is relatively trivial and far more exciting calculations occur when dealing with furnaces and associated equipment. The approach is exactly the same, however, as illustrated in the following simple example.

A A units

Material	°F	Heat above datum (BTU)		Heat above datum (BTU)	°F	Material
CaO	60	O		24 626 000	188	Slurry
Na₂CO₃ solution	110	9 760 000				
Heat of reaction		4 907 300	Causticizer			
Heat supplied		10 558 700		600 000		Losses (estimated)
Total		25 226 000		25 226 000		

SI units

Material	K	Heat above datum [kJ]		Heat above datum [kJ]	K	Material
CaO	288	O		26 165 000	355	Slurry
Na₂CO₃ solution	316	10 407 400				
Heat of reaction		5 600 000	Causticizer			
Heat supplied		10 790 600		633 000		Losses (estimated)
Total		26 798 000		26 798 000		Total

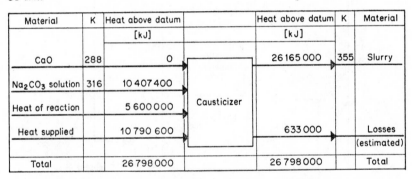

FIGURE 1–3. Heat balance for Example 1–3. Basis: 1 shift of 8 hours [2.88 ks]. Datum: 60°F [288 K]

Example 1–4

A fertilizer is dried in a rotary dryer by a parallel flow of hot flue gas produced in an oil-fired furnace. The following information has been noted on a test run:

throughput of dry fertilizer	5 ton/hr	[1.41 kg/s]
initial moisture content	15% w/w	
final moisture content	5% w/w	
temperature of fertilizer ex-dryer	180°F	[355 K]
consumption of fuel oil	142 lb/hr	[0.0179 kg/s]
net calorific value of fuel oil	16 000 BTU/lb	[37 200 kJ/kg]

heat losses from dryer
(radiation plus sensible heat
loss in dry flue gas)   200 000 BTU/hr   [58.6 kW]

latent heat of vaporization of
water   1090 BTU/lb   [2535 kJ/kg]

mean specific heat of flue gas   0.275 BTU/lb °F   [1.15 kJ/kg °F]

mean specific heat of dried
fertilizer   0.24 BTU/lb °F   [1.003 kJ/kg °F]

air–fuel ratio   30 lb/lb   [30 kg/kg]

Draw up a heat balance for the furnace and the dryer and calculate the thermal efficiency of the furnace.

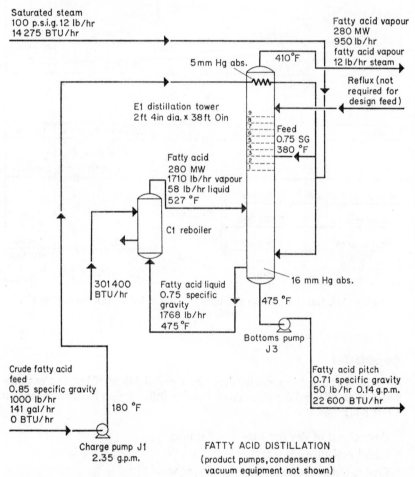

FIGURE 1–4. Part of a typical process flowsheet (British Units)

The flow arrangement for the system is shown in Figure 1–4. It will be assumed that the wet fertilizer, the combustion air and the oil enter the system at 60°F [288 K] and this will be taken as the datum temperature.

Heat into the plant

(i) Wet fertilizer at 60°F [288 K]: enthalpy = 0.

(ii) Combustion air at 60°F [288 K]:

weight of air = 30 × 142 = 4260 lb/hr
[= 30 × 0.0179 = 0.537 kg/s]
enthalpy = 0

(iii) 142 lb/hr [0.0179 kg/s] oil at 60°F [288 K]:

sensible heat = 0

combustion heat = 142 × 1600 = 2 272 000 BTU/hr
[= 0.0179 × 37 220 = 666 kW]

Heat leaving the plant

(i) Heat in dried fertilizer:

weight of dry fertilizer = 5 × 2240 = 11 200 lb/hr [1.41 kg/s]

weight of water at inlet = 11 200 × 15/100 = 1680 lb/hr
[= 1.41 × 15/100 = 0.212 kg/s]

weight of water at outlet = 11 200 × 5/100 = 560 lb/hr
[= 1.41 × 5/100 = 0.071 kg/s]

heat in dry fertilizer leaving the plant
= 11 200 × 0.24 (180 − 60) = 322 560 BTU/hr
[= 1.41 × 1.003 (355 − 288) = 94.75 kW]

heat in water remaining in fertilizer
= 560 × 1.0 (180 − 60) = 67 200 BTU/hr
[= 0.071 × 4.18 (335 − 288) = 19.88 kW]

total heat = (322 560 + 67 200) = 389 760 BTU/hr
[(94.75 + 19.88) = 114.63 kW]

(ii) heat loss in water vapour in flue gas:

weight of water evaporated = (1680 − 560) = 1120 lb/hr
[= (0.212 − 0.071) = 0.141 kg/s]

Assuming vaporization takes place at the outlet temperature of the fertilizer, 180°F [355 K], then

latent heat in flue gas = 1120 × 1090 = 1 220 800 BTU/hr
[= 0.141 × 2535 = 357 kW]

sensible heat in water vapour
= 1120 × 1.0 (180 − 60) = 134 400 BTU/hr
[= 0.141 × 4.18 (355 − 288) = 39.48 kW]

total heat = (1 220 800 + 134 400) = 1 355 200 BTU/hr
[= (357 + 39.48) = 396.48 kW]

(iii) radiation losses plus heat in dry flue gas: this was given as 200 000 BTU/hr [58.6 kW].

Heat balance

Heat leaving dryer:

	BTU/hr	[kW]
dried fertilizer	389 760	[114.63]
water vapour	1 355 200	[396.48]
radiation + dry flue gas	200 000	[58.60]
Total	1 944 960	[569.71]

As the enthalpy of the wet fertilizer is 0, this amount of heat must enter with the flue gas from the furnace. For the furnace therefore:

	BTU/hr	[kW]
heat in:		
air	0	[0]
oil	2 272 000	[666]
heat out:		
flue gas	1 944 960	[596.71]
losses (by difference)	327 040	[69.29]

The efficiency of the furnace is thus

$$= 1\,944\,960 \times 100/2\,272\,000 \; [596.71 \times 100/666] = 85.6\%$$

1.4 Flowsheets

The flowsheet is essentially a road map of the process, and it pictorially and graphically identifies the chemical process steps in their proper sequence. Sufficient detail should be included in the diagrams in order that a proper mechanical interpretation may be made of the chemical requirements. Several types of flowsheet may be prepared for a plant, depending on the complexity of the process and the important examples are summarized in the following.

1.4.1 Block Diagrams

Block diagrams are used for illustrating preliminary or basic concepts only, and in essence each block describes what is to be done rather than how it is to be carried out. Block diagrams are used in process surveys, research summaries and for generally 'talking out' a processing idea, especially at the process development and evaluation stage. A typical example is shown in Figure 1–5. A similar concept is that of the pictorial flow diagram, which incorporates small sketches of the proposed plant in place of blocks and is especially useful in dealings with senior management and for publicity material.

1.4.2 Process Flowsheets

Perhaps the most important flowsheet in plant design is the process flowsheet, which is used to present the heat and mass balance of the process together with a reasonably detailed indication of the operating conditions,

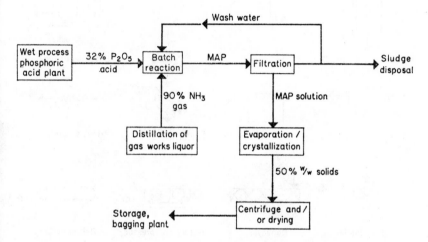

FIGURE 1–5. Block diagram for the manufacture of monammonium phosphate

including the flowrates, temperatures and pressures for each basic item of process equipment or processing step. Where possible, consumption of services should be incorporated including items such as steam, water, air, fuels, refrigeration, and circulating oils. A typical example of part of a process flowsheet is shown in Figure 1–4. This type of diagram is complimentary to the tables of data discussed in section 1.4 under heat and mass balances and does not replace them. In order to avoid lengthy descriptions of the operation involved, symbols are incorporated on process flowsheets to indicate the plant item concerned. The principle is similar to that used in electrical engineering, where symbols for a condenser, rheostat and so on are well known. Unfortunately the symbols used in flowsheets of chemical plant are not standardized and in addition to those proposed by the British Standards Institution and engineering bodies, many companies have their own variations. Some useful and generally accepted symbols are shown in Figure 1–6. Where a fairly simple plant is involved, it may be possible to include the overall dimensions of larger equipment items, if these are known at this stage, though such information is usually included on the engineering flowsheet.

2

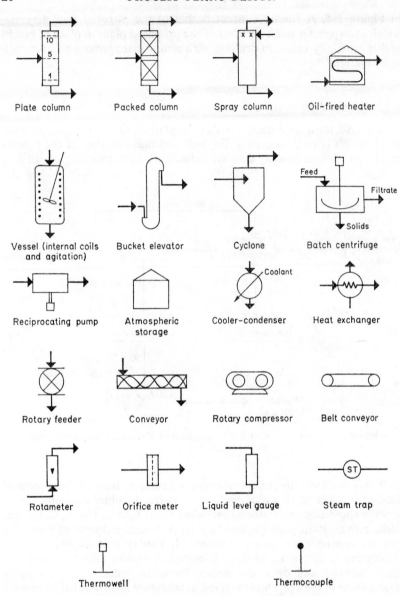

Plate column   Packed column   Spray column   Oil-fired heater

Vessel (internal coils   Bucket elevator   Cyclone   Batch centrifuge
and agitation)

Reciprocating pump   Atmospheric   Cooler-condenser   Heat exchanger
                      storage

Rotary feeder   Conveyor   Rotary compressor   Belt conveyor

Rotameter   Orifice meter   Liquid level gauge   Steam trap

Thermowell   Thermocouple

FIGURE 1–6. Typical flowsheet symbols

1.4.3 Engineering Flowsheets

These are usually prepared when the design is completed and are used to
coordinate all the data from the drawings of individual plant items,
which must be prepared separately. The flowsheet usually takes the form
of a scale elevation of the plant, not necessarily laid out in the proposed
form, but with the plant items at their intended elevations above the ground
datum. It is conventional to show pipes with flanges, heat exchangers with

TABLE 1–2. TYPICAL INFORMATION INCLUDED ON ENGINEERING FLOWSHEETS

Item number	P2	C1	E2	E7	V11
Description	Pump	Column	Reboiler	Condenser	Drum
Function	Reflux to C1	D.A.A. separation	Vapour supply	Refrigerant condensation	Reflux storage
Dimensions (ft)	—	36 × 8 dia.	14 × 2 dia.	20 × 1.5 dia.	105 ft^3
Material of construction	Mild Steel (MS)/Rubber lined	MS	MS	MS	MS
Design pressure (lb/in.^{2} g)	14.7	380 mm Hg	380 mm Hg	14 at	380 mm Hg
Design temperature (°F)	100	260	100	140	100
Steam pressure (lb/in.^{2} g)	—	—	125	—	—
Cooling water (lb/hr)	—	—	—	55 700	—
Power (hp)	5	—	—	—	—
Instruments	—	—	—	—	—
Hazards	—	Pneumatic Fire	Fire	Pneumatic Fire	Fire
Drawing number	—	CE/C1/28511	CE/E2/3279	CE/E7/25766	CE/V11/70937

supports, flanges and nozzles and so on, each item being given a code number. Various companies have their own ideas on this, but E for exchangers, C for columns, V for vessel and so on seems to be generally accepted. Below the drawing of the plant, information relating to each plant item is tabulated, and typical examples are included in Table 1–2. Closely allied to the engineering flowsheets are mechanical flow diagram or piping flowsheets. These include all pipe sizes, size and type of valves, pipe fittings, etc., and are necessary where this information, which is

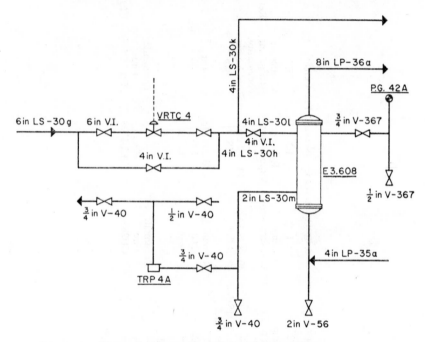

FIGURE 1–7. Part of a typical piping flow diagram

required by mechanical, electrical and instrument engineers, is too complicated to be included in the process flowsheet. No attempt is made at this stage to indicate pipe layout (see Chapter 11) and in many cases this activity has been superceded by the use of models of the plant. A portion of a typical piping flow diagram is shown in Figure 1–7 and useful symbols for incorporation in such diagrams in Figure 1–8. In such diagrams it is conventional to number the various pipelines and branches as an aid to clarity and also in locating lines once the plant is completed. Again, there are a multitude of systems, though the following coding is fairly widespread:

nominal pipe size/material code/sequence number

For example, '2-Cl-6a'—represents a 2 in. diameter, number 6a, carrying chlorine, whilst '4-S150-21' refers to pipe 21, which is 4 in. diameter carrying steam at 150 lb/in.^{2} g. Other materials codes are fairly obvious and a matter of personal reference. In numbering pipes, letters are often used to indicate branches.

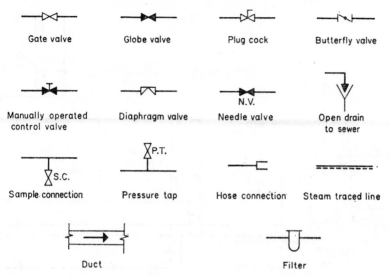

| Gate valve | Globe valve | Plug cock | Butterfly valve |

| Manually operated control valve | Diaphragm valve | Needle valve N.V. | Open drain to sewer |

| Sample connection S.C. | Pressure tap P.T. | Hose connection | Steam traced line |

Duct   Filter

FIGURE 1–8. Symbols used in piping flow diagrams

Included on the piping flow diagram are the instruments to be specified. These are indicated by appropriate symbols and the following typical abbreviations:

RTC	recording temperature controller
RFM	recording flowmeter
ILC	indicating level controller
ORFM	orifice for recording flowmeter
PG	pressure gauge
HPA	high pressure alarm

In most designs, diagrams of instrumentation lines, power supplies and so on have to be prepared, though, in general, these are a specialist function and beyond the scope of the plant designer. Nevertheless, instruments and control lines should be presented on the piping flow diagram and specified on the drawings of the plant items together with full mechanical details.

1.4.4 Other Diagrams

There are many other types of diagram illustrating the data connected with plant design, and the use of these depends a great deal on the complexity of the process and the policy of a particular organization. In many plant designs, utility flowsheets or diagrams are prepared not only for clarity in presentation, but with a view to optimization of the system and reducing consumption both of material and energy. Such diagrams incorporate details of services such as air, water, steam, heat transfer media together with vents, purges, safety relief, blowdown and so on. A typical example is shown in Figure 1–9. Such diagrams are prepared when the amount of detail is too great to include on the process flowsheet.

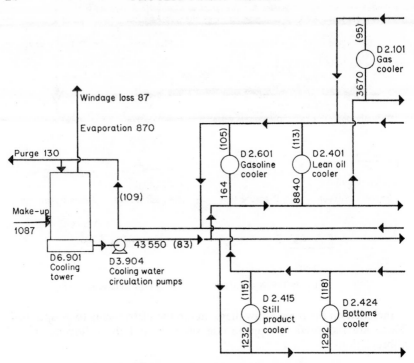

FIGURE 1–9. Part of a typical utility flowsheet (flow in gal/min, temperature in °F)

A particularly clear way of illustrating heat and mass flows through a process is by means of a Sankey diagram. In such diagrams, a stream is represented by two parallel lines whose distance apart is equivalent to the heat or mass flow of the stream. A simple example is shown in Figure 1–10 where an oil-fired heater is used to heat a heat transfer medium. This is pumped to an evaporator in which steam is generated; the heat transfer fluid returning to the heater. In the Sankey diagram, the basis is 100 parts of heat per unit time in the fuel oil. It is graphically obvious that, based on the thermal value of the oil, the heater is 50 per cent efficient, and that 75 per cent of the heat in the oil appears in the steam. One must question the value of circulating 50 per cent of the heat in the recycle, as this represents not only inefficient operation but capital tied up unnecessarily.

Such diagrams are especially useful and illuminating in displaying cash flows in a process. Although difficult to draw up in all but very simple processes, instead of displaying mass and energy flows, the actual moneys invested in the material and energy are shown. In this way one can determine the contribution of labour to the various processing steps, the significance of steam leaks and radiation losses, the amount of capital tied up in the process materials, in storage tanks, hoppers, recycle loops and so on. The evaluation of such data, whilst expensive, is absolutely vital in optimizing not only the plant but the entire processing operation, and such an exercise can pay significant dividends in the long term.

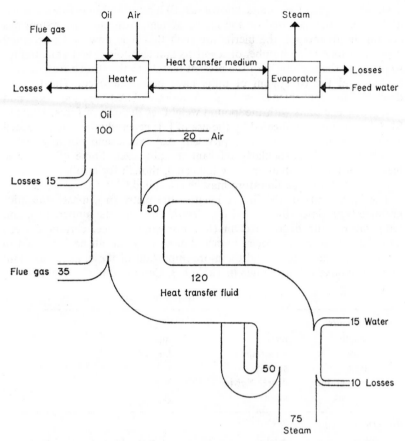

FIGURE 1–10. Representation of heat flows. Basis: 100 parts of heat/unit time in the fuel oil

1.5 Units

The chemical engineer often finds that the data he uses are expressed in a great variety of different units, and it is vital that he should be able to express quantities in a common system of units and also convert from one system to another. Such activities are basic to plant design and it is felt appropriate to give some consideration to this important topic in this introductory chapter.

Most of the physical properties evaluated in the laboratory are usually expressed in the C.G.S. system, whilst the dimensions of plant, its through-put and operating parameters are expressed in some form of general engineering units or special units, which have their origin in the history of the particular industry. For example, the C.G.S. unit of dynamic viscosity is the 'poise' or 'g/cm s'. In engineering units this becomes 'lb/ft hr' and in the oil industry the usual unit is 'Redwood secs', being the time taken for a standard volume of oil to flow through a specified orifice, under a given set of conditions. In an attempt to overcome a confusing situation such as this, the International Organization for Standardization have adopted a standard system of units known as SI units (Système International d'Unités), which is based on the metric system. Although the system has certain disadvantages, the merits are such that it is now being accepted by an ever increasing number of scientists and technologists. In an attempt to promote the usefulness of the present text, both the British Engineering System and the SI system of units have been employed. The former, described for convenience as Anglo-American (A.A.) uses the 'foot' and 'hour' as basic units with the 'pound weight' as the unit of force. Units in SI units have been indicated by the use of [ ]; generally the values quoted are direct conversions of the equivalent A.A. value and not any metric standard—this is particularly relevant to tube sizes. Some of the basic features of the SI system are now indicated, though for a fuller treatment various articles in the literature may be consulted.[2]-[6]

The basic units of the SI are length—the metre (m), mass—the kilogramme (kg), time—the second (s), electric current—the ampere (A), and temperature—the degree Kelvin (K). For convenience, certain derived units have been given special names though they can all be expressed in terms of the basic units, and the more important of these, as far as plant design is concerned are shown in Table 1–3. Certain of the basic units may

TABLE 1–3 THE SI SYSTEM OF UNITS

(a) *Basic units*

length	metre	m
mass	kilogramme	kg
time	second	s
temperature	degree Kelvin	K
current	ampere	A

(b) *Derived units*

force	newton	N	$1\ N = 1\ kg\ m/s^2$
work, energy, heat	joule	J	$1\ J = 1\ N\ m = 1\ kg\ m^2/s^2$
power	watt	W	$1\ W = 1\ J/s = 1\ kg\ m^2/s^2$
electric potential	volt	V	$1\ V = 1\ W/A = 1\ kg\ m^2/s^3\ A$
frequency	hertz	Hz	$1\ Hz = 1\ cycle/s$
electric resistance	ohm	Ω	$1\ \Omega = 1\ V/A = 1\ kg\ m^2/s^3\ A^2$

(c) *Important prefixes for unit multiples and sub multiples*

10^{-6}	micro	μ
10^{-3}	milli	m
10^{-2}	centi	c
10^3	kilo	k
10^6	mega	M

(d) *Common physical constants*

gravitation acceleration, $g = 9.807\ m/s^2$

gas constant, $R = 8.314\ kJ/kmol\ K$

Stefan-Boltzmann constant, $\sigma = 5.67 \times 10^{-8}\ W/m^2\ K^4$

gas molecular volume, 1 kmol occupies 22.41 m^3 at $1.013 \times 10^5\ N/m^2$ and 273.15 K (s.t.p.)

be of an inconvenient size in certain cases, though the use of multiplying prefixes shown in the same table overcomes this problem. The SI system is coherent in that when unit quantities are multiplied or divided, no numerical values are involved. For example, g does not appear unexpectedly as in a gravitational system of units, but only when the force of gravity is actually involved. Thus the weight of a mass of m kg is a force of mg N, where g is the acceleration due to gravity. There is little significance in the fact that 1 N is approximately the weight of an apple!

Conversion factors for some common units are shown in the Appendix and their uses are illustrated by the following example.

Example 1–5

Calculate the film coefficient of heat transfer for acetone condensing on the outside of a tube bundle using the following data:
At a film temperature of 167°F [348 K],

thermal conductivity, $k = 0.095$ BTU/hr ft² °F/ft

density, $\rho = 0.75 \times 62.4 = 46.7$ lb/ft³

latent heat, $\lambda = 155$ BTU/lb

viscosity, $\mu = 0.009$ cP $= 0.009 \times 2.42 = 0.0218$ lb/ft hr

number of tubes $= 250$, hence $j = \sqrt{250} = 15.8$

temperature drop across film, $\Delta T_f = 15$°F

tube outer diameter $= \frac{3}{4}$ in. $= 0.0625$ ft

The coefficient should be calculated in both British (A.A.) and SI units and the result checked.

The appropriate relationship is

$$h_0 = 0.72((k^3\rho^2 g\lambda)/(jD_0\mu\Delta T_f))^{0.25}$$

Thus, in British units,

$$h_0 = 0.72 \left\{ \frac{0.095^3 \times 46.7^2 \times (32 \times 3600^2) \times 155}{(15.8 \times 0.0625 \times 0.0218 \times 15)} \right\}^{0.25}$$

$$= 561 \text{ BTU/hr ft}^2 \text{ °F}$$

In SI units, first check the units:

$$\left(\frac{k^3\rho g\lambda}{jD_0\mu\Delta T_f} \right)^{0.25}$$

$$= \left\{ \left(\frac{W}{mK}\right)^3 \left(\frac{kg}{m^3}\right)^2 \left(\frac{J}{kg}\right) \left(\frac{m}{s}\right) (m)^{-1} \left(\frac{Ns}{m^2}\right)^{-1} \left(K^{-1}\right) \right\}^{0.25}$$

or in basic units,

$$\left(\frac{kg^3 m^6 kg^2 \ m^2 m^2 m^2 s^2}{s^9 m^3 K^3 m^6 s^2 kg \ s^2 \ m \ s \ kg \ m \ K} \right)^{0.25} = (kg^4/s^{12}K^4)^{0.25}$$

$$= kg/s^3 K = W/m^2 \ K$$

From the Appendix,

$$k = 0.095 \times 1.731 = 0.1645 \text{ W/m K}$$
$$\rho = 46.7 \times 16.02 = 749 \text{ kg/m}^3$$
$$g = 9.807 \text{ m/s}^2 \qquad \text{(Table 1--3)}$$
$$\lambda = 155 \times 2.326 \times 10^3 = 351 \times 10^3 \text{ J/kg}$$
$$j = 15.8$$
$$D_0 = 0.0625 \times 0.305 = 0.0191 \text{ m}$$
$$\mu = 0.009 \times 0.1/10^2 = 0.009 \times 10^{-3} \text{ Ns/m}^2$$
$$\Delta T_f = 15 \times 0.556 = 8.35 \text{ K}$$

Thus, the heat transfer coefficient becomes

$$h_0 = 0.72 \left\{ \frac{0.1645^3 \times 749^2 \times 9.807 \times 357 \times 10^3}{15.8 \times 0.0191 \times 0.009 \times 10^3 \times 8.35} \right\}^{0.25}$$

$$= 3140 \text{ W/m}^2 \text{ K}$$

Converting the previous value,

$$h_0 = 561 \times 5.68 = 3170 \text{ W/m}^2 \text{ K}$$

A final point, worthy of note in using SI units, is that certain common physical properties, for example the specific heat of water, are no longer unity. In this case, the value is 4.187 kJ/kg K, though as a bonus, the density of air is approximately 1 kg/m^3, the specific heat of air 1 kJ/kg K, the density of water 10^3 kg/m^3 and the latent heat of vaporization of water about 2 MJ/kg.

REFERENCES

1. Ross, T. K. *An Introduction to Chemical Engineering*. London: Pitman, 1953.
2. 'The Use of SI Units', *Pub. PD 5686*. London: British Standards Institution, 1967.
3. Anderton, P. and Bigg, P. H. *Changing to the Metric System*. London: H.M.S.O., 1965.
4. Ede, A. J. *Int. J. Heat. Mass. Trans.* 1966, **9**, 837.
5. Bigg, P. H. *Brit. J. App. Phys.*, 1964, **15**, 1243.
6. Bigg, P. H. *Chem. in Brit.* 1963, **87**, 407.
7. Mullin, J. W. *Chem. Engr.* (London) 1967, No. 211, CE 176.

Part 2: Process Plant Design

Part 2: Process Plant Design

Chapter 2: Cost Estimation and Optimization

2.1 Introduction

In Chapter 1, the factors governing the selection of a process for the manufacture of a given chemical were considered, and the importance of designing for a reasonable return on capital invested and hence a viable commercial operation was discussed. Although other factors such as plant location, availability of raw materials and services may be of considerable importance, the prime factor in evaluating any process is the cost of producing unit weight of the end product. In order to obtain this information, it is necessary to prepare preliminary designs of plant and equipment for each process so that the most economic and usually that with the highest profitability may be selected. Once this has been carried out, it is necessary to produce the most efficient and economic design for the chosen process— a procedure which involves optimizing each section of plant and also smaller items such as pipework and pumps. It is in this operation that the designer's skill and experience is so valuable. Throughout both these operations, that is process evaluation and optimization of plant design, a knowledge of cost estimation procedures is vital, and it is for this reason that the topic is discussed at an early stage in the present text.

It is convenient at the outset to distinguish between detailed cost estimates and approximate costing. The former are usually carried out by specialists in a cost estimating department in the preparation of a final tender. Such estimates are relatively accurate and are based on the detailed plant design in which all the equipment is sized, the pipework layouts have been prepared and the instruments are completely specified.

In this chapter, only approximate cost estimates are considered as these are quite adequate for process evaluation and optimization procedures. Although such estimates are based on the major features of the system and the more important design variables, it is possible, from a knowledge of the proposed plant location, a sketch of the process flow sheet, the size of the major items of equipment and the service requirements, to estimate capital and operating costs to about ±15 per cent. It is feasible, of course, to achieve a greater accuracy with more time and effort, though this is rarely justified in the initial stages of design preparation.

In using cost estimate procedures, it should be noted that a rapid estimation is vital in the early stages of any design, though any uncertainty in the data results in incorrect optimization. The use of cost correlations and factored estimates can reduce the time involved and are generally sufficiently accurate.

It may be said that cost estimation has been somewhat neglected as a part of the art of plant and process design, and it is only in recent years that it has been developed on a rational basis. There is still a desperate need for an improvement in techniques and the availability of essential data. In outlining the methods available, it has been necessary to draw heavily on the few but excellent texts on the topic, and these should be consulted as the following sections represent only a brief introduction to the subject and are in no way a comprehensive survey.

2.2 Cost of Major Items of Equipment

The first stage in preparing a cost estimate for a plant is to evaluate the cost of the major items such as vessels, columns, heat exchangers, pumps and filters. It is convenient at this stage to exclude extras such as insulation, erection and process piping, as these can be estimated by the factored

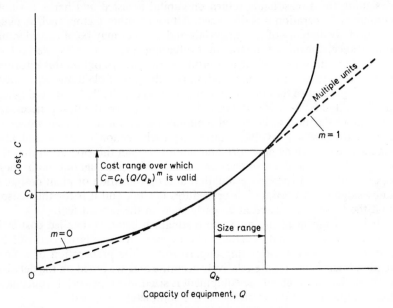

FIGURE 2-1. Cost correlations for major plant items

estimate procedure from the total cost of the major items as discussed in a later section.

Perhaps the simplest method of obtaining this information, and sometimes the only method where no data is available, is to compute the cost from the weight of the plant item:

$$C = FWc \qquad (2\text{--}1)$$

where W is the weight of the unit in lb [kg]

        c is the cost of the material of construction in £/lb [£/kg]

        F is a factor allowing for the cost of fabrication, delivery, etc.

It is, of course, possible to obtain prices for standard items of equipment such as pumps from the stock list of the appropriate manufacturer, but where an item of special design is required, a quotation must be sought from the supplier. This may entail considerable effort, for example in the preparation of drawings, and is not particularly efficient, especially where the time factor is critical. In assessing quotations it is important to note possible discounts and also to distinguish between a 'firm price' and those containing an escalation clause which permits increases in the price due to rises in the cost of materials or labour.

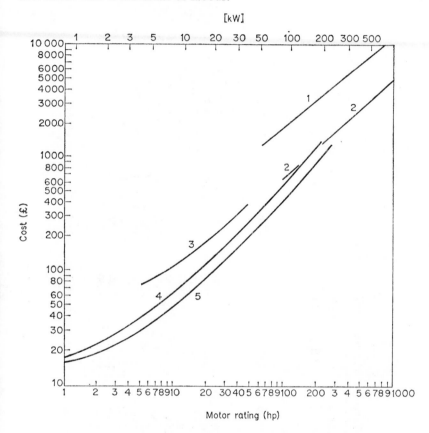

1. Back pressure steam turbines with reduction gear
2. Simple impulse turbines (no reduction gear) speed ≈3000 r.p.m. [50 Hz]
3. Electric motor FLP foot mounted 6 pole (970 r.p.m.) [16 Hz]
4. Electric motor TEFC foot mounted 6 pole (970 r.p.m.) [16 Hz]
5. Electric motor TEFC foot mounted 4 pole (1450 r.p.m) [24 Hz]

FIGURE 2–2. Electric motors and turbines—typical variation in cost

The most convenient way of determining the costs of major items for process evaluation is to make use of cost correlations, in which the logarithm of capital cost of an item is plotted against the logarithm of a capacity parameter, such as volume, surface area, throughput or power rating. Such a plot is shown diagrammatically in Figure 2–1.

In theory, the cost of a unit with zero capacity should be nil and the cost then rises exponentially as the capacity is increased, ultimately tending to infinity simply because it is impossible to fabricate a single unit above a certain limiting capacity. In practice, there is always some cost even at

TABLE 2–1. TYPICAL EQUIPMENT COST CORRELATIONS[1],[4]

Item	Base size Q_b		Base cost C_b	Size range		m
	(A.A.)	[SI]	£ (1970)	(A.A.)	[SI]	
Blower (7.5 lb/in.²) [50 kN/m²]	1400 ft³/min	0.66 m³/s	4870	1400–6000 ft³/min	0.66–2.80 m³/s	0.35
Air compressor	240 hp	180 kW	56 500	240–2000 hp	180–1500 kW	0.29
Plate and frame filter	10 ft²	1 m²	570	10–300 ft²	1–30 m²	0.58
Heat exchangers						
Shell and tube	50 ft²	4.6 m²	950	50–300 ft²	4.6–27.6 m²	0.48
Finned tube	700 ft²	64 m²	3800	700–3000 ft²	64–290 m²	0.58
Reboiler	400 ft²	37 m²	2875	400–600 ft²	37–56 m²	0.25
Centrifugal pumps	10 hp	7.5 kW	925	10–25 hp	7.5–18.6 kW	0.68
	25 hp	18.6 kW	2540	25–100 hp	18.6–75.0 kW	0.86
Storage tanks	300 gal	1.4 m³	170	300–1400 gal	1.4–6.5 m³	0.66
Process towers	6000 lb	2720 kg	2050	6000–20 000 lb	2720–9320 kg	0.79
	20 000 lb	9320 kg	5370	20 000–30 000 lb	9320–136 200 kg	0.71
Reactors	50 gal	0.25 m³	710	50–300 gal	0.25–1.5 m³	0.41
Pressure vessel	300 gal	1.5 m³	1480	300–2800 gal	1.5–12.5 m³	0.69
	800 gal	3.6 m³	1230	300–20 000 gal	3.6–90 m³	0.54

(These prices do not include delivery or installation and a factor should be allowed for these items.)

zero capacity—usually a value allowing for such items as overheads, administration, etc., and the slope of the curve, m, then increases from 0 to 1, at which point it is more practical and economic to use multiple units of the same size. Between these two limits, there is a certain range of capacity

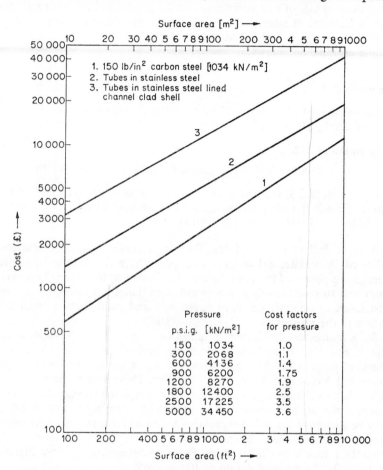

FIGURE 2–3. Approximate costs of tubular heat exchangers. Floating head, 16 ft [4.88 m] tubes

Q for which m is essentially constant and the cost is conveniently correlated by a relation of the form

$$C = C_b(Q/Q_b)^m \qquad (2–2)$$

where C_b and Q_b are the cost and capacity of a unit of basic size. Typical values of these base components with the corresponding exponent m are given in Table 2–1, which is based on information given in the literature.[1]–[4]

A great deal of this type of information is drawn from the past experience of a company and is often not generally available. There is, however, a reasonable amount published in the technical journals, some of which devote a section of each issue to this vital information service. One particularly valuable source of data is a booklet *A Guide to Capital Cost Estimation*[5] and Figures 2–2 and 2–3 represent typical plots presented in this publication. The cost data quoted in this chapter are only intended to serve as examples of cost correlation, and a great deal of further information is included in later chapters relating to particular equipment items.

The use of equation 2–2 and data in Table 2–1 may be illustrated by a simple example as follows:

Example 2–1

What is the cost of a shell and tube heat exchanger with a heat transfer area of 200 ft^2 [18.6 m^2]?

In equation 2–2,

$$C = 950(200/50)^{0.48} [= 950(18.6/4.6)^{0.48}] = 1850 \text{ £}(1970)$$

From Figure 2.3, the cost of a floating head tubular exchanger of area 200 ft^2 [18.6 m^2] with stainless steel tubes 16 ft [4.88 m] long operating at 150 lb/in.^{2} [1.034 MN/m^2] is 2100 £(1970).

A great deal of published cost data is presented in units of currency followed by a date, \$(1959) for example, where the date indicates the purchasing power of the pound sterling or dollar at that time. This varies from year to year (usually a downwards trend) and it is necessary to use a cost index to upgrade prices of equipment and plant from a given time datum to the present.

Such indices are published as tables or graphs in *The Economist* and *The Engineering News Record Magazine*, and of particular relevance to chemical plant costs, *Nelson's Refinery Index*, *Chemical Engineering Cost Index* and *Marshall and Steven's Cost Index*. These indices are not applicable solely to chemical plant items, but to complete plants and to a lesser extent materials and services. The reduced purchasing power of capital may be a decisive factor in process evaluation and it is of particular importance in assessing depreciation of a given unit. Typical values extrapolated from data quoted in *Chemical and Process Engineering*[6],[7] are included as Table 2–2 based on a datum of 100 in 1965.

TABLE 2–2. TYPICAL COST INDEX VARIATION[6],[7]

1958	84	1965	100
1959	87	1966	107.7
1960	88	1967	105.5
1961	90	1968	110.0
1962	92	1969	115.0
1963	93	1970	124.1
1964	96		

2.3 Costs of Complete Plants

In the preliminary screening of alternative processes, a knowledge of the cost of a complete plant is very useful, and some such data is available in

TABLE 2–3. TYPICAL COST DATA FOR COMPLETE PLANTS[1],[3]

Product	Size (A.A.)	Size [SI]	Cost £(1970)	Size range (A.A.)	Size range [SI]	Exponent
Acetylene (from natural gas)	10 ton/day	0.12 Mg/ks	1 409 000	10–100 ton/day	0.12–1.20 Mg/ks	0.73
Ammonia (from natural gas)	100 ton/day	1.2 Mg/ks	2 113 000	100–300 ton/day	1.20–3.60 Mg/ks	0.63
Carbon dioxide	50 ton/day	0.60 Mg/ks	148 800	50–300 ton/day	0.60–3.60 Mg/ks	0.67
Chlorine (electrolytic)	5000 ton/year	0.015 Mg/ks	2 820 000	5000–30 000 ton/year	0.015–0.09 Mg/ks	0.38
Ethanol	1300 ton/year	0.065 Mg/ks	930 000	1300–104 000 ton/year	0.065–5.20 Mg/ks	0.60
Ethylene	11 000 ton/year	0.33 Mg/ks	1 860 000	11 000–330 000 ton/year	0.33–9.90 Mg/ks	0.71
Hydrochloric acid	10 ton/day	0.12 Mg/ks	63 400	10–20 ton/day	0.12–0.24 Mg/ks	0.60
Hydrogen (from natural gas)	10^6 ft^3/day	300 m^3/ks	296 000	10^6–10^7 ft^3/day	300–3000 m^3/ks	0.57
Nitric acid	50 ton/day	0.60 Mg/ks	592 000	50–300 ton/day	0.60–3.60 Mg/ks	0.55
Oxygen	20 ton/day	0.24 Mg/ks	304 000	20–100 ton/day	0.25–1.25 Mg/ks	0.56
Phosphoric acid (54% P_2O_5)	50 ton/day	0.60 Mg/ks	592 000	50–300 ton/day	0.60–3.60 Mg/ks	0.63
Sulphuric acid (contract)	50 ton/day	0.60 Mg/ks	211 870	50–1000 ton/day	0.60–12.0 Mg/ks	0.70

the literature. In the complete absence of any information, a very rough indication is given by the multiple of the output of final product and the value of the product per unit weight. Fortunately, a great deal of cost data on complete plants can be correlated by a relation of the form shown in equation (2–2), and typical information for use in this equation, again based on the work of Baumann,[1] is shown in Table 2–3. This provides investment costs with an accuracy of 30 per cent and it will be noted that the exponent m is virtually constant and equal to 0.6. This is an indication of the validity of the 'six-tenths rule' which states that the ratio of the costs of two plants producing the same product is proportional to the ratio of their capacities raised to the power of 0.6, or

$$C_2/C_1 = (Q_2/Q_1)^{0.6} \qquad (2\text{–}3)$$

Example 2–2

What is the increase in the cost of a 50 ton/day [0.60 Mg/ks] contact sulphuric acid plant, if its output is doubled?
In equation (2–3),

$$C_2/211\ 870 = (100/50)^{0.6}$$

$$[C_2/211\ 870 = (1.20/60)^{0.6}]$$

$$\therefore C_2 = 321\ 200\ \pounds(1970)$$

Hence an increase of $(321\ 200 - 211\ 870) = \pounds109\ 330$, which is 52 per cent of the original cost.

The fact that the cost is not doubled indicates the economic advantages in designing very large processing units—a principle which is valid as long as duplication of plant items is unnecessary. It should be remembered, however, that the losses are proportionately larger when a large plant fails, and in a hazardous and unreliable process it may be advantageous to construct a number of smaller plants.

A much more accurate method of determining the investment in a complete plant is the factored estimate procedure in which the total plant cost is extrapolated from the cost of major items of equipment. This is a most useful technique which, by careful selection of appropriate factors, can provide results within an accuracy of 10 per cent. The method is eminently suited to development of cost equations required for optimization operations. Suggested factors are given in Table 2–4 from which

$$I_F = I_E(1 + \sum_i f_i)(1 + \sum_i f_i') \qquad (2\text{–}4)$$

or $$I_F = I_E \cdot f_L \qquad (2\text{–}5)$$

where I_F is the fixed investment cost of a complete plant,
I_E is the cost of the major items of processing equipment,
f_i are factors for the cost of piping, instruments, etc.,
f_i' are factors for indirect costs, such as contractors, fees.

TABLE 2–4. FACTORS FOR ESTIMATING TOTAL PLANT COSTS

Installed cost of process equipment	I_E

Typical
Experience Factors as Fractions of I_E

Process piping, f_1 solids processing		0.07–0.10
mixed processing		0.10–0.30
fluids processing		0.30–0.60
Instrumentation, f_2 little automatic control		0.02–0.05
some automatic control		0.05–0.10
complex automatic control		0.10–0.15
Buildings, f_3 outdoor		0.05–0.20
indoor and outdoor		0.20–0.60
indoor		0.60–1.00
Facilities, f_4 minor addition		0–0.05
major addition		0.05–0.25
new site		0.25–1.00
Outside lines, f_5 with existing plant		0–0.05
separated unit		0.05–0.15
scattered unit		0.15–0.25

TOTAL PHYSICAL COST	$I_E(1 + \sum_i f_i)$

Experience Factors as Fractions of Physical Cost

Engineering and Construction, f_1'		
straightforward		0.20–0.35
complex		0.35–0.50
Size factor, f_2' large		0–0.05
small		0.05–0.15
experimental		0.15–0.35
Contingencies, f_3' firm process		0.10–0.20
subject to change		0.20–0.30
tentative		0.30–0.50

TOTAL PLANT COST	$I_F = I_E(1 + \sum_i f_i)$
	$(1 + \sum_i f_i)$

The factor f_L is known as the Lang factor and for most plants this is approximately equal to a value of 3. It is of course possible to approach the complete plant cost with greater accuracy by considering more detail and breaking the various factors down still further, though this is not generally justified. Further information on typical factors is given in the literature.[8]–[9] It is usual to add a factor to the calculated I_F, say 10 per cent, for contingencies as illustrated in the following example.

Example 2–3

What is the total cost of the synthesis section of an ammonia plant with a throughput of 30 ton/day [0.36 Mg/ks]?

Using the techniques outlined in section 2.2 and data in this and other chapters, total cost of major items of equipment, $I_E = £168\ 500$. Thus, direct costs are:

f_1 insulation	0.15
f_2 installation	0.15
f_3 piping	0.60
f_4 foundations	0.10
f_5 buildings	0.07
f_6 structures	0.06
f_7 electrical	0.10
f_8 painting and clean-up	0.07
f_9 instrumentation	0.05

$$1.35 = \sum_i f_i$$

Indirect costs are:

f_1' contractors' overheads	0.30
f_2' engineering fee	0.13
f_3' contingencies	0.13

$$0.56 = \sum_i f_i'$$

Hence   total investment $= 168\ 500\ (1 + 1.35)(1 + 0.56)$

$$= £618\ 000$$

From Table 2–3, the cost of a complete plant for the production of ammonia from natural gas with an output of 30 ton/day [0.36 Mg/ks] is given by

$$I_F = 2\ 113\ 000\ (30/100)^{0.63}$$
$$[\ = 2\ 113\ 000\ (0.36/1.20)^{0.63}]$$
$$= £988\ 000$$

which leaves £370 000 for the plant needed for preparation of the synthesis gas.

2.4 Investment in Auxiliary Services

Whilst section 2.3 describes the computation of complete plant costs, it will be appreciated that, for any complex of chemical plants, additional costs are incurred in providing ancillary services such as roads, canteens, laboratories and machine shops. The cost of these services may be quite substantial and they must be taken into account when assessing the cost of a new plant. There are essentially two ways of dealing with the problem depending on whether the proposed plant is an addition to an existing plant or a plant at a completely new site—generally known as a 'grass roots project'. In the former case it is convenient to charge the new plant a proportion of the cost of existing services at the site based on the amount of the services to be consumed. In the case of a 'grass roots' scheme and large additions to existing sites, the whole of the auxiliary services cost must be included in the investment required for the new plant. Table 2–5 shows some typical auxiliary costs as a percentage of the cost of a complete plant. This list, which is based on the data of Baumann[1] is not comprehensive and reference should be made to other sources of information.

TABLE 2–5. INVESTMENT IN AUXILIARY SERVICES

	% of total installed plant cost	
	range	mean
Auxiliary buildings	3.0–9.0	5.0
Steam supply and distribution	2.6–6.0	3.0
Water supply including cooling	0.4–3.7	1.8
Product storage	0.7–2.4	1.8
Electrical sub-station	0.9–2.6	1.5
Water distribution and treatment	0.3–3.1	1.5
Communications	0.1–0.3	0.2
Roads	0.2–1.2	0.6

2.5 Estimation of Manufacturing Costs

Previous sections have dealt with computing the investment required to install plant for a given process and the cost of running the plant is now considered. This cost has essentially three parts:

(i) costs proportional to investment,
(ii) costs proportional to the production rate,
(iii) costs proportional to the labour requirements.

The first of these includes items such as maintenance, insurance, fire protection, security and administration costs and these usually amount to 10–20 per cent of the total investment cost of the plant each year. The second part is by far the greatest proportion of the manufacturing costs and includes the cost of raw materials and, where relevant, catalysts and utilities such as steam, power and water. In addition, smaller items such as licence fees are included at this stage. The consumption of utilities and chemicals are usually estimated from mass and energy balances, and prices of raw materials are available in trade journals such as *Chemical Age*, although it may be possible to reduce these by special purchase agreements. In process evaluation, complete mass and energy balances may not be available for a particular plant and in this case data on the consumption of utilities such as that complied by Aries and Newton[3] is most useful. Some values are shown in Table 2–6 with an indication of typical costs. Where applicable, the cost of packaging and transport should be included in the manufacturing costs at this stage.

In many processes, the cost of labour is decreasing because of increased utilization and installation of automatic control. For large-scale processes involving mainly liquids and gases, the labour costs are usually between 5 per cent and 15 per cent of the total manufacturing costs, though where a great deal of solids handling is involved this figure may approach 25 per cent.

TABLE 2–6. CONSUMPTION AND COST OF UTILITIES

(a) *Consumption*

Plant	Electricity (kWh/lb product)	[MJ/kg product]	Steam (lb/lb product) [kg/kg product]	Water (gal/lb product)	[kg/kg product]
Acetic acid from acetylene	0.210	1.67	3.10	43.0	430
Catalytic cracker	0.001	0.01	0.10	1.2	12
Oxygen	0.220	1.75	2.20	13.0	130
Phosphoric acid (54% P_2O_5)	0.052	0.40	0.70	7.3	73
Contact sulphuric acid	0.015 [1 MJ = 0.28 kWh]	0.12	0.10	2.2	22

(b) *Cost*

Steam	400 lb/in.^{2} [2750 kN/m^2]	0.50–0.75 £/1000 lb	[1.1–1.7 £/Mg]
	100 lb/in.^{2} [675 kN/m^2]	0.25–0.75 £/1000 lb	[0.6–1.7 £/Mg]
Electricity		0.005–0.01 £/kWh ($\frac{1}{2}$–1 p/kW h)	[1.1–2.1 £/1000 MJ]
Water	Cooling	0.05–0.13 £/1000 gal	[0.011–0.027 £/Mg]
	Process	0.10–0.25 £/1000 gal	[0.02–0.05 £/Mg]

Example 2–4

Estimate the cost of manufacturing 30 ton/day [0.36 Mg/ks] oxygen on a plant using the air liquefaction process.

(i) *Cost proportional to investment*

Using the data in Table 2–3, the total investment in plant is given by

$$I = 304\,000\,(30/20)^{0.56} = 382\,000\ £(1970)$$
$$[= 304\,000\,(0.36/0.24)^{0.56} = 382\,000\ £(1970)]$$

Taking the cost of administration, maintenance, insurance, etc. as say 15 per cent of the total investment, then the cost proportional to investment is

$$0.15 \times 382\,000 = 57\,300\ £/\text{year}$$
$$[57\,300/31.5 = 1\,820\ £/\text{Ms}]$$

(ii) *Cost proportional to production*

Raw materials: air, cost = nil.
Utilities:

electricity   0.220 kWh/lb product   cost = 0.220 × 0.008
(from Table 2–6)   = 0.001 76 £/lb

[1.75 MJ/kg product   cost = 1.75 × 1.8/1000
= 0.003 20 £/kg]

steam   2.20 lb/lb product   cost = 2.20 × 0.5/1000
                                                  = 0.001 10 £/lb

               [2.20 kg/kg product   cost = 2.20 × 1.1/1000
                                                  = 0.002 31 £/kg]

water   13.0 gal/lb product   cost = 13.0 × 0.10/1000
                                                  = 0.001 30 £/lb

               [130 kg/kg product   cost = 130 × 0.02/1000
                                                  = 0.002 60 £/kg]

Hence total cost of utilities = 0.0416 £/lb O_2 [0.008 11 £/kg O_2]
                              = 0.004 16 × 30 × 2240 × 365
                              = 102 050 £/year
                              [= 0.008 11 × 0.36 × 10^3 × 10^3
                              = 3240 £/Ms]

(iii) *Cost proportional to labour*

Taking the labour costs as say, 10 per cent of the total manufacturing costs, then cost of labour

$$= (102\,050 + 57\,300) \times 0.10/(1 - 0.10) = 17\,706 \text{ £/year}$$
$$[= (1820 + 3240) \times 0.10/(1 - 0.10) = 562 \text{ £/Ms}]$$

Hence, total manufacturing costs

$$= (17\,706 + 57\,300 + 102\,050) = 177\,060 \text{ £/year}$$
$$[= (1820 + 3240 + 562) = 5622 \text{ £/Ms}]$$

or, per unit mass of product,

$$177\,060/(365 \times 30 \times 2240) = 0.0072 \text{ £/lb}$$
$$[5622/(10^3 \times 0.36 \times 10^3) = 0.0156 \text{ £/kg}]$$

2.6 Estimation of the Total Cost of Product

In the previous section the manufacturing costs involved in production have been considered, and it is important to realize that these may represent but a mere fraction of the price at which the product must be sold in order to break even, let alone show a workable profit. No charge has been allowed for the cost of building the production unit, for example. This is probably the major item which must be added to the manufacturing costs, together with the costs involved in transport, rates, administration, advertising, offices and so on. The costing of these items is of major importance, and for the purposes of a first order estimate of product costs they are grouped together as overheads. The usual way of assessing these is to take 80–90 per cent of the labour costs as a guide, though of course the amount involved will vary with the company and factors, such as plant location, proximity of markets, packaging and so on. Where there is uncertainty in assessing overheads a fairly hefty contingency item should be included in the estimates.

There are many ways of estimating the most important item, capital charges, i.e. the cost of building the plant. The financing of chemical plant, or indeed any large project, is a matter for experts, albeit with often little chemical engineering background and yet, in preliminary costing of alternatives, three simple procedures may be adopted:

(i) The simplest technique, and the one most suited to first order optimization techniques, is to assume that the total cost of the installed plant is paid back in equal annual increments throughout the life of the plant together with simple interest on the total sum involved. For example, if the installed cost of a plant is £1 000 000 and it is to be paid for in 7 years, finance being available at 8 per cent p.a., then

$$\text{annual capital repayment} = 1\ 000\ 000/7 = 142\ 860\ \text{£/year}$$
$$\text{interest} = 1\ 000\ 000 \times 8/100 = 80\ 000\ \text{£/year}$$
$$\text{total} = \text{£}222\ 860/\text{year}$$

or $(7 \times 222\ 860) = \text{£}1\ 560\ 020$ during the life of the plant.

The number of years over which the cost of a plant is paid off often bears little relationship to the actual life of the plant or the period over which it is operated. A more important consideration is the length of time for which there is a demand for the product or for which it can be sold at an economical and profitable price. Other factors may also be involved, such as the type of product, the plant output, the plant operating conditions and the materials of construction.

(ii) Where it is likely that the demand for a product may fall off during the life of the plant, for example where one is dealing with an expanding field where new and better processes and indeed products are being developed, it is convenient to pay back a larger proportion of the capital in the earlier years of the depreciation period. The easiest way of doing this is to pay back a fraction of $2(n - x + 1)/(1 + n)n$ of the total capital in the xth year of a total depreciation period of n years. For example, using the previous data, $n = 7$; thus in the first year $(x = 1)$ the fraction to be repaid is

$$2(7 - 1 + 1)/(8 \times 7) = 7/28 \text{ or } \text{£}250\ 000.$$

In the second year, the fraction is 6/28 and so on. The total repaid and the annual interest remain the same as before, though in this case the total cost of the product decreases during the depreciation period.

The amounts paid with methods (i) and (ii) are compared in Figure 2–4, which illustrates the greater rate of pay-off in the second case. In practically all optimization calculations, whether involving a piece of plant or a complete production unit, for simplicity it is usual to assume linear depreciation, and it is this method which is used in later sections of this text.

(iii) A more profitable way of depreciating the plant cost involves amortization. In simple terms, a sum is invested each year at compound interest and the total of these sums plus the accrued interest is used to pay the total sum involved at the end of the depreciation period. Thus if the total capital invested in the plant is £P and money can be invested at

i per cent compound interest, then the minimum amount to be invested each year, £R, to produce £P in n years is given by

$$R = Pi/((1 + i)^n - 1) \text{ £/year} \qquad (2\text{-}6)$$

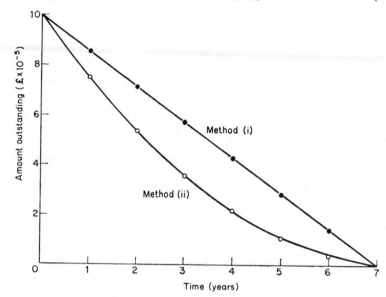

FIGURE 2-4. Repayment of capital

As before interest on the loan must be paid each year for the n years. Thus with the data in (i) and (ii) and assuming that a compound interest of 6 per cent after tax can be obtained on moneys invested, then the sum to be set aside each year is

$$R = 1\ 000\ 000 \times 0.06/((1 + 0.06)^7 - 1)$$
$$= 60\ 000/(0.505) = \text{£118 812/year}$$

or a total of $(7 \times 118\ 812) = \text{£832 872}$ during the depreciation period. Taking the interest on the loan as 8 per cent simple as before,

interest = 80 000 £/year or £560 000 over the 7-year period.

Thus the total repayments amount to £1 392 872 compared to £1 560 000 using methods (i) and (ii)—a saving of £167 128 or £23 875/year, which may be a significant proportion of the cost of product and may even determine whether or not the product can be sold at a profit. As an example of amortization, the manufacturing costs obtained in Example 2-4 will be extended to obtain the cost per unit weight of product.

Example 2-5

What is the total cost of producing unit weight of oxygen in the plant considered in Example 2-4?

The total manufacturing costs were computed as 177 060 £/year [5622 £/Ms].

Capital repayments. The total investment was 382 000 £(1970). Assuming this amount is borrowed at 8 per cent simple interest and the plant is written off after 8 years, then

$$\text{annual interest} = 382\,000 \times 8/100 = £30\,560/\text{year}$$

If money can be invested at $6\frac{1}{2}$ per cent compound interest, then

$$\text{amortization payments} = (382\,000 \times 6.5/100)/((1.065)^8 - 1)$$
$$= 24\,830/0.6538$$
$$= £37\,978/\text{year}$$

Hence, total capital repayments = £68 538/year   [2170 £/Ms]

Taking the overheads as 85 per cent of the labour costs, this is equivalent to

$$17\,706 \times 85/100 = 15\,050 \text{ £/year} \quad [562 \times 85/100 = 477 \text{ £/Ms}]$$

Thus we now have the major cost components involved in producing the oxygen:

manufacturing	177 060 £/year
capital repayments	68 538 £/year
overheads	15 050 £/year
Total	260 648 £/year [8280 £/Ms]

This is equivalent to

$$260\,648/(365 \times 30 \times 2240) = 0.0106 \text{ £/lb}$$
$$[8280/(10^3 \times 0.36 \times 10^3) = 0.0230 \text{ £/kg}]$$

2.7 Profitability and Investment Appraisal

Apart from certain additional costs which are essentially peculiar to a particular company, such as advertising, sales and distribution costs, it is the total cost of product which must be carefully examined to ascertain whether or not the material can be sold at a reasonable and acceptable profit, normally expressed as a percentage of the capital invested in the project. If this is not the case then the plant must be redesigned to make better use of the capital invested, either by way of more efficient use of services and materials or by greater utilization of optimization techniques. If, bearing in mind the time and cost involved in redesign of plant and equipment, the product cost is still unattractive, then it may be that the whole process must be rethought and, in extreme cases, the validity of the entire project re-examined.

One way of evaluating the likely profit margin is to compare the total cost of product with the current market price and if possible the market demand. Where a new product is involved it is necessary to carry out market surveys, and in any event it is important to determine for how long a product can be sold at a satisfactory profit. As discussed previously, this period is often far more significant in assessing capital repayments than the time over which a plant may be operated.

The question of what is an acceptable return on capital is very much dependent on the policy of any one company, and although 20 per cent is a generally accepted minimum, it has been suggested that a large chemical organization will not invest in a new process unless it is possible to sell the product for less than half the current market price.

The viability of investment lies, sad to say, mainly in the hands of the economist and the financial expert, and it is important to realize that an increase of $\frac{1}{2}$ per cent in the bank rate has probably far more effect on the profitability of a project than an increase of, say, 10 per cent in the efficiency of a distillation column. Nevertheless, it is worth considering two basic examples which serve to illustrate the principles involved.

Example 2-6

A plant is to be fabricated from material X, although it has been suggested that either material Y or Z may prove a more attractive investment. If a 20 per cent return on investment is required, which material should be specified?

Material of construction	X	Y	Z
Cost of installed plant ($£I$)	45 000	65 000	60 000
Estimated life of plant (years)	8	12	20
Maintenance costs as a percentage of I ($£$/year)	0.05	0.035	0.025

Assuming that the manufacturing costs and overheads are the same in each case, the important variables are depreciation and maintenance.

For material X,

depreciation = 45 000/8 = £5630/year
maintenance = 0.05 × 45 000 = £2250/year } total = £7880/year

For material Y,

depreciation = 65 000/12 = £5420/year
maintenance = 0.035 × 65 000 = £2275/year } total = £7695/year

For material Z,

depreciation = 60 000/20 = £3000/year
maintenance = 0.025 × 60 000 = £7500/year } total = £4500/year

Thus comparing materials Y and Z with X,

Y: extra profit = (7880 − 7695) = £185/year

extra investment = (65 000 − 45 000) = £20 000

$$\text{return on additional investment} = \frac{185 \times 100}{20\,000} = 1\% \text{ p.a.}$$

Z: extra profit = (7880 − 4500) = £3350/yr

extra investment = (60 000 − 45 000) = £15 000

$$\text{return on additional investment} = \frac{3350 \times 100}{1500} = 22.3\% \text{ p.a.}$$

Thus material Y is quite unrealistic as an alternative to X, and if a return on investment of 20 per cent is acceptable material Z would be specified.

Example 2–7

23 500 lb/hr [2.96 kg/s] of a process fluid is evaporated at 340°F [444 K] in a tubular unit using steam at 250 lb/in² [1720 kN/m²]. Unfortunately, the mode of operation permits the build up of scale inside the tubes equivalent to a resistance of 0.01 hr ft² °F/BTU [0.001 76 m² K/W] and this has to be removed periodically at an annual cost of £8500. It is proposed to retube the exchanger with alloy tubes and it is estimated that due to the resulting reduction in scale formation, equivalent to a resistance of 0.007 hr ft² °F/BTU [0.001 23 m² K/W] the annual cleaning costs will be £1750. If the present unit can be operated for five years, for how long must the modified unit be operated to produce the same annual profit?

Data

latent heat of process fluid = 470 BTU/lb [1093 kJ/kg]

value of process fluid = 0.7 £/1000 lb [1.55 £/Mg]

outside film coefficient (steam) = 1500 BTU/hr ft² °F [8.52 kW/m²K]

inside film coefficient (process fluid) = 250 BTU/hr ft² °F [1.42 kW/m²K]

installed cost of existing evaporator, $I = 44 \, A^{0.8}$ £ [$I = 295 A^{0.8}$ £]

installed cost of modified evaporator, $I = 119 \, A^{0.8}$ £ [$I = 797 \, A^{0.8}$ £]

where A ft² [m²] is the heat transfer area,

fixed charges = 40% of installed cost

cost of steam = £0.75/1000 lb [£1.65/1000 kg]

interest rate on capital = 8%

Existing unit

$$\text{heat load, } Q = 2.35 \times 10^4 \times 4.70 \times 10^2$$
$$= 11.02 \times 10^6 \text{ BTU/hr}$$
$$[= 2.96 \times 1093 = 3235 \text{ kW}]$$

The resistances are

$$1/h_o = 0.000\,67 \qquad\qquad [0.117\,37]$$
$$1/h_i = 0.004\,00 \qquad\qquad [0.704\,23]$$
$$\text{scale} = \underline{0.010\,00} \qquad\qquad [\underline{1.760\,00}]$$
$$1/U = 0.014\,67 \text{ hr ft}^2 \text{ °F/BTU} \qquad [2.582 \text{ m}^2 \text{ K/kW}]$$

Hence $U = 68.2$ BTU/hr ft² °F   [0.387 kW/m² K]

Also

$$\Delta t_m = 401 - 340 = 61°F$$
$$[= 478 - 444 = 34 \text{ K}]$$

and so the heat transfer area,

$$A = 11.02 \times 10^6/(61 \times 68.2) = 2660 \text{ ft}^2$$
$$[= 3235/(34 \times 0.387) = 245.9 \text{ m}^2]$$

and the installed cost of exchanger,

$$I = 44 \times 2660^{0.8}$$
$$[= 295 \times 245.9^{0.8}] = £22\,000$$

Hence

$$\left. \begin{array}{l} \text{depreciation} = 22\,000/5 = 4400 \text{ £/year} \\ \text{interest} = 81/100 = 1760 \text{ £/year} \\ \text{fixed costs} = 401/100 = 8800 \text{ £/year} \end{array} \right\} \text{Total} = 149 \text{ £/year}$$

$$\begin{aligned} \text{cost of steam} &= (11.02 \times 10^6 \times 0.75 \times 24 \times 365)/(825 \times 1000) \\ &\quad [= (3235 \times 1.65 \times 3600 \times 24 \times 365)/(1920 \times 1000)] \\ &= 87\,800 \text{ £/year} \end{aligned}$$

$$\begin{aligned} \text{value of product} &= (23\,500 \times 24 \times 365 \times 0.7)/1000 \\ &\quad [= (2.95 \times 3600 \times 24 \times 365 \times 1.55)/1000] \\ &= 144\,900 \text{ £/year} \end{aligned}$$

$$\begin{aligned} \text{annual profit} &= 144\,900 - (87\,800 + 8500 + 14\,960) \\ &= 33\,640 \text{ £/year} \end{aligned}$$

With modified tubing

The resistances are

$$\begin{array}{ll} 1/h_o = 0.000\,67 & [0.117\,37] \\ 1/h_i = 0.004\,00 & [0.704\,23] \\ \text{scale} = 0.007\,00 & [1.230\,00] \\ \hline 1/U = 0.011\,67 \text{ hr ft}^2 \text{ °F/BTU} & [2.051 \text{ m}^2 \text{ K/kW}] \end{array}$$

Hence $\quad U = 85.7 \text{ BTU/hr ft}^2 \text{ °F} \qquad [0.488 \text{ kW/m}^2 \text{ K}]$

$$\begin{aligned} \text{heat load, } Q &= 85.7 \times 2660 \times 61 = 1.39 \times 10^7 \text{ BTU/hr} \\ &\quad [= 0.488 \times 245.9 \times 34 = 4080 \text{ kW}] \end{aligned}$$

$$\begin{aligned} \text{cost of steam} &= (1.39 \times 10^7 \times 0.75 \times 24 \times 365)/(825 \times 1000) \\ &\quad [= (4080 \times 1.65 \times 3600 \times 24 \times 365)/(1920 \times 1000)] \\ &= 110\,900 \text{ £/year} \end{aligned}$$

$$\begin{aligned} \text{value of product} &= (1.39 \times 10^7 \times 24 \times 365 \times 0.7)/(470 \times 1000) \\ &\quad [= (4080 \times 3600 \times 24 \times 365 \times 1.55)/(1093 \times 1000)] \\ &= 182\,200 \text{ £/year} \end{aligned}$$

The installed cost of the exchanger,

$$I = 119 \times 2660^{0.8}$$
$$[= 797 \times 245.9^{0.8}] = £59\,500$$

Hence depreciation $= 59\,500/n$ £/year

$$\text{interest} = 8I/100 = 4760 \text{ £/year}$$

$$\text{fixed costs} = 0.4I/100 = 23\,800 \text{ £/year}$$

annual profit, $33\,640 = 182\,200 - (110\,900 + 1750$

$$+ 59\,500/n + 4760 + 23\,800)$$

and so $\qquad n = 8.1$ years

A value of 20 per cent has been suggested as the minimum acceptable return on capital, and yet in theory this value (r_{min}) should be the highest rate of interest which could be earned by the capital if invested elsewhere. If on costing a process, the calculated rate of return is less than this rate of interest, it is clearly not worthwhile proceeding with the venture but more profitable to invest for the higher return elsewhere.

In practice the value of r_{min} is somewhat higher than the rate obtainable elsewhere as it incorporates a factor to allow for the effects of risk and inflation. Since the mid 1960s inflation has been proceeding at an ever increasing rate, and whilst a value of r_{min} of 10 per cent a few years ago was acceptable, as mentioned previously, the corresponding value in 1971 is more likely to lie in the range of 15–20 per cent.

In the case of selection of a piece of equipment the simple 'cost and return' basis is used for comparison:

$$O_1 + r_{min} \cdot C_1 \leqslant O_2 + r_{min} C_2 \qquad (2\text{--}6)$$

where $\qquad O_1$, O_2 are the operating costs of items 1 and 2,

$\qquad\qquad C_1$, C_2 are the capital costs of items 1 and 2,

$\qquad\qquad r_{min}$ is the minimum acceptable rate of return.

Equation (2–6) as written for the case where item 1 is economically preferable to item 2 and the example shown below should serve to illustrate the method.

Example 2–8

Considering that four pieces of equipment A, B, C and D are capable of performing the same duty and given their capital and operating costs as in the table below, show how variation of the value of r_{min} affects the choice of an economic selection.

TABLE 2-7. DATA FOR ECONOMIC SELECTION (EXAMPLE 2–8)

Item	Capital cost	Power cost	Maintenance	Total operating costs
A	£10 000	£800	£1200	£2000
B	£15 000	£300	£500	£800
C	£20 000	£100	£100	£200
D	£25 000	£50	£50	£100

If values of r_{min} are selected as 10, 15 and 20 per cent, Table 2–8 may then be calculated.

TABLE 2–8. VALUES OF r_{min} (EXAMPLE 2–8)

Item	$r_{min} = 10\%$		$r_{min} = 15\%$		$r_{min} = 20\%$	
	$r_{min}C$	$O + r_{min}C$	$r_{min}C$	$O + r_{min}C$	$r_{min}C$	$O + r_{min}C$
A	£1000	£3000	£1500	£3500	£2000	£4000
B	£1500	£2300	£2250	£3050	£3000	£3800
C	£2000	£2200	£3000	£3200	£4000	£4200
D	£2500	£2600	£3750	£3850	£5000	£5100

If $r_{min} = 10$ per cent, item C has clearly the lowest total cost on this simple 'cost + return' basis. If, however, r_{min} increases above 12 per cent, the choice changes from C to B and item B remains the economic favourite to the chosen limit of 20 per cent. If r_{min} were to increase further, item A would eventually become the choice at $r_{min} = 23.5$ per cent.

Consider now the example of a plant costing £800 000 to build which gives a product X. For the same capital outlay a different plant can be erected to produce an alternative product Y. Conditions are such that each plant will only be run for five years and then both will be scrapped. The profits in each of the five years obtained from selling X and Y are shown in Table 2–9 and a choice is to be made as to which plant should be built.

TABLE 2–9. ANNUAL PROFITS

Year	1	2	3	4	5	TOTAL
Product X	100 000	200 000	300 000	400 000	500 000	£1 500 000
Product Y	500 000	400 000	300 000	200 000	100 000	£1 500 000

In this example, the capital outlay, total profit and peak year profit are all equal, the difference being in the years in which the profits were made. In practice, the case of product Y would always be chosen because predictions of early year's profits are most likely to be accurate. Also this is when the peak profits arise and the effects of inflation are more predictable in the near future. If the cash flows are plotted as in Figure 2–5, it will be seen that the pay-off time for product Y is under two years but for product X is $3\frac{1}{2}$ years.

Each project can be considered to give a certain return in each year, allowing 20 per cent depreciation in each case. Thus in the first year:

for product Y, return $= \dfrac{\text{profit} - \text{depreciation}}{\text{total capital}}$

$$= \left(\frac{500\,000 - 160\,000}{800\,000} \right) \times 100 = 42.5\%$$

for product X, return $= \left(\dfrac{100\,000 - 160\,000}{800\,000} \right) \times 100 = -7.5\%$

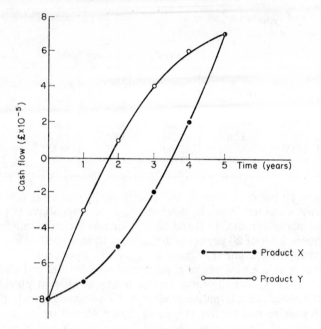

FIGURE 2–5. Cumulative cash flow diagram

In the second year:

$$\text{for Y, return} = \left(\frac{400\ 000 - 160\ 000}{800\ 000}\right) \times 100 = 30\%$$

$$\text{for X, return} = \left(\frac{200\ 000 - 160\ 000}{800\ 000}\right) \times 100 = 5\%$$

Similarly for the remaining years the returns are 17.5 per cent, 5 per cent and −7.5 per cent for Y and 17.5 per cent, 30 per cent and 42.5 per cent for X. On this basis projects are similar.

The average return on the average capital employed is

$$\frac{1000\ \{(500 - 160) + (400 - 160) + (300 - 160) + (200 - 160) + (100 - 160)\}/5}{1000\ \{800 + (800 - 160) + (640 - 160) + (480 - 160) + (320 + 160)\}/5}$$

$$= 29.1\% \text{ for both processes}$$

Another way of looking at the two projects is to say that each requires a loan of £800 000 to be repaid with interest at an unknown rate r at the end of each year by means of the profits. This concept is illustrated in Figure 2–6 opposite and forms the fundamental basis of discounted cash flow.

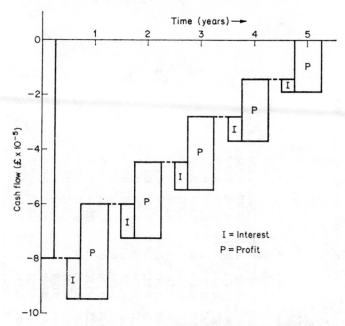

FIGURE 2-6. Diagrammatic representation of discounted cash flow

Consider Product Y. At the end of year 1,

$$\text{interest} = £800\,000r$$
$$\text{original capital} = £800\,000$$
$$\text{profit} = £500\,000$$

The amount outstanding at the end of year 1 is therefore $= 800\,000$ $(1 + r) - 500\,000$.

In a similar way, at the end of year 2 the amount outstanding is

$$800\,000(1 + r)^2 - 500\,000(1 + r) - 400\,000$$

Repeating this procedure to the end of year 5 when the project ends, and since the amount outstanding is to be zero, the following equation may be derived:

$$0 = 800\,000(1 + r)^5 - 500\,000(1 + r)^4 - 400\,000(1 + r)^3$$
$$- 300\,000(1 + r)^2 - 200\,000(1 + r) - 100\,000$$

or

$$0 = -800\,000 + \frac{500\,000}{(1 + r)} + \frac{400\,000}{(1 + r)^2} + \frac{300\,000}{(1 + r)^3} + \frac{200\,000}{(1 + r)^4} + \frac{100\,000}{(1 + r)^5}$$

$$(2\text{-}7)$$

from which the unknown rate of interest r may be calculated as approximately 35 per cent.

TABLE 2-10. DISCOUNT FACTORS. THIS TABLE SHOWS THE PRESENT VALUE OF UNITY DISCOUNTED FOR DIFFERENT NUMBERS OF YEARS AND AT DIFFERENT RATES OF DISCOUNT

Year	Rate of discount 5%	10%	12%	14%	15%	16%	17%	18%	19%	20%
0	1.000	1.000	1.000	1.000	1.000	1.000	1.000	1.000	1.000	1.000
1	0.952	0.909	0.893	0.877	0.870	0.862	0.855	0.847	0.840	0.833
2	0.907	0.826	0.797	0.769	0.756	0.743	0.731	0.718	0.706	0.694
3	0.864	0.751	0.712	0.675	0.658	0.641	0.624	0.609	0.593	0.579
4	0.823	0.683	0.636	0.592	0.572	0.552	0.534	0.516	0.499	0.482
5	0.784	0.621	0.567	0.519	0.497	0.476	0.456	0.437	0.419	0.402
6	0.746	0.564	0.507	0.456	0.432	0.410	0.390	0.370	0.352	0.335
7	0.711	0.513	0.452	0.400	0.376	0.354	0.333	0.314	0.296	0.279
8	0.677	0.457	0.404	0.351	0.327	0.305	0.285	0.266	0.249	0.233
9	0.645	0.424	0.361	0.308	0.284	0.263	0.243	0.225	0.209	0.194
10	0.614	0.386	0.322	0.270	0.247	0.227	0.208	0.191	0.176	0.162
11	0.585	0.350	0.287	0.237	0.215	0.195	0.178	0.162	0.148	0.135
12	0.557	0.319	0.257	0.208	0.187	0.168	0.152	0.137	0.124	0.112
13	0.530	0.290	0.229	0.182	0.163	0.145	0.130	0.116	0.104	0.093
14	0.505	0.263	0.205	0.160	0.141	0.125	0.111	0.099	0.083	0.078
15	0.481	0.239	0.183	0.140	0.123	0.108	0.095	0.084	0.074	0.065
16	0.458	0.218	0.163	0.123	0.107	0.093	0.081	0.071	0.062	0.054
17	0.436	0.198	0.146	0.103	0.093	0.080	0.069	0.060	0.052	0.045
18	0.416	0.180	0.130	0.095	0.081	0.069	0.059	0.051	0.043	0.038
19	0.396	0.164	0.116	0.083	0.070	0.060	0.051	0.043	0.037	0.031
20	0.377	0.149	0.104	0.073	0.061	0.051	0.043	0.037	0.031	0.026

In an analogous manner for product X, the following equation may be derived:

$$0 = -800\,000 + \frac{100\,000}{(1 + r)} + \frac{200\,000}{(1 + r)^2} + \frac{300\,000}{(1 + r)^3} + \frac{400\,000}{(1 + r)^4} + \frac{500\,000}{(1 + r)^5}$$

when the rate of interest is found to be approximately 19 per cent.

This technique demonstrates clearly the importance of the years in which the profits were earned and shows the desirability of early returns in preference to those accruing later. The factors $1/(1 + r)$, $1/(1 + r)^2$, $1/(1 + r)^3$, etc. are known as discount factors, and are tabulated in Table 2–10 to facilitate discounted cash flow calculations.

To avoid the necessity of solving an nth order equation arising from discounting over n years, it is common practice to insert the appropriate value of r_{min} in the right-hand side of equation (2–7) and calculate the value of the resulting expression. The resulting figure is known as the 'present worth' or 'net present value' (N.P.V.) of the project, and it is used to select quickly and simply between alternatives.

If $r_{min} = 20\%$, for product Y,

present worth

$$= -800\,000 + \frac{500\,000}{1.20} + \frac{400\,000}{(1.20)^2} + \frac{300\,000}{(1.20)^3} + \frac{200\,000}{(1.20)^4} + \frac{100\,000}{(1.20)^5}$$

$$= +£203\,000$$

Similarly for product X, present worth $= -£12\,700$.

To select between projects it is only necessary to decide upon the minimum acceptable rate of return and calculate the present worth of each project, the one with the highest resulting value being chosen, i.e. the process for product Y.

Whilst this treatment of the technique has been elementary, the procedure is similar for more complex situations. It is essential that *all* receipts and costs resulting from the project should be considered. The cash flow in any year will equal the earnings, i.e. profit before depreciation plus any cash receipts, including grants, less capital expenditure and tax payable. It is likely that estimated earnings will fall over the life of the project as efficiency and certainty of estimation will tend to reduce with time. The next example takes into account the effect of a government grant and taxation.

Example 2–9

It is proposed to build a small plant on an existing site in a development area. The capital cost of the project is £5000 and the level of available grant aid is 45 per cent. The plant would be run for ten years and then scrapped to realize 5 per cent of its original value. If the minimum acceptable return on capital is 20 per cent, should the project proceed if the estimated gross profits are as shown in column D of Table 2–11?

Table 2–11 shows how the project is assessed on discounted cash flow (D.C.F.) techniques. The capital expenditure is made in years 0 and 1 and grant aid is received in the following years. Gross profits build up to a

TABLE 2–11. EXAMPLE OF A D.C.F. CALCULATION

A Year	B Capital expenditure	C Development grant	D Gross profit	E Tax allowances	F Taxable profit D − E	G Tax payable 0.40F	H Cash flow (B+C+D) − G	J Discount factors 15%	K Net present worth 15% J × H	L Discount factors 20%	M Net present worth 20% L × H
0	−3000	—	—	—	—	—	−3000	1.000	−3000	1.000	−3000
1	−2000	1350	1000	1000	0	—	+350	0.870	+304	0.833	+291
2	—	900	1500	800	700	0	2400	0.756	1820	0.694	1670
3	—	—	2000	640	1360	290	1710	0.658	1130	0.579	990
4	—	—	1800	512	1268	544	1256	0.572	722	0.482	605
5	—	—	1500	410	1090	514	986	0.497	490	0.402	396
6	—	—	1200	328	872	436	764	0.432	330	0.335	256
7	—	—	1100	262	838	349	751	0.376	282	0.279	210
8	—	—	1000	210	790	335	665	0.327	208	0.233	155
9	—	—	500	168	332	316	184	0.284	52	0.194	36
10	—	—	200	134	66	133	67	0.247	17	0.162	11
11	+250	—	—	286	−286	26	224	0.215	48	0.135	30
12	—	—	—	—	—	−114	114	0.187	21	0.112	13
								TOTAL	£2424	TOTAL	£1663

peak in year 3 and then fall. Tax allowances of 20 per cent of the initial capital cost in year 1 are permitted and these are reduced in successive years to 80 per cent of the allowance in the previous year. The tax allowance in year 11 represents a balancing charge between the net total capital expenditure from column B and the sum of the tax allowances claimed for the years 1 to 10 inclusive. Tax is payable a year after it was earned and the current (1971) rate of 40 per cent is assumed. Hence, the cash flows and the net present values may be calculated for rates of return equal to 15 per cent and 20 per cent. The value of the N.P.V. at 20 per cent is +£1663 which shows that this would be an excellent investment and should be started as soon as possible.

This treatment of D.C.F. techniques has been, of necessity, brief. For a fuller account reference may be made to more comprehensive, specialized texts.[10]

2.8 Optimization

Optimization is, in essence, design in order to attain maximum profit by way of the most effective utilization of capital. With most items of plant this operation may be relatively straightforward, though where a complete plant unit is involved the problem is far more complex. To take a very simple example, consider the transport of fluid along a pipeline. It is fairly obvious that the bigger the pipe diameter, the greater is the cost per unit length, and hence from the point of view of capital charges the smallest diameter possible would be specified. However, with reduction in diameter for a given throughput, the pressure drop, and hence the operating costs, are increased and a larger diameter would be specified. Between these two extremes there is obviously some value of the pipe diameter at which the total costs, i.e. capital plus operating costs are a minimum value and this is the optimum pipe diameter. Similar, though obviously more complex considerations, may be applied to other plant items and several calculations of this type are included in the appropriate sections of this book.

Although in a plant each item of equipment may have been optimized along the lines suggested above, it does not follow that the complete plant makes the best use of capital, nor is it designed for the most economic and efficient operation. Many other factors such as heat recovery, economics in utilization of services and recycling of process streams must be considered, and perhaps a complete new set of process conditions evolved. This topic merits a detailed treatment far beyond the scope of this book, and indeed it is in many ways impossible to generalize in such a complex problem.

Equally as important as optimization of plant items and process variables is the way in which a plant is operated especially where the efficiency of a particular operation is a function of time. For example, for how long should an evaporator be run before the tubes are descaled, and when should a catalyst be renewed? Obviously this is again a complex question and very much dependent on the particular plant and to some extent on company policy. The following examples are intended as an introduction to this topic.

Example 2–10

Sea water is pumped through a 200 ft [61 m] line at 70 000 gal/hr [0.088 m^3/s], the power requirements being given by

$$w = 0.09 \, V^3/d^5 \text{ hp } [= 6.89 \, V^3/d^5 \text{ kW}]$$

where V = volumetric flow rate (ft^3/s)[m^3/s],

d = pipe diameter (ft) [m].

What is the optimum pipe diameter using mild steel and stainless steel and which material would be specified if a 20 per cent return on capital is acceptable?

Data

	mild steel	*stainless steel*
cost of piping (£/ft) [£/m]	$1.2d^{0.6}$ [$8.04d^{0.6}$]	$2.2d^{0.6}$ [$14.7d^{0.6}$]
length of service (years)	4	10
purchase cost of piping (£)	a	b
installed cost (£)	$a + 1500$	$b + 3000$
maintenance cost (£/year)	a	$0.06b$
cost of electricity (£/kWh)	0.005	0.005

(i) *Optimum diameter for mild steel*

profit = value of product − (operating costs + fixed costs + overheads)
or $p = v - (o + f + e) = v - c'$.

Value of product, v: v is the value of 70 000 gal/hr [0.088 m^3/s] sea water which is constant and quite independent of the pipe diameter, d.

Operating costs, o:

$$\text{pumping costs} = (0.08 \times 3.12^3/d^5) \times 24 \times 365 \times 0.005 \times 0.745$$
$$[= (6.89 \times 0.088^3/d^5) \times 24 \times 365 \times 0.005]$$
$$= 79.6/d^5 \text{ £/year } [0.207/d^5 \text{ £/year}]$$
$$\text{maintenance costs} = 1.2d^{0.6} \times 200 = 240d^{0.6} \text{ £/year}$$
$$[= 61 \times 8.04d^{0.6} = 490d^{0.6} \text{ £/year}]$$
$$\text{total operating costs, } o = 240d^{0.6} + 79.6/d^5 \text{ £/year}$$
$$[= 490d^{0.6} + 0.207/d^5 \text{ £/year}]$$

Fixed costs, f:

$$\text{installed cost} = 240d^{0.6} + 1500 \quad [= 490d^{0.6} + 1500]$$
$$\text{depreciation} f = 60d^{0.6} + 375 \text{ £/yr } [= 123d^{0.6} + 375 \text{ £/yr}]$$

Overheads, e: these are constant and independent of the pipe diameter, d. Hence

$$\text{total costs, } c' = 79.6/d^5 + 300d^{0.6} + 375 \text{ £/year} \qquad (2\text{–}8)$$
$$[= 0.207/d^5 + 613d^{0.6} + 375 \text{ £/year}]$$
$$\text{total investment, } I = 240d^{0.6} + 1500 \text{ £} \qquad (2\text{–}9)$$
$$[= 490d^{0.6} + 1500 \text{ £}]$$

The optimum pipe diameter is that at which the return on capital invested is 20 per cent, i.e. when $dp/dI = 0.20$. Now

$$p = v - c'$$

thus

$$\frac{dp}{dI} = \frac{dv}{dI} - \frac{dc'}{dI}$$

and as v is independent of d, hence $dv/dI = 0$ and

$$\frac{dp}{dI} = 0 - \frac{dc'}{dd} \cdot \frac{dd}{dI}$$

or

$$-\frac{dc'}{dd} \cdot \frac{dd}{dI} = 0.20 \qquad (2\text{--}10)$$

From (2–8),

$$\frac{dc'}{dd} = -\frac{398}{d^6} + \frac{180}{d^{0.4}} \left[= -\frac{1.04}{d^6} + \frac{368}{d^{0.4}} \right]$$

From (2–9)

$$\frac{dI}{dd} = \frac{144}{d^{0.4}} \left[= \frac{294}{d^{0.4}} \right]$$

Therefore substituting in (2–10),

$$\left(\frac{398}{d^6} - \frac{180}{d^{0.4}} \right) \left(\frac{d^{0.4}}{144} \right) = 0.20$$

$$\left[\left(\frac{1.04}{d^6} - \frac{368}{d^{0.4}} \right) \left(\frac{d^{0.4}}{294} \right) = 0.20 \right]$$

from which $d = 1.122$ ft [0.341 m].

(ii) *Optimum diameter for stainless steel*

This calculation is made in exactly the same way as for mild steel (and provides a useful exercise for the student) giving an optimum pipe diameter, $d = 1.29$ ft [0.392 m]. The data is summarized in the following section.

(iii) *Comparison of materials*

	mild steel	stainless steel
optimum pipe diameter (ft) [m]	1.122 [0.341]	1.290 [0.392]
purchase cost of pipe (£)	257	472
installed cost of pipe (£)	1757	3472
running costs:		
maintenance (£/year)	257	28
depreciation (£/year)	439	347
power (£/year)	45	22
	741	397

Thus, comparing stainless steel with mild steel,

extra profit from using stainless steel = (741 − 397) = £344/year

additional investment required = (3472 − 1757) = £1715

which represents a return of 344 × 100/1715 = 20.1 per cent per annum. This is just beyond the accepted minimum and hence stainless steel would be specified.

Example 2–11

A chemical Y is produced in a process involving the vapour phase catalytic oxidation of a feedstock X. The performance of the catalyst decreases with time and periodically the plant is shut down and the catalyst is reactivated by steam—an operation which takes 24 hours [86.4 ks]. When in production the consumption of X is 1000 lb/hr [0.126 kg/s] and any unreacted material is sent to waste. What is the optimum operating period between shutdowns?

Data

cost of X = 200 £/ton [197 £/1000 kg]

value of Y = 270 £/ton X converted [266 £/1000 kg X converted]

cost of reactivation: steam £15, labour £120, fixed costs £85.

Catalyst activity:

time (days) [ks]	0 0	10 864	20 1728	30 2592	40 3456
Percentage conversion of X	99.4	98.9	98.2	97.6	97.0

Assuming that the costs given are the only ones affecting the calculation, then

profit = value of product (v) − {cost of feed (F') + cost of shutdown (s)}

Considering these in turn,

Value of product, v

During the operating period, $24 \times 1000nz/2240$ tons of X are consumed [$126nz$ kg] where n days [ks] is the length of the operating period, z is the *mean* conversion of X → Y over n days [ks]. Thus

$$\text{value of product}, v = 270 \times 1000 \times 24nz/2240 \text{ £}$$
$$[= 266 \times 126nz/1000 \text{ £}]$$

The total period of the cycle is $(n + 1)$ days [$n + 86.4$ ks] and for the length of the cycle,

$$v = 270 \times 1000 \times 24nz/\{2240(n + 1)\} = 2.89 \times 10^3 nz/(n + 1)\text{£/day}$$
$$[= 266 \times 126nz/\{1000(n + 86.4)\} = 33.5nz/(n + 86.4)\text{£/ks}]$$

Cost of feed, F'

Cost of feed

$$24 \times 1000n \times 200/2240 \text{ £ } [= 126n \times 197/1000 \text{ £}]$$

or $F' = 24 \times 1000n \times 200/\{2240(n + 1)\} = 2.14 \times 10^3 \, n/(n + 1) \text{ £/day}$

$$[= 126n \times 197/\{1000(n + 86.4)\} = 24.8n/(n + 86.4) \text{ £/ks}]$$

Cost of shutdown, s

Total cost = £(15 + 120 + 85) = £220

Hence $s = 220/(n + 1)$ £/day

$$[= 220/(n + 86.4) \text{ £/ks}]$$

Profit, P

The profit over the period of the cycle is given by

$P = 2.89 \times 10^3 nz/(n + 1) - \{2.14 \times 10^3 n/(n + 1) + 220/(n + 1)\}$ £/day

$[= 33.5nz/(n + 86.4) - \{24.8n/(n + 86.4) + 220/(n + 86.4)\}$ £/ks]

$= (10^3 n (2.89z - 2.14) - 220)/(n + 1)$ £/day

$[= (n(33.5z - 24.8) - 220)/(n + 86.4)$ £/ks]

This equation may be used to calculate p as a function of n as shown in Table 2–12. The data are plotted in Figure 2–7, from which the maximum profit is seen to be obtained when the operating period is 31 days [2680 ks].

TABLE 2–12. CALCULATION OF OPERATING PERIOD

Operating period n (days)	[ks]	Mean % conversion over n days [ks]^z	Profit p (£/day)	[£/ks]
10	864	0.9915	639.2	7.40
20	1728	0.989	673.3	7.79
30	2592	0.985	677.1	7.83
40	3456	0.982	675.6	7.81

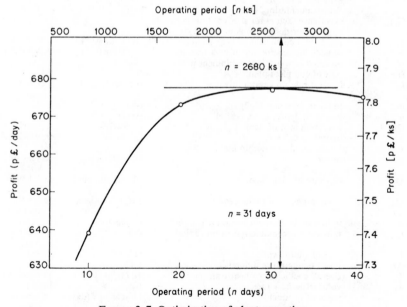

FIGURE 2–7. Optimization of plant operation

Optimization brings to bear the skills of both the engineer and to some extent the economist on the complexities of plant design and the importance of accuracy, and the return on the effort employed cannot be stressed too highly. It has been possible here only to introduce the topic and to illustrate one or two basic ideas, though further examples of the optimization of equipment items are included in other chapters. It has been suggested that the optimization of complete plants is of a complex nature, but it should be noted that profitability equations of the types used in Examples 2-10 and 2-11 provide a sound basis for tackling such problems and the advent of the computer has rendered this form of cost optimization commonplace throughout the design and contracting organizations.

2.9 Nomenclature

A	heat transfer area	(ft^2)	[m^2]
a	purchase cost of mild steel piping	(£)	[£]
b	purchase cost of stainless steel piping	(£)	[£]
C	cost of plant item	(£)	[£]
C_1, C_2	cost of complete plant of capacity Q_1, Q_2	(£)	[£]
C_b	cost of plant item of capacity Q_b	(£)	[£]
c	cost of material of construction	(£/lb)	[£/kg]
c'	total cost of production	(£/year)	[£/year or £/ks]
d	nominal pipe diameter	(ft)	[m]
e	cost of overheads	(£/year)	[£/year]
F	factor in equation (2-1)	(—)	[—]
F'	cost of feedstock	(£/day)	[£/ks]
f	fixed costs	(£/year)	[£/year]
f_1, f_2	experience factors for piping, instrumentation, etc.	(—)	[—]
f'_1, f'_2	experience factors for size, contingencies, etc	(—)	[—]
f_L	Lang factor defined in equation (2-5)	(—)	[—]
h_i	inside film coefficient of heat transfer	(BTU/hr ft^2 °F)	[kW/m^2 K]
h_o	outside film coefficient of heat transfer	(BTU/hr ft^2 °F)	[kW/m^2 K]
I_E	installed cost of plant equipment items	(£)	[£]
I_F	cost of complete plant	(£)	[£]
i	interest rate on capital invested	(per cent/year)	[per cent/year]
m	index in equation (2-2)	(—)	[—]
n	length of depreciation period	(year)	[year]
O_1, O_2	operating cost of items 1, 2, etc.	(£/year)	[£/year]
o	operating costs	(£/year)	[£/year]
P	total capital invested in plant	(£)	[£]
p	profit	(£/year)	[£/year or £/ks]
Q	capacity of plant item	(ft^2, hp, lb, etc.)	[m^2, kW, kg, etc.]
Q_o	basic capacity of plant item	(ft^2, hp, lb, etc.)	[m^2, kW, kg, etc.)
R	minimum investment for amortization	(£/year)	[£/year]
r_{min}	minimum acceptable return on capital	(per cent/year)	[per cent/year]
s	cost of shutdown	(£/day)	[£/ks]
Δt_m	mean temperature difference	(°F)	[K]
U	overall coefficient of heat transfer	(BTU/hr ft^2 °F)	[kW/m^2 K]
V	volumetric flow rate	(ft^3/s)	[m^3/s]
v	value of product	(£/year or £/day)	[£/year or £/ks]
W	weight of plant item	(lb)	[kg]
w	power required for pump	(hp)	[kW]
x	given number of years within the depreciation period	(years)	[years]
z	mean conversion efficiency of catalyst during n days	(—)	[—]

REFERENCES

1. Baumann, H. C. *Fundamentals of Cost Engineering in the Chemical Industry.* New York: Reinhold, 1964.
2. Chilton, C. H. *Cost Engineering in the Process Industries,* New York: McGraw-Hill Book Co., 1960.
3. Aries, R. E. and Newton, R. D. *Chemical Engineering Cost Estimation.* New York: McGraw-Hill Book Co., 1955.
3a. Guthrie, K. M. *Chem. Eng.*, 24 March 1969, **76**, 114.
4. Hackney, J. W. *Control and Management of Capital Projects.* New York: Wiley, 1965.
5. *A Guide to Capital Cost Estimation.* London: The Institution of Chemical Engineers, 1970.
6. Anon. *Chem. Proc. Eng.* 1964, **45**, 152.
7. Anon. *Chem. Proc. Eng.* 1970, **51**, 59.
8. Baumann, H. C., Hirsh, J. H. and Glazier, E. M. *Chem. Eng.* 1960, **56**, 37.
9. Waddell, R. M. *Chem. Eng. Prog.* 1961, **57**, 51.
10. Alfred, A. M. and Evans, J. B. *Discounted Cash Flow.* London: Chapman and Hall, 1967.

Chapter 3: Heat Exchange Equipment

3.1 Introduction

Apart from perhaps fluid flow and the transport of process materials, the transfer of heat between two process streams is the most commonly encountered operation in process plant design. Indeed one can think of very few processes in which heat, or for that matter cold, is not transferred at some time or other. In addition to the heat evolved or absorbed when a chemical reaction is carried out, the majority of unit operations involve the transfer of heat in some way—an operation which is carried out in a variety of equipment designs, some involving direct contact and others indirect. Drying, for example, may be effected by direct heat transfer from hot flue gases to the process material, whereas in a pipe-furnace on a crude distillation unit, heat is transferred from the hot flame gases mainly by radiation to the process material, which is contained within pipes. This chapter is mainly concerned with equipment in which heat is transferred primarily by convection from one fluid to another; the fluids being separated by a wall through which the heat is transferred. Such equipment takes many forms, of which the shell and tube type is the most common. A shell and tube unit is generally known as a heat exchanger if neither fluid evaporates or condenses. A condenser (or heater) is a shell and tube unit in which one fluid condenses, and an evaporator (or cooler) is a unit in which one fluid is evaporated. All three units are basically similar in design.

The design of these types of unit may be undertaken in four basic stages:

(i)   The approximate sizing of the unit is estimated, based on an assumed overall coefficient of heat transfer.

(ii) For the preliminary configuration, an accurate calculation is made of the film coefficients and hence the design overall coefficient. This should correspond to the assumed value in (i). If this is not the case, a further value is assumed and the calculations repeated.

(iii) From the design coefficient, the area for heat transfer is computed and calculations are then made to decide the exact number and layout of tubes, baffles and so on. Such calculations are usually based on optimization procedures taking into account factors such as water velocity and optimum outlet temperature for example.

(iv) Once the design has been completed and the sizes of all components obtained, the mechanical design of the unit may be carried out and hence the final specification drawn up.

The stages listed will be illustrated by worked examples later in this chapter, which is concerned in the main with step (iii) of the above procedure. In step (i), the initial estimate of the likely overall coefficient is largely a matter of experience, though some guide is given by Table 3–1,

TABLE 3–1. APPROXIMATE FILM COEFFICIENTS[1]

	Film coefficient h_i or h_o	
	(BTU/ hr ft² °F)	[W/m² K]
No change of state		
water	300–2000	1700–11 400
gases	3–50	17–280
organic solvents	60–500	340–2800
oils	10–120	57–680
Condensation		
steam	1000–3000	5700–17 000
organic solvents	150–500	850–2800
light oils	200–400	1140–2280
heavy oils	20–50	114–280
ammonia	500–1000	2800–5700
Evaporation		
water	800–2000	4540–11 400
organic solvents	100–300	570–1700
ammonia	200–400	1140–2280
light oils	150–300	850–1700
heavy oils	10–50	57–280

which lists approximate film coefficients.[1] The value of the approximate overall coefficient is then given by

$$U = 0.7 \, h_i h_o/(h_i + h_o) \qquad (3\text{–}1)$$

Example 3–1

What is the estimated value of the overall coefficient of heat transfer when acetone is condensed on the outside of tubes through which cooling water is passed in a shell and tube unit?

From Table 3–1,

$$h_i = 1000 \, [5700]$$

$$h_o = 250 \, [1420]$$

and hence in equation (3–1),

$$U = 0.7 \times 1000 \times 250/(1000 + 250) = 140 \text{ BTU/hr ft}^2 \text{ °F}$$

$$[= 0.7 \times 5700 \times 1420/(5700 + 1420) = 795 \text{ W/m}^2 \text{ K}]$$

The relationships and methods of calculating film coefficients in stage (ii) are well documented in the standard texts on the subject and will not be discussed here.[2]–[4] Special mention is made of Holland's book, *Heat Transfer*, which is uniform with the present volume.[5] The main part of this chapter is devoted to stage (iii) of the design calculation, and stage (iv), mechanical design, is considered in some detail in Chapter 7. Heat exchangers of the shell and tube type are designed according to the various engineering codes of practice and these should be consulted especially in connection with stage (iv) of the calculation.[6]–[7]

3.2 General Description of Shell and Tube Units

Of the many varied types of shell and tube exchangers, three important examples are shown in Figure 3–1, which is based on B.S. 1500 [8]. In (a) the unit has fixed tube plates, which is a cheap and relatively simple form of construction. It is virtually impossible, however, to clean the outside of the tubes and this type of unit is limited to applications where the shell

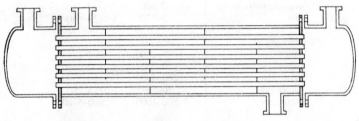

(a) Fixed tube plate exchanger

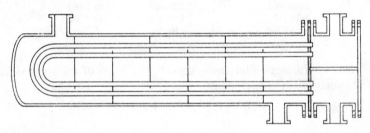

(b) U–tube exchanger

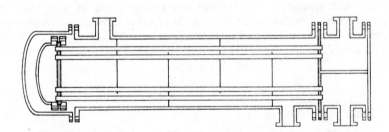

(c) Floating head type exchanger

FIGURE 3–1. Main types of shell and tube heat exchangers

side fluid is relatively clean. Figure 3–1(b) shows the U-tube form of construction. This again is relatively simple and the design allows for differential thermal expansion between the shell and the tubes. The entire tube bundle is easily removed and hence cleaning the outside of the tubes is quite straightforward. Because of the bend it is difficult to clean the inside of the tubes, and this type of unit is limited to situations where the tube side fluid is clean. The floating head design shown in Figure 3–1(c) also allows for differential expansion and both the shell and tubes are readily accessible for cleaning, although the construction is relatively expensive. The floating head type of exchanger is used widely throughout the chemical and petroleum industries and the component parts of a two tube-side pass, single shell-side pass are shown in Figure 3–2. This is

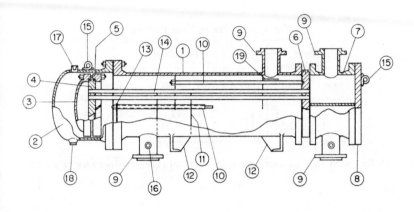

1. Shell	8. Channel cover	14. Tube
2. Shell cover	9. Branch	15. Lifting lug
3. Floating head cover	10. Tie rod and spacer	16. Test connection
4. Floating tube plate	11. Cross baffle or	17. Vent connection
5. Clamp ring or hook bolt	tube support plate	18. Drain connection
6. Fixed tube plate	12. Saddle	19. Impingement plate
7. Channel	13. Floating head support	

FIGURE 3–2. Component parts of a shell and tube heat exchanger—Floating head type[6]

largely self explanatory, though the following notes on the important components are included as a guide to layout detail:

(i) *Tube Bundle*

The tubes are usually expanded into the tube plate, thus providing a seal against leakage under normal operating conditions. The bundle consists of the tubes, the tube plates (and floating head) together with baffles, tube support plates, tie rods and spacers. The layout of holes in the tube plate must be such as to permit easy cleaning both inside and outside the tubes. The tube pitch is the minimum centre-to-centre distance between adjacent tubes, and either square or triangular pitch is employed. The former enables the tubes (on the shell side) to be cleaned more easily and the pressure drop is less on the shell side, when the fluid flows at right angles to the tube axis. Triangular layout gives a larger area of heat transfer per unit volume of exchanger. Typical pitches for various tube sizes are shown in Table 3–2. Tubes are not laid out symmetrically and it is usual to omit tubes directly under the shell inlet nozzle to minimize the contraction effect. In condensers it is necessary to fit impingement plates under the shell nozzle and space must be left for this purpose. The tube count is the total number of tubes in the tube layout. Further consideration of tube layout is given in Chapter 7.

TABLE 3–2. TYPICAL TUBE PITCHES

Tube outer diameter		Tube Pitch			
		Square		Triangular	
(in.)	[cm]	(in.)	[cm]	(in.)	[cm]
$\frac{3}{4}$	1.91	1	2.54	$\frac{15}{16}$ or 1	2.38 or 2.54
1	2.54	$1\frac{1}{4}$	3.17	$1\frac{1}{4}$	3.17

(*N.B.:* The SI units are direct conversions and not the equivalent metric tube size.)

(ii) *Baffles*

In order to obtain economic values of the outside film coefficient, the fluid on the shell side should flow within the turbulent regime, that is with a reasonably high linear velocity. This is achieved by installing baffles perpendicular to the shell axis, thus deflecting the fluid so that it flows at right angles to the tube bundle. The velocity is then increased by reducing the baffle spacing, bearing in mind the need to avoid excessive pressure drop. Baffles are normally 25 per cent segmental cut, that is the height of the baffle is (0.75 × shell inner diameter). Other designs employ 'disc and doughnut' shaped baffles, where alternate baffles form firstly an annulus at the shell periphery and then an aperture at the shell axis. The spacing between baffles is known as the baffle pitch, and this is usually between 20 and 100 per cent of the shell inner diameter.

The shell inlet nozzles are fitted with impingement plates to reduce the velocity effect of the fluid and where the inlet fluid is a vapour-liquid mixture, the plates are usually perforated to minimize erosion problems. Impingement plates are not necessary in reboilers, where the inlet velocity is less than 3 ft/s [1 m/s].

(iii) *Pass Partition Plates*

Pass partition plates are fitted in the shell and in the channel and floating head cover, to force the shell and tube side fluids to travel from one of the exchangers to the other, several times; one traverse is known as a pass. The exchanger shown in Figure 3–2 has no shell side plate and a single partition plate in the channel, thus giving two tube side passes. The unit is therefore described as a one shell-side, two tube-side pass or a 1:2 exchanger. The plates required in the channel and floating head cover to give various tube side passes are shown in Figure 3–3.

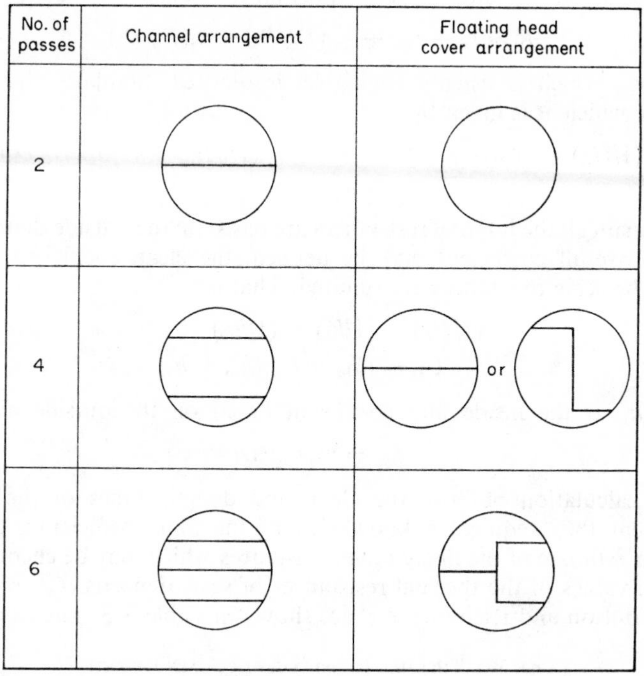

No. of passes	Channel arrangement	Floating head cover arrangement
2		
4		or
6		

FIGURE 3–3. Partition plates required for various numbers of tube-side passes

3.3 Heat Transfer Considerations

3.3.1 Overall Coefficient of Heat Transfer

The basic relationship for all calculations of heat transfer rates is

$$Q = U_D A \Delta t_m \qquad (3\text{–}2)$$

where Q is the amount of heat transferred per unit time (BTU/hr) [W],

U_D is the design value of the overall coefficient of heat transfer (BTU/hr ft² °F) [W/m² K],

A is the area available for heat transfer (ft²) [m²],

Δt_m is the logarithmic mean temperature difference (°F) [K].

The overall coefficient is the reciprocal of the total resistance to heat transfer ($R = 1/U_D$), and in the case of shell and tube units, this is made up of five components:

r_o = the fluid film on the outside of the tube (hr ft² °F/BTU) [m² K/W]

r_{do} = the dirt or scale on the outside of the tube (hr ft² °F/BTU) [m² K/W]

r_m = the tube wall ($r_m = x/k$) (hr ft² °F/BTU) [m² K/W]

r_{di} = the dirt or scale on the inside of the tube (hr ft² °F/BTU) [m² K/W]

r_i = the fluid film on the inside of the tube (hr ft² °F/BTU) [m² K/W]

Thus, basing all the internal resistances on the outside diameter,

$$R = r_o + r_{do} + r_{di}(D_o/D_i) + r_i(D_o/D_i) \qquad (3\text{-}3)$$

where r_m, which is usually negligible, is omitted. Similarly the overall design coefficient is given by

$$(1/U_D) = (1/h_o) + (1/h_{do}) + (1/h_{di})(D_i/D_o) + (1/h_i)(D_i/D_o) \qquad (3\text{-}4)$$

where again all the internal resistances are based on the outside diameter. A further overall coefficient may be defined, the clean coefficient, U_C, in which the scale resistances are omitted. That is,

$$(1/U_C) = (1/h_o) + (1/h_{io})$$

or

$$U_C = (h_{io} \times h_o)/(h_{io} + h_o) \qquad (3\text{-}5)$$

where h_{io} is the inside film coefficient based on the outside diameter,

$$h_{io} = (h_i D_o)/D_i$$

The calculation of both the clean and design values of the overall coefficient thus requires a knowledge of the film coefficients, together with an estimate of the likely scale resistances which may be encountered. Typical values of the thermal resistances of scale deposits (r_{do}, r_{di}) taken from Coulson and Richardson[1] are shown in Table 3-3. The calculation

TABLE 3-3. THERMAL RESISTANCES OF SCALE DEPOSITS[1]

	r_{do}, r_{di}	
	(hr ft^2 °F/BTU)	[m^2 K/W]
Water ($u = 3$ ft/s [1 m/s], $T< 120$ °F [320 K]) distilled	0.0005	0.000 09
sea	0.0005	0.000 09
clear river	0.0012	0.000 2
untreated cooling tower	0.0033	0.000 6
treated cooling tower	0.0015	0.000 3
hard well	0.0033	0.000 6
Steam good quality, oil free	0.0003	0.000 05
poor quality, oil free	0.0005	0.000 09
exhaust from reciprocating engines	0.001	0.000 2
Liquids treated brine	0.0015	0.000 3
organics	0.001	0.000 2
fuel oils	0.006	0.000 1
tars	0.01	0.002
Gases air	0.0015–0.003	0.000 3–0.000 5
solvents	0.0008	0.000 015

of film coefficients for various fluids and operating conditions forms the major part of the standard works of reference on heat transfer[2]–[5] and will not be considered in any detail here, other than a brief summary of the most usual system, where two fluids are passed through the exchanger, neither undergoing a change in state.

The exact form of the correlation for the film coefficient depends on the fluid flow regime of the system as defined by the Reynolds Number, Re, where

$$Re = DG/\mu \tag{3-6}$$

where D is the inside diameter of the tube (ft) [m],
G is the mass flowrate of the fluid (lb/hr ft^2) [kg/s m^2]
μ is the viscosity of the fluid at the temperature of the film (lb/ft hr) [Ns/m^2]

When Re < 2100 the flow is streamline, and when Re > 10^4 the flow is turbulent. Having resolved which flow regime predominates, the following correlations may then be used for evaluating the film coefficient:

(i) *Inside tubes.* For streamline flow, the Sieder and Tate relation may be used:

$$(h_i D/k) = 1.86\{(DG_t/\mu)(C_p\mu/k)(D/L)\}^{0.33}(\mu/\mu_w)^{0.14} \tag{3-7}$$

For turbulent flow,

$$(h_i D/k) = 0.027(DG_t/\mu)^{0.8}(C_p\mu/k)^{0.33}(\mu/\mu_w)^{0.14} \tag{3-8}$$

(ii) *Shell side.* For the shell side film coefficient, Kern[2] recommends

$$(h_o D_e/k) = 0.36(D_e G_s/\mu)^{0.55}(C_p\mu/k)^{0.33}(\mu/\mu_w)^{0.14} \tag{3-9}$$

In this correlation, D_e is the equivalent mean diameter, which is defined as $D_e = 4$ (free area)/(wetted perimeter) and typical values for various tube arrangements are included in Table 3-4; the free area being shown as the

TABLE 3-4. VALUES OF D_e FOR VARIOUS TUBE LAYOUTS

Tube outer diameter		Tube pitch		D_e	
(in.)	[cm]	(in.)	[cm]	(in.)	[cm]
$\frac{3}{4}$	1.91	1 □	2.54 □	0.95	2.41
$\frac{3}{4}$	1.91	$\frac{15}{16}$ △	2.38 △	0.55	1.40
$\frac{3}{4}$	1.91	1 △	2.54 △	0.73	1.85
1	2.54	$1\frac{1}{4}$ □	3.17 □	0.99	2.52
1	2.54	$1\frac{1}{4}$ △	3.17 △	0.72	1.83

□ = tubes on square pitch
△ = tubes on triangular pitch

shaded area in Figure 3-4. In equation (3-9), G_s is the fluid mass velocity on the shell side given by

$$G_s = m_s/a_s \text{ (lb/ft}^2 \text{ hr) [kg/m}^2\text{s]}$$

where m_s = mass flow of fluid (lb/hr) [kg/s]
a_s = cross area for flow (ft^2) [m^2]

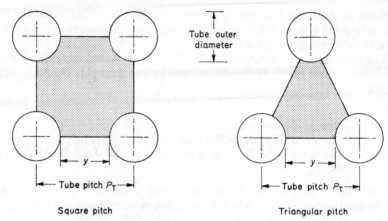

Square pitch   Triangular pitch

FIGURE 3-4. Definition of free area in calculating D_e

The cross area for flow is defined as

$$a_s = (D_s By)/(P_T 144) \text{ (ft}^2) \tag{3-9a}$$
$$[= (D_s By \times 10^{-6})/P_T] \text{ [m}^2]$$

where D_s = shell inner diameter (in.) [mm],
$\quad B$ = baffle spacing (in.) [mm],
$\quad y$ = clearance between tubes, defined in Figure 3-4 (in.) [mm],
$\quad P_T$ = tube pitch (in.) [mm].

Thus it is now possible for a given configuration and set of operating conditions to calculate h_o and h_{io} from equations (3-9) and either (3-7) or (3-8), and hence the value of U_C from equation 3-5. The addition of suitable scale resistances from Table 3-3 permits the computation of the design overall coefficient U_D by means of equation (3-4).

The same method of calculation may be adopted for assessing the performance of an existing exchanger. From a knowledge of the operating conditions and flowrates, h_o and h_{io} may be obtained and hence U_C, the overall coefficient for the exchanger, assuming operation under perfectly clean conditions. The actual coefficient U_D is obtained from equation (3-2) and the actual scale resistance from

$$R_d = (U_C - U_D)/U_C U_D \text{ (ft}^2 \text{ hr } °F/BTU) \text{ [m}^2 \text{ K/W]} \tag{3-10}$$

If this is greatly in excess of the sum of the two appropriate values taken from Table 3-3, then the operation of the unit is uneconomic and imminent cleaning or possible retubing is indicated.

3.3.2 Heat Transfer Area

The second term in equation (3-2), the area available for heat transfer, is based on the outside diameter of the tubes and is given by

A = (no of tubes) × (tube length between inner faces of tube sheets)

× (external surface area per unit length of tube).

The area per unit length of various tube sizes is shown in Table 3–5. Where the film coefficient is controlling on one side of the tubes there are advantages to be gained in specifying finned tubes, as illustrated in the following example.

TABLE 3–5. SURFACE AREA OF TUBES

Tube outer diameter		Surface area/ unit length	
(in.)	[cm]	(ft²/ft)	[m²/m]
$\frac{3}{4}$	1.91	0.1963	0.0599
$\frac{7}{8}$	2.22	0.2297	0.0701
1	2.54	0.2618	0.0930
$1\frac{1}{4}$	3.17	0.3272	0.0998

Example 3–2

Air is heated in the shell of a shell and tube exchanger in which steam condenses in the tubes. What is the likely effect on the overall coefficient of doubling the outside surface area?

From Table 3–1, typical film coefficients are

h_{io} (condensing steam) = 1500 BTU/hr ft² °F [8500 W/m² K]

h_o (air) = 10 BTU/hr ft² °F [60 W/m² K]

Hence in equation (3–5),

$U_C = (1500 \times 10)/(1500 + 10) = 9.93$ BTU/hr ft² °F

$[= (8500 \times 60)/(8500 + 60) = 59.6$ W/m² K]

Assuming the air velocity remains unchanged, increasing the surface area by a factor of 2 effectively doubles the value of h_o, and $h_o = 20$ BTU/hr ft² °F [120 W/m² K]. Thus,

$U_C = (1500 \times 20)/(1500 + 20) = 19.73$ BTU/hr ft² °F

$[= (8500 \times 120)/(8500 + 120) = 118.3$ W/m² K]

Thus with a twofold increase in the overall coefficient the size of the exchanger could be halved, although the saving must be offset against the extra cost of installing finned tubes and also higher operating costs due to the increased pressure drop.

3.3.3 Mean Temperature Difference

The third term in equation (3–2) is the temperature driving force, Δt_m defined as

$$\Delta t_m = \frac{(T_1 - t_2) - (T_2 - t_1)}{2.3 \log_{10}\{(T_1 - t_2)/(T_2 - t_1)\}} \qquad (3\text{–}11)$$

where the various temperatures are as indicated in Figure 3–5. Equation (3–11) applies only where true counterflow exists, and therefore it is

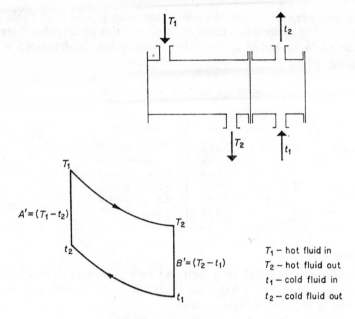

Δt_m = logarithmic mean of A' and B'

$$= \frac{(A'-B')}{\log_e (A'/B')}$$

for true counterflow systems only

FIGURE 3–5. Mean temperature difference

unapplicable to multipass heat exchangers in which the shell-side fluid continuously passes back and forth across the tube bundle. For use in equation (3–2), therefore, Δt_m as given by equation (3–11) is multiplied by a factor F to allow for the deviation from true counterflow. This is given by

$$F = \frac{(r^2 + 1)\log_{10}\{(1 - p)/(1 - pr)\}/(r - 1)}{\{2/p - 1 - r + \sqrt{(r^2 + 1)}\}/\{2/p - 1 - r - \sqrt{(r^2 + 1)}\}} \quad (3–12)$$

where $r = (T_1 - T_2)/(t_2 - t_1)$ and $p = (t_2 - t_1)/(T_1 - t_1)$.

Fortunately, equation (3–12) is available in the literature[1],[2] as a series of charts, thus facilitating its use. The above equations make the assumption that the exchanger is perfectly insulated and that the overall coefficient, the mass velocities and the specific heats remain constant. When the factor F is less than 0.75, the use of a particular heat exchanger is not recommended and an alternative layout must be chosen.

In calculating film coefficients, the physical properties of the fluids are required at the temperature of the film; this changes as the fluid passes through the exchanger and average fluid temperatures may be obtained from the following approximate equations:

For the *hot* fluid,

$$T_A = T_2 + F_c(T_1 - T_2)$$

For the *cold* fluid,

$$t_A = t_1 + F_c(t_2 - t_1)$$

where F_c is the caloric fraction, values of which are available in Kern.[2]

Similarly, it is often necessary to calculate the temperature of the tube wall, especially where large temperature differences or heat sensitive materials are involved and this is given by the following equations:

For the hot fluid in the tubes,

$$t_w = t_A + (h_{io}/(h_{io} + h_o))(T_A - t_A) = T_A - (h_o/(h_{io} + h_o))(T_A - t_A)$$

For the hot fluid in the shell,

$$t_w = t_A + (h_o/(h_{io} + h_o))(T_A - t_A) = T_A - (h_{io}/(h_{io} + h_o))(T_A - t_A).$$

In parallel flow and counterflow exchangers, $(T_2 - t_2)$ is called 'approach' and where $t_2 > T_2$, $(t_2 - T_2)$ is called 'temperature cross'. This is to be avoided wherever possible.

At this stage, the application of equation (3-2) will now be illustrated by a worked example.

Example 3-3

A horizontal shell and tube exchanger is to be designed to condense 60 000 lb/hr [756 kg/s] isobutane at 120 lb/in.^{2} [827 kN/m^2], using cooling water on the tube side. The water is available at 80°F [300 K] and is required in another process at 102.5°F [312.5 K]. $\frac{3}{4}$ in. outer diameter [19.05 mm], 14 s.w.g. [2.11 mm wall] tubes are to be used on 1 in. [25.4 mm] square pitch and the total scale resistance may be taken as 0.003 ft^2 hr °F/BTU [0.000 53 m^2 K/W]. What size of exchanger should be specified?

(i) *Preliminary Calculation*

The latent heat of condensation of isobutane at 120 lb/in.^{2} [827 kN/m^2] is 118.2 BTU/lb [274.9 kJ/kg]. Hence

Heat load, $Q = 6 \times 10^4 \times 1.182 \times 10^2 = 7.092 \times 10^6$ BTU/hr

$$[= 7.56 \times 2.749 \times 10^2 = 2.078 \times 10^3 \text{ kJ/s}$$

$$= 2.078 \times 10^6 \text{ W}]$$

Allowing a 10 per cent overload,

$$Q = 7.80 \times 10^6 \text{ BTU/hr} \qquad [2.285 \times 10^6 \text{ W}]$$

Rise in water temperature $= (102.5 - 80) = 22.5°F$

$$[= (312.5 - 300) = 12.5 \text{ K}]$$

and the hence cooling water flow rate,

$$w = (7.8 \times 10^6)/(1.0 \times 22.5) = 3.47 \times 10^5 \text{ lb/hr}$$

$$[= (2.285 \times 10^3)/(4.187 \times 12.5) = 43.7 \text{ kg/s}]$$

The equilibrium condensation temperature of isobutane is 146°F [337 K] and hence

$$\Delta t_m = \frac{(146 - 102.5) - (146 - 80)}{2.3 \log_{10} (146 - 102.5)/(146 - 80)} = 54.3°F$$

$$[= \frac{(337 - 312.5) - (337 - 300)}{2.3 \log_{10} (337 - 312.5)/(337 - 300)} = 30.2 \text{ K}]$$

(In condensation, $T_1 = T_2$ and no correction factor F is necessary).

Assuming an overall coefficient $U_D = 90$ BTU/hr ft^2 °F [510 W/m^2 K], based on data given in Table 3–1, the area required for heat transfer is

$$A = Q/U_D \cdot \Delta t_m$$
$$= (7.80 \times 10^6)/(90 \times 54.3) = 1596 \text{ ft}^2$$
$$[= (2.285 \times 10^6)/(510 \times 30.2) = 148.3 \text{ m}^2]$$

The area per unit length of tube is 0.1963 ft^2/ft [0.0599 m^2/m] (Table 3–5) and hence the total length of tubing required is

$$1596/0.1963 = 8130 \text{ ft} \ [148.3/0.0599 = 2478 \text{ m}]$$

Adopting a standard tube length of 16 ft [4.88 m], (Chapter 7), the total number of tubes required is

$$8130/16 \ [2478/4.88] = 508$$

The volumetric flowrate of cooling water is

$$3.47 \times 10^5/(62.4 \times 3600) = 1.54 \text{ ft}^3/\text{s}$$
$$[43.7/10^3 = 43.7 \times 10^{-3} \text{ m}^3/\text{s}]$$

Assuming a water velocity of (say) 2.5 ft/s [0.7 m/s], area for flow required/pass is

$$1.54/2.5 = 0.616 \text{ ft}^2$$
$$[(43.7 \times 10^{-3})/0.7 = 0.0572 \text{ m}^2]$$
$$\text{area for flow/tube}^{(2)} = 1.86 \times 10^{-3} \text{ ft}^2$$
$$[= 1.729 \times 10^{-4} \text{ m}^2]$$
$$\text{Hence number of tubes/pass} = 0.616/(1.86 \times 10^{-3})$$
$$[= 0.0572/(1.729 \times 10^{-4})] = 331$$

Thus the 508 tubes required will be arranged in 2 passes of 254 each. This smaller number of tubes/pass will increase the inside water velocity and hence h_1. The nearest standard size to this configuration (see Chapter 7) is a 29 in. internal diameter [0.737 m] shell which can accommodate 526 tubes arranged in 2 passes.

The preliminary design is therefore 526 tubes each 16 ft long [4.88 m] arranged in 2 passes in a 29 in. internal diameter shell [0.737 m].

(ii) *Calculation of Overall Coefficient*

On the *tube* side, for cooling water, the coefficient is given by

$$h_1 = 150\,(1 + 0.011t_a)u/^{0.8}d_1{}^{0.2}$$
$$[= 1063\,(1 + 0.002\,93t_a)u^{0.8}/d_1{}^{0.2}] \qquad (3\text{-}13)$$

where t_a = mean water temperature (°F) [K],

u = water velocity (ft/s) [m/s],

d_1 = tube internal diameter (in.) [m].

In this case the water velocity,

$$u = (1.54/(1.86 \times 10^{-3} \times 526/2)) = 3.15\ \text{ft/s}$$
$$[= (43.7 \times 10^{-3}/(1.729 \times 10^{-4} \times 526/2)) = 0.96\ \text{m/s}]$$
$$d_1 = 0.584\ \text{in.}\ [1.48 \times 10^{-2}\ \text{m}]$$

Hence $h_1 = 150\,(1 + 0.011 \times 91)3.15^{0.8}/0.584^{0.2} = 835\ \text{BTU/hr ft}^2\ °\text{F}$

$[= 1063\,(1 + 0.002\,93 \times 306)0.96^{0.8}/(1.48 \times 10^{-2})^{0.2}$

$\qquad\qquad\qquad\qquad\qquad = 4741\ \text{W/m}^2\ \text{K}]$

$h_{1o} = 835 \times 0.584/0.75 = 650\ \text{BTU/hr ft}^2\ °\text{F}$

$[= 4741 \times 1.48 \times 10^{-2}/(1.91 \times 10^{-2}) = 3674\ \text{W/m}^2\ \text{K}]$

On the *shell* side, the appropriate relation for condensation of a vapour is

$$h_o = 0.72\{(k^3\rho^2g\lambda)/(jd_o\mu\Delta T_t)\}^{0.25}\ \text{BTU/hr ft}^2\ °\text{F} \qquad (3\text{-}14)$$
$$[= 4.09\{(k^3\rho^2g\lambda)/(jd_o\mu\Delta T_t)\}^{0.25}\ \text{W/m}^2\ \text{K}]$$

where k, ρ and μ are the physical properties of the film,

g is the gravitational constant (ft/hr²) [m/s²]

λ is the latent heat of condensation (BTU/lb) [J/kg]

j is $\sqrt{}$(total number of tubes)

$$\Delta T_f = \Delta t_m \cdot \frac{1 - (1/h_{1o} + \text{scale resistance})}{1/U_D}$$

Thus in this case, taking physical properties from the literature,[1],[2]

$$h_o = 0.72 \left\{ \frac{(0.075^3 \times 31.8^2 \times 4.17 \times 10^6 \times 118.2)}{(526 \times (0.75/12) \times 0.329 \times 24.1)} \right\}^{0.25}$$

$$= 142\ \text{BTU/hr ft}^2\ °\text{F} \qquad [806\ \text{W/m}^2\ \text{K}]$$

Hence in equation (3–5),

$$U_C = (650 \times 142)/(650 + 142) = 116.5\ \text{BTU/hr ft}^2\ °\text{F}$$
$$[= (3674 \times 806)/(3674 + 806) = 661.5\ \text{W/m}^2\ \text{K}]$$

In equation (3–10),

$$U_D = U_C/(R_d U_C + 1)$$
$$= 116.5/(0.003 \times 116.5 + 1) = 86.3\ \text{BTU/hr ft}^2\ °\text{F}$$
$$[= 661.5/(0.000\,53 \times 661.5 + 1) = 490\ \text{W/m}^2\ \text{K}]$$

(iii) *Specification of Exchanger*

The calculated value of the design coefficient, 86.3 BTU/hr ft^2 °F [490 W/m^2 K] is sufficiently close to the assumed value of 90 BTU/hr ft^2 °F [510 W/m^2] for the preliminary design to be specified as the final design. Were this not the case, the calculation would have to be reworked, based on a different assumed value of the coefficient in the initial stage. It now remains to check that the area proposed is adequate and to calculate the overload factor which is incorporated in the design.

$$\text{Area required} = (7.80 \times 10^6)/(86.3 \times 54.3) = 1665 \text{ ft}^2$$
$$[= (2.285 \times 10^6)/(490 \times 30.2) = 155 \text{ m}^2]$$
$$\text{Area available} = 526 \times 16 \times 0.1963 = 1652 \text{ ft}^2$$
$$[= 526 \times 4.88 \times 0.0599 = 153 \text{ m}^2]$$

This is sufficiently close for the design to be acceptable, bearing in mind the 10 per cent design factor which was allowed in the initial calculation. The design heat load is thus

$$Q = 86.3 \times 1652 \times 54.3 = 7.74 \times 10^6 \text{ BTU/hr}$$
$$[= 490 \times 153 \times 30.2 = 2.27 \times 10^6 \text{ W}]$$

and the weight flow of isobutane which can be handled is

$$7.74 \times 10^6/118.2 = 65\ 480 \text{ lb/hr } [8.25 \text{ kg/s}]$$

a design factor of 9.1 per cent.

The design specification is thus 526 tubes, each 16 ft [4.88 m] long arranged in 2 tube-side passes in a 29 in. [0.737 m] inner diameter shell.

3.4 Pressure Drop in Shell and Tube Exchangers

The film coefficients on both the shell and tube sides and hence the rate of heat transfer depend on the fluid velocity. The higher the velocity, however, the greater is the pressure drop through the exchanger, and in practical terms there is a maximum velocity which may be used; this is fixed by the maximum permissible pressure drop. The pressure drop in both the shell and tubes may be calculated from relations given by Kern.[2]

(i) *Shell Side*

$$\Delta P_s = (fG_s^2 D_s L)/(4.35 \times 10^9 D_e s \phi_s B) \ (\text{lb/in.}^2) \qquad (3\text{–}15)$$
$$[= (fG_s^2 D_s L)/(2 \times 10^6 D_e s \phi_s B)] \ [\text{kN/m}^2]$$

where G_s = mass velocity in the shell (lb/ft^2 hr) [kg/m^{2}s] (obtained by dividing the mass flow by the flow area, a_s, given in equation (3–9a),

D_s = shell inner diameter (ft) [m],

D_e = equivalent diameter of shell (ft) [m] (Table 3–4),

s = specific gravity of the fluid (—) [—],

$\phi_s = (\mu/\mu_w)^{0.14}$ (—) [—],

L = tube length (ft) [m],

B = baffle spacing (ft) [m],

f = friction factor (ft^2/in.^{2}) [m^2/m^2].

obtained from Kern[2] or the approximate relationship $f = 0.013 \ \text{Re}^{-0.2} \text{ft}^2/\text{in.}^2 \ [= 1.87 \ \text{Re}^{-0.2} \text{ m}^2/\text{m}^2]$, which applies where Re is in excess of 500.

(ii) *Tube Side*

On the tube side, the pressure drop consists of two components, the friction losses in the tubes and the loss due to change in direction. Kern[2] recommends the following equation:

$$\Delta P_t = (fG_t^2Ln)/(5.22 \times 10^{10}D_t s\phi_t) + (4nu^2 \times 62.4s)/(2 \times 144g) \text{ (lb/in.}^2)$$

$$[= (fG_t^2Ln)/(2 \times 10^6 D_t s\phi_t) + 2.26nu^2s] \text{ [kN/m}^2] \qquad (3\text{-}16)$$

where G_t = mass velocity in the tubes (m_t/a_t) (lb/ft² hr) [kg/m² s]
$\quad a_t = Na'/n$ (ft²) [m²],
$\quad N$ = total number of tubes (—) [—],
$\quad n$ = number of tube-side passes ( ) [—],
$\quad a'$ = cross area of flow/tube (ft²) [m²],
$\quad D_t$ = tube diameter (ft) [m],
$\quad u$ = fluid velocity (ft/s) [m/s],
$\quad f$ = friction factor (ft²/in.²) [m²/m²].

Values of f are obtained from Kern[2] or the approximate relationship

$$f = 0.005 \text{ Re}^{-0.33} \text{ ft}^2/\text{in.}^2 \text{ } [= 0.72 \text{ Re}^{-0.33} \text{ m}^2/\text{m}^2]$$

which applies for exchanger tubes where the Reynolds Number is in excess of 1000.

Example 3-4

What is the tube-side pressure drop for the exchanger specified in Example 3-3?

In this case,

flow of water = 3.47×10^5 lb/hr [= 43.7 kg/s]

area of flow/tube = 1.86×10^{-3} ft² [= 1.729×10^{-4} m²]

Hence area for flow/pass = $1.86 \times 10^{-3} \times 526/2 = 0.489$ ft²

$$[= 1.729 \times 10^{-4} \times 526/2 = 0.0454 \text{ m}^2]$$

mass velocity, $G_t = 3.47 \times 10^5/0.489 = 7.10 \times 10^5$ lb/ft² hr

$$[= 43.7/0.0454 = 963 \text{ kg/m}^2 \text{ s}]$$

$$\text{Re} = D_t G_t/\mu = (0.584/12)(7.10 \times 10^5)/1.96$$

$$[= (1.48 \times 10^{-2} \times 963)/0.81 \times 10^{-3}]$$

$$= 1.76 \times 10^4$$

friction factor, $f = 0.005 \times (1.76 \times 10^4)^{-0.33} = 0.000\,23$ ft²/in.²

$$[= 0.72 \times (1.76 \times 10^4)^{-0.33} = 0.033 \text{ m}^2/\text{m}^2]$$

For water at the temperatures under consideration it will be assumed that $\phi_t = 1.0$. Hence

$$\Delta P_t = \frac{0.000\,23 \times (7.10 \times 10^5)^2 \times 16.0 \times 2}{5.22 \times 10^{10} \times (0.584/12) \times 1.0 \times 1.0}$$
$$+ 4 \times 2 \times 3.15^2 \times 62.4 \times 1.0/2 \times 144 \times 32.2$$

$$= (1.456 + 0.532) = 1.988, \text{ say } 2\,\text{lb/in.}^2$$

$$[= \frac{0.033 \times 963^2 \times 4.88 \times 2}{2 \times 10^6 \times 1.48 \times 10^{-2} \times 1.0 \times 1.0}$$
$$+ 2.26 \times 2 \times 0.96^2 \times 1.0$$

$$= 10.03 + 3.67 = 13.70\,\text{kN m}^2]$$

3.5 Optimum Operating Conditions

In designing shell and tube units, there are many instances where the operating conditions are not precisely defined. For example, in Example 3–3 the required duty is the condensation of 60 000 lb/hr [7.56 kg/s] of isobutane and the amount of cooling water used, and for that matter the outlet water temperature, are of secondary importance. When such parameters are not defined further degrees of freedom are apparent, and these enable the design of the exchanger to be optimized. Consider these two parameters.

3.5.1 Water Outlet Temperature

In a condenser or a cooler it is fairly obvious that either a large quantity of cooling water may be used, when the temperature rise will be small (thus enabling possible re-use of the water in other units), or the water consumption may be kept to a minimum, in which case the rise in water temperature will be much greater. (Normally the maximum permissible water temperature is 120°F [322 K] as scaling tends to increase at this point.) In the first case, a large logarithmic mean temperature difference will result in a smaller exchanger for a given coefficient and duty, and hence the depreciation and maintenance charges will be less. Against this, the costs of water and the pumping costs will be high. Where a small water flow is employed the reverse situation is the case—high capital charges and low operating costs. Between these two extremes there is obviously an optimum value of the water outlet temperature, which should be incorporated into the design. The total annual cost,

$$C = (\text{cost of water} + \text{pumping costs}) + (\text{capital charges})$$

The first part of this cost, the operating costs, is directly proportional to the water flow rate, m lb/hr [kg/s] and the second part is a function of the heat transfer area. Thus from equation (3–2) and Figure 3–5,

$$Q = mC_p(t_2 - t_1) = UA\,\Delta t_m$$

or $\qquad m = Q/C_p(t_2 - t_1), \qquad A = Q/U\,\Delta t_m$

Then operating costs $= C_w \theta Q/C_p(t_2 - t_1)$ £/year

where C_w is the cost of supplying and pumping water (£/lb) [£/kg]

$\qquad \theta$ is the operating time (hr/year) [s/yr],

and $\qquad$ capital charges $= C_F\, Q/(U\,\Delta t_m)$ £/year where C_F is the annual fixed charge/unit area (£/ft²) [£/m²]. Thus

$$C = (C_w \theta Q)C_p(t_2 - t_1)$$
$$+ C_F Q/\{U((T_1 - t_2) - (T_2 - t_1))/\log_e(T_1 - t_2)/(T_2 - t_1)\}$$

Assuming U is constant and writing $\Delta t_1 = (T_2 - t_1)$, this equation may be differentiated and the optimum condition is then obtained when $dC/dt_2 = 0$. Equating the respective parts,

$$\frac{U\theta C_w}{C_F C_p} \left\{ \frac{T_1 - t_2 - \Delta t_1}{t_1 - t_2} \right\}^2 = \log_e \frac{T_1 - t_2}{\Delta t_1} - \left\{ 1 - \frac{1}{(T_1 - t_2)/\Delta t_1} \right\} \quad (3\text{-}17)$$

Colburn[10] has plotted this equation and his curve is shown in Figure 3-6.

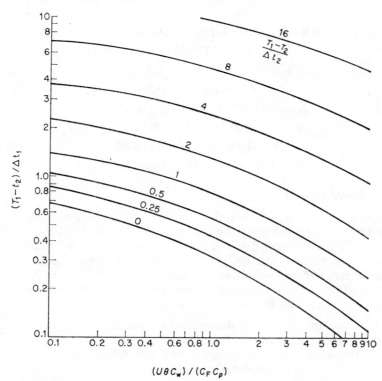

FIGURE 3-6. Determination of optimum water outlet temperature[10]

Example 3-5

A hydrocarbon fraction is to be cooled from 160 to 130°F [344 to 327 K] in a shell and tube unit using cooling water available at 65°F [291 K]. The plant is operated 8000 hr/year [28.8 Ms/year] and the water cost including pumping is £0.2/1000 gal [£0.044/m³]. The installed cost of the heat exchanger is £4/ft² [£43.1/m²] and it is to be depreciated over 7 years. If the interest, maintenance and fixed charges amount to 15 per cent of the installed cost per annum, what is the optimum water outlet temperature?

From Equation 3–1 and Table 3–1, the overall coefficient will be approximately $U = 40$ BTU/hr ft^2 °F [227 W/m^2 K]. The water cost, C_w is given by

$$C_w = 0.2/(1000 \times 10) = £2.0 \times 10^{-5}/\text{lb}$$
$$[= 0.044/10^3 = £4.40 \times 10^{-5}/\text{kg}]$$

The capital charges C_F are given by

$$C_F = (4 \times 0.15) + (4/7) = 1.17 \text{ £/ft}^2 \text{ year}$$
$$[= (43.1 \times 0.15) + (43.1/7) = 12.63 \text{ £/m}^2 \text{ year}]$$

Hence $(U\theta C_w)/(C_F C_p) = (40 \times 8000 \times 2.0 \times 10^{-5})/(1.17 \times 1.0)$
$$[= (227 \times 28.8 \times 10^6 \times 4.40 \times 10^{-5})/$$
$$(12.63 \times 4.187 \times 10^3)]$$
$$= 5.47$$

$$(T_2 - T_1)(\Delta t_1) = (160 - 130)/(130 - 65)$$
$$[= (344 - 327)/(327 - 291)] = 0.46$$

Thus, from Figure 3–6,

$$(T_1 - t_2)/\Delta t_1 = 0.21$$

Hence $T_1 - t_2 = 0.21 \times (130 - 65) = 13.7°\text{F}$
$$[= 0.21 \times (327 - 291) = 7.6 \text{ K}]$$
$$t_2 = (160 - 13.7) = 146.3°\text{F}$$
$$[= (344 - 7.6) = 336.4 \text{ K}]$$

3.5.2 Water Velocity

In condensers an important variable is the cooling water velocity, and again an optimum value may be calculated. With high velocities, a high film coefficient is obtained, which results in a low heat transfer area and hence a cheaper installed cost. This must be offset against the increased cost of pumping and supplying the water. At low water velocities the reverse situation is the case—low running costs with a high capital investment. The water velocity may be varied by:

(i) increasing the amount drawn from the mains,
(ii) recycling some of the water leaving the unit. This in turn affects the water outlet temperature and any optimization procedure involves the evaluation of two parameters.
(iii) variation in tube layout and particularly the number of tubes per pass.

It should be realized, however, that there is a limit to manipulation in this respect if it is desired to specify an exchanger of standard size with corresponding savings in manufacturing costs, and also the addition of further tube passes adds considerably to the cost of an exchanger of a given area.

The variation of fluid velocity and the evaluation of optimum conditions is by no means restricted to water flowing through tubes and the same basic principles apply to air, gases and a multitude of process materials—indeed wherever the operating conditions are incompletely specified. With liquids, velocities through the tubes are usually in the range 3–5 ft/s [0.9–1.52 m/s], and reasonable gas velocities are 50–100 ft/s [15–30 m/s]. The calculation of an optimum water velocity is now illustrated by a worked example.

Example 3–6

Calculate the optimum water velocity and water outlet temperature for the condenser specified in Example 3–3. Assume a heat load of 7.8 × 10⁶ BTU/hr [2.285 × 10⁶ W] and that the installed cost of £1.50/ft² [£16.15/m²] is depreciated over 10 years, the interest and fixed costs being 5 per cent p.a. Assume also that the water is returned to a ring main and hence no charge for consumption is involved and that the cost of power is £0.0063/kWh.

In this calculation, there are two components of the total annual cost, the capital charges and the running costs. The steps in the calculation are as follows:

(i) *Capital Charges*

A series of five water velocities u are chosen in the range 1.0–5.0 ft/s, [0.305–1.53 m/s] and three values of t_2, the outlet water temperature— 95, 100, 105°F [308, 311, 314 K]—and hence the following calculation is made 15 times. For one value of t_2 and one value of u, the inside film coefficient h_1 is obtained from equation 3–13 and hence h_{1o} from

$$h_{1o} = h_1 D_1 / D_o.$$

The outside film coefficient is then obtained from equation 3–14 which requires a value for the total number of tubes. This is obtained by calculating the total water flow from the duty and t_2, and from the value of u knowing that two passes are to be used.

Assuming a total scale resistance of 0.003 ft² hr °F/BTU [0.000 53 m² K/W], the design overall coefficient is obtained from equation (3–4) and hence the heat transfer area required from equation (3–2), bearing in mind that Δt_m is a function of the chosen water outlet temperature t_2. The installed cost of the exchanger is then £1.5A [£16.15A] and the annual capital charges are £(1.5A × (10 + 5)/100) [£16.15A × 0.15]. The calculated values for the 15 sets of conditions, together with the intermediate steps, are shown in Tables 3–6 and 3–6a.

(ii) *Running Costs*

From the heat transfer area calculated above the total length of tubing is A/0.1963 [A/0.0598]. The total number of tubes is known as is the number of tubes/pass to give the chosen velocity u, and hence the tube length (not necessarily a standard size) may be calculated and then the ratio L/D_1. The Reynolds Number is calculated in the usual way and hence the ratio $(R/\rho u^2)$ may be obtained from friction charts available in the literature.[1] The various head losses are then:

$$\text{friction loss} = 8(R/\rho u^2)(L/D_1)(u^2/2g) \quad \text{(ft) [m] (per pass)}$$
$$\text{head lost due to bends} = 8u^2/2g \qquad \text{(ft) [m]}$$
$$\text{velocity head} = u^2/2g \qquad \text{(ft) [m]}$$

The height of the exchanger above datum is assumed to be 4 ft [1.22 m].

The sum of these is the total head lost. The power required for pumping is then:

$$\frac{1}{550} \times \text{head lost (ft)} \times \text{mass flow of water (lb/s) (hp)}$$

$$[10^{-3} \times \text{head lost [m]} \times \text{mass flow of water [kg/s]} \times g[\text{m/s}^2] \text{ kW}]$$

This is then multiplied by (100/pump efficiency)—the latter being taken as 90 per cent. The cost is then calculated assuming operation for 8000 hr/yr and noting that 1 hp hr = 0.745 kWh—the cost of electricity being £0.0063/kWh. The cost of water is neglected in this problem. The running costs and hence the total cost per annum is obtained and the results for the 15 sets of conditions are shown in Tables 3-6 and 3-6a.

TABLE 3-6. OPTIMIZATION OF CONDENSER (BRITISH UNITS)

Cooling water velocity, u (ft/s)	1	2	3	4	5

Outlet temperature 95°F, m = 145 lb/s, Δt_m = 57.5°F

Number of tubes	1250	625	416	312	250
h_1 (BTU/hr ft^2 °F)	327	457	813	1010	1185
h_0 (BTU/hr ft^2 °F)	129	141	148	154	158
U (BTU/hr ft^2 °F)	69.1	77.4	87.0	97.2	96.0
A (ft^2)	1955	1750	1560	1464	1412
Capital charges (£/yr)	440	395	350	330	317
L/D_1	128	218	302	383	461
Friction losses (ft)	0.065	0.358	0.98	2.06	3.71
Total head loss (ft)	4.205	4.916	6.24	8.00	11.25
Pump power (hp)	1.112	1.299	1.645	2.11	2.95
Pumping costs (£/yr)	46.0	53.6	68.2	87.5	123.0
Total costs (£/yr)	486	429	418	417	440

Outlet temperature 100°F, m = 108 lb/s, Δt_m = 55.2°F

Number of tubes	930	470	312	234	182
h_1 (BTU/hr ft^2 °F)	331	462	812	1025	1200
h_0 (BTU/hr ft^2 °F)	134	146	154	159	165
U (BTU/hr ft^2 °F)	69.9	78.8	90.0	94.6	98.7
A (ft^2)	2030	1780	1560	1490	1430
Capital charges (£/yr)	457	400	352	336	322
L/D_1	176	310	406	578	614
Friction losses (ft)	0.089	0.51	1.32	2.78	5.16
Total head loss (ft)	4.229	5.07	6.58	9.02	12.65
Pump power (hp)	0.83	0.996	1.291	1.77	2.48
Pumping costs (£/yr)	34.5	41.5	53.7	73.5	103.0
Total costs (£/yr)	492	441	406	409	423

TABLE 3-6. OPTIMIZATION OF CONDENSER (BRITISH UNITS)—(contd.)

Cooling water velocity, u (ft/s)	1	2	3	4	5

Outlet temperature 105°F, m = 86.6 lb/s, Δt_m = 52.5°F

Number of tubes	746	373	249	187	149
h_1 (BTU/hr ft^2 °F)	340	475	845	1050	1238
h_o (BTU/hr ft^2 °F)	138	150	158	164	169
U (BTU/hr ft^2 °F)	71.1	81	91.6	97.7	100.1
A(ft^2)	2090	1840	1620	1520	1480
Capital charges (£/yr)	470	415	364	342	333
L/D_1	213	400	527	665	810
Friction losses (ft)	0.108	0.658	1.71	3.36	6.52
Total head loss (ft)	4.25	5.22	6.87	9.83	14.5
Pump power (hp)	0.27	0.823	1.08	1.545	2.28
Pumping costs (£/yr)	27.4	34.0	44.9	64.1	94.5
Total costs (£/yr)	497	449	409	406	428

TABLE 3-6a. OPTIMIZATION OF CONDENSER (SI UNITS)

Cooling water velocity, u [m/s]	0.305	0.610	0.915	1.22	1.53

Outlet temperature 308 K, m = 65.8 kg/s, Δt_m = 31.9 K

Number of tubes	1250	625	416	312	250
h_1 [W/m^2 K]	1857	2596	4618	5737	6731
h_o [W/m^2 K]	733	801	841	875	897
U [W/m^2 K]	393	440	494	552	545
A [m^2]	182	163	145	136	131
Capital charges (£/yr)	440	395	350	330	317
L/D_1	128	218	302	383	461
Friction losses [m]	0.020	0.109	0.299	0.628	1.132
Total head loss [m]	1.283	1.500	1.903	2.440	3.431
Pump power [kW]	0.830	0.969	1.227	1.574	2.201
Pumping costs (£/yr)	46.0	53.6	68.2	87.5	123.0
Total costs (£/yr)	486	429	418	417	440

Outlet temperature 311 K, m = 49.0 kg/s, Δt_m = 30.7 K

Number of tubes	930	470	312	234	182
h_1 [W/m^2 K]	1880	2624	4612	5822	6816
h_o [W/m^2 K]	761	829	875	903	937
U [W/m^2 K]	397	448	571	537	561
A [m^2]	189	165	145	138	133
Capital charges (£/yr)	457	400	352	336	322
L/D_1	176	310	406	578	614
Friction losses [m]	0.027	0.156	0.403	0.848	1.57
Total head loss [m]	1.290	1.55	2.01	2.751	3.860
Pump power [kW]	0.62	0.743	0.963	1.320	1.850
Pumping costs (£/yr)	34.5	41.5	53.7	73.5	103.0
Total costs (£/yr)	492	441	406	409	423

TABLE 3–6a. OPTIMIZATION OF CONDENSER (SI UNITS)—(contd.)

Cooling water velocity, u [m/s]	0.305	0.610	0.915	1.22	1.53

Outlet temperature 314 K, $m = 39.3$ kg/s, $\Delta t_m = 29.2$ K

Number of tubes	746	373	249	187	149
h_i [W/m² K]	1931	2698	4800	5694	7032
h_o [W/m² K]	784	852	897	932	960
U [W/m² K]	404	460	520	555	569
A [m²]	194	171	151	141	138
Capital charges (£/yr)	470	415	364	342	333
L/D_i	213	400	527	665	810
Friction losses [m]	0.033	0.201	0.522	1.03	1.99
Total head loss [m]	1.296	1.592	2.095	3.00	4.423
Pump power [kW]	0.201	0.614	0.806	1.153	1.700
Pumping costs (£/yr)	27.4	34.0	44.9	64.1	94.5
Total costs (£/yr)	497	449	409	406	428

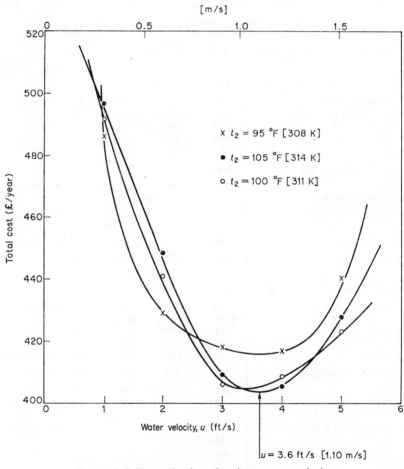

FIGURE 3–7. Determination of optimum water velocity

These data are plotted in Figure 3-7, from which it will be seen that the cost is a minimum, when the water velocity is 3.6 ft/s [1.10 m/s] and the outlet water temperature is 105°F [314 K]. It now remains to recalculate the exchanger based on these conditions and the nearest standard size will be specified.

This type of calculation obviously lends itself to processing by means of a computer and in this event much more complicated factors can be considered, including possible recirculation of materials, mixing of streams, materials of construction, price fluctuations and so on. Indeed programmes are available whereby the whole of a heat exchanger chain, containing say 50 units, can be optimized both in relation to performance and design, taking into account the interrelated functions of adjacent plant and equipment. The example worked here is greatly simplified and yet it serves to illustrate some basic concepts of optimization techniques which will be extended in other sections of the present text.

3.6 Evaporator Design

3.6.1 Evaporator Types

A great number of chemical processes involve the transfer of heat to a boiling liquid, that is evaporation, at some stage or other, whether it be in a reboiler in a distillation column train or the concentration of solutions prior to crystallization. Evaporation involves fundamental problems in handling as well as in transfer and economy of heat, and it is consideration of these requirements that has led to the existing diversity of evaporator designs. In basic terms, an evaporator consists of a vessel of suitable size, fitted with connections for feed liquor and products and with some heating device, such as steam coils or tubes, installed by which the solvent (usually aqueous) is boiled off.

Perry[10] has provided a useful classification of evaporator types which is summarized as follows:

(a) *Solar Heat*

This is confined to the evaporation of sea water and saline waters in deserts. The evaporation of 1 in. [2.54 cm] water corresponds to the evaporation of 113 tons/acre [28.4 kg/m²] of pond surface.

(b) *Direct Fired Heating*

The most important type is the steam boiler though, in general, equipment in this class has never been standardized. Radiation may be the most important factor, but where waste flue gases are used the temperature is too low for radiation to be important and the gas-film coefficient is controlling.

(c) *Jacketed Apparatus*

Usually small batches are involved, particularly in the food industry where such kettles are open and made of copper or aluminium. Coefficients in the range 575–1575 BTU/hr ft^2 °F [3260–8940 W/m^2 K] have been obtained depending on the condition of the surface and the temperature difference.[11]

(d) Tubular Heating Surfaces

(i) *Horizontal tubes.* This unit consists of a vertical cylindrical shell to which are attached two steam chests, connected by the horizontal tube bundle which is surrounded by the process liquor. Tubes are usually $\frac{7}{8}$–1$\frac{1}{4}$ in. [22.2–31.8 mm] diameter and 4–15 ft [1.22–4.6 m] long, and the maximum heat transfer area for one unit is around 5000 ft^2 [465 m^2]. Horizontal tube evaporators should not be used for viscous liquids or scale is likely to be encountered.

(ii) *Vertical tubes.* The standard vertical tube evaporator consists of a vertical shell with two horizontal tube sheets, a downcomer usually being provided in the centre of the tube bundle or callandria. This design is widely used and suitable for situations where solid material or scale may be deposited. Tubes may be 1$\frac{1}{4}$–3 in. [31.8–76.2 mm] in diameter and 2.5–6 ft [0.76–1.83 m] long with up to 20 000 ft^2 [1860 m^2] heat transfer surface in one unit. Many variations of this design are used, including conical or dished bottoms and eccentric, multi, external or annular downcomers.

In the Kestner natural-circulation or climbing-film evaporator, there is no downcomer and all the feed is discharged after passing up through the tubes, although recirculation is possible. Tubes may be 1$\frac{1}{4}$–2 in. [31.8–50.8 m] diameter and 12–20 ft [3.7–6.1 m] long. Forced circulation evaporators consist of vertical tubes through which fluids are pumped with a positive velocity, resulting in very high heat transfer coefficients. They may be used for the evaporation of clear liquor or, where a salt receiver is incorporated into the circulation system, for fluids from which solids are deposited. Forced circulation evaporators are used for evaporation of viscous liquors; where expensive materials are used in fabrication of the unit; where the available temperature difference is low; and where solids tend to grow on the tubes. Tubes are usually $\frac{3}{4}$–1 in. [19.1–25.4 mm] diameter and 8–15 ft [2.44–4.58 m] long.

(iii) *Inclined tubes.* There are a variety of additional designs incorporating tubular construction including inclined tubes, which have high heat transfer coefficients, but are not suitable where solid or scale formation is likely. Hairpin tubes and coils are also used; the latter for the production of distilled water.

The choice of an evaporator for a particular duty is largely one of tradition rather than a definite technical reason, and several different types may be found performing exactly the same service with apparently equal satisfaction. In general, horizontal tube units are best suited to non-foaming, non-viscous liquors, where solids or scale are not deposited. Vertical tube evaporators are the most versatile of all types and are especially suited to very viscous liquids where solids are likely to form. Although forced circulation evaporators are more expensive, they are particularly successful with foaming or viscous fluids which have a tendency to scale, and the entrainment losses are low. There is little detailed information available on evaporator costs, though relative costs of various types, including all the accessories associated with the installation, are shown in Figure 3–8.[12]

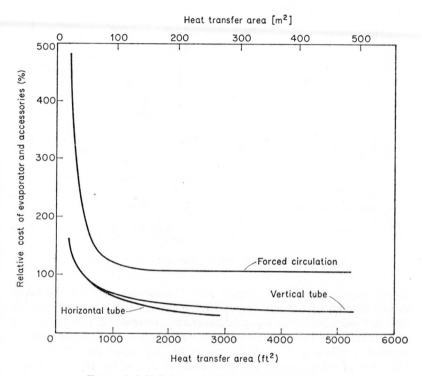

FIGURE 3–8. Relative cost of evaporator installations

3.6.2 Evaporator Operation

(i) *Multiple effects*

In a single effect evaporator, the vapour driven off from the feed liquor is condensed and no heat is recovered. Evaporators may be coupled in series, however, so that the vapour from one unit is used as the heating medium for the next, with the process material passing either in co-current flow (forward feed) or counterflow (backward feed) to the vapours. Such an arrangement is known as multiple effect operation. The purpose of multiple effect evaporation is to reduce the steam consumption for a given amount of evaporation as, ignoring heat losses and boiling point elevation for the moment, 1 lb [0.45 kg] of steam will evaporate 1 lb [0.45 kg] of vapour in one effect, 2 lb [0.90 kg] of vapour in two effects and so on. It is fairly obvious, however, that in passing from single to multiple effect operation the saving in steam costs must be offset by increased capital expenditure on equipment, and the choice of the number of effects is one of optimization. The behaviour of the various parameters involved is

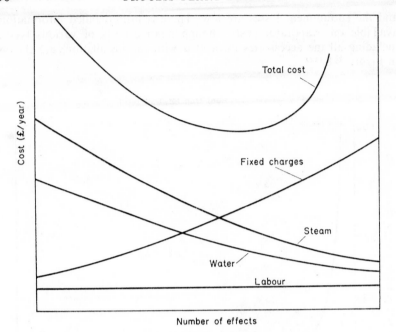

Number of effects

FIGURE 3–9. Optimum number of effects in evaporation

shown diagrammatically in Figure 3–9. The labour costs may be taken as independent of the number of effects, though the fixed costs, which include repairs, maintenance and depreciation, increase with the number of effects as shown. The cost of condenser water and steam decrease with the number of effects and the total cost passes through a minimum value. This type of calculation is rarely carried out except for an entirely new process as the number of effects in the various chemical industries are now relatively standardized, as indicated in Table 3–7.

TABLE 3–7. TYPICAL MULTIPLE EFFECT EVAPORATORS

Material	Number of effects	Liquor feed arrangements
Salt	4	Parallel
Caustic soda	2–3	Backward
Sugar	5–6	Forward

The calculation of multiple effect evaporators is largely one of heat balance, as illustrated by the following example which is based on the data of Meisler.[2]

Example 3–7

50 000 lb/hr [6.3 kg/s] of a solution at 100°F [311 K] is to be concentrated from 10 per cent to 50 per cent solids in a forward-feed triple effect

evaporator which has equal heat transfer surfaces in each effect. Steam is available at 26.7 lb/in.^{2}a [184 kN/m^2] and the last effect is operated under vacuum at 1.95 lb/in.^{2}a [13.4 kN/m^2]; cooling water being available for the condenser at 85°F [303 K]. Assuming a negligible boiling point rise, a mean specific heat of 1.0 BTU/lb °F [4.18 kJ/kg K], that the condensate leaves each effect at its saturation temperature and negligible heat losses, calculate (a) the steam consumption, (b) the heating surface per effect, (c) the condenser water requirement.

The overall coefficients for each effect are $U_1 = 600$ BTU/hr ft^2 °F [3407 W/m^2 K], $U_2 = 250$ [1420 W/m^2 K] and $U_3 = 125$ BTU/hr ft^2 °F [710 W/m^2 K].

$$\text{Total solids in feed} = 0.10 \times 50\,000 = 5000 \text{ lb/hr}$$
$$[= 0.10 \times 6.3 = 0.63 \text{ kg/s}]$$
$$\text{Total product} = 5000 \times 100/50 = 10\,000 \text{ lb/hr}$$
$$[= 0.63 \times 100/50 = 1.26 \text{ kg/s}]$$
$$\text{Total evaporation} = 50\,000 - 10\,000 = 40\,000 \text{ lb/hr}$$
$$[= 6.30 - 1.26 = 5.04 \text{ kg/s}]$$

As a first approximation it will be assumed that, with equal heating surfaces in each effect, the pressure difference between each effect will be equal, i.e.

$$\text{pressure drop/effect} = (26.70 - 1.95)/3 = 8.25 \text{ lb/in.}^2$$
$$[= (184.0 - 13.4)/3 = 56.8 \text{ kN/m}^2]$$

From steam tables information relating to vapour conditions in each effect may be obtained, as shown in Table 3–8. The steam to the first effect may be approximated by

$$\text{(total evaporation)}/(F' \times \text{number of effects})$$

TABLE 3–8. VAPOUR CONDITIONS IN EACH EFFECT (EXAMPLE 3–7)

Steam	Pressure		Temperature		Latent heat	
	(lb/in.^{2} a)	[kN/m^2]	(°F)	[K]	(BTU/lb)	[kJ/kg]
Steam to first effect	26.70	184.0	244	391	949	2207
Steam to second effect	18.45	127.1	224	379	961	2235
Steam to third effect	10.20	70.3	194	363	981	2282
Vapour to condenser	1.95	13.4	125	325	1022	2377

a = absolute pressure

where F' depends on the temperature of the feed. Where the feed enters at its boiling point $F' = 0.75$, though in this case 0.70 is more appropriate as the feed enters at 25°F [14 K] below its boiling point, 125°F [325 K]. Thus,

$$\text{steam to first effect} = 40\,000/(0.70 \times 3) = 19\,050 \text{ lb/hr}$$
$$[= 5.04/(0.70 \times 3) = 2.40 \text{ kg/s}]$$

Consider each effect in turn.

1st effect

$$\text{heat in steam} = 19\,050 \times 949 = 18\,078\,450 \text{ BTU/hr}$$
$$[= 2.40 \times 2207 = 5297 \text{ kW}]$$
$$\text{heat to feed} = 50\,000 \times 1.0(224 - 100)$$
$$= 6\,200\,000 \text{ BTU/hr}$$
$$[= 6.30 \times 4.18(379 - 311) = 1817 \text{ kW}]$$
$$\text{heat available for evaporation} = (18\,078\,450 - 6\,200\,000)$$
$$= 11\,878\,450 \text{ BTU/hr}$$
$$[= (5297 - 1817) = 3480 \text{ kW}]$$
$$\text{liquor evaporated} = (11\,878\,450/961) = 12\,361 \text{ lb/hr}$$
$$[= 3480/2235 = 1.56 \text{ kg/s}]$$
$$\text{liquid transferred to 2nd effect} = (50\,000 - 12.361) = 37\,639 \text{ lb/hr}$$
$$[= 6.30 - 1.56 = 4.74 \text{ kg/s}]$$

2nd effect

$$\text{heat from first effect} = 11\,878\,450 \text{ BTU/hr } [= 3480 \text{ kW}]$$
$$\text{heat from flash evaporation} = 37\,639 \times 1.0\,(224 - 194)$$
$$= 1\,129\,170 \text{ BTU/hr}$$
$$[= 4.74 \times 4.18\,(379 - 363) = 331 \text{ kW}]$$
$$\text{heat available for evaporation} = (11\,878\,450 + 1\,129\,170)$$
$$= 13\,007\,620 \text{ BTU/hr}$$
$$[= 3480 + 331 = 3811 \text{ kW}]$$
$$\text{liquor evaporated} = 13\,007\,620/981 = 13\,260 \text{ lb/hr}$$
$$[= 3811/2282 = 1.67 \text{ kg/s}]$$
$$\text{liquor transferred to 3rd effect} = 37\,639 - 13\,260 = 24\,379 \text{ lb/hr}$$
$$[= 4.74 - 1.67 = 3.07 \text{ kg/s}]$$

3rd effect

$$\text{heat from second effect} = 13\,007\,620 \text{ BTU/hr } [= 3811 \text{ kW}]$$
$$\text{heat from flash evaporation} = 24\,379 \times 1.0(194 - 125)$$
$$= 1\,682\,151 \text{ BTU/hr}$$
$$[= 3811 + 493 = 4304 \text{ kW}]$$
$$\text{liquor evaporated} = 14\,689\,771/1022 = 14\,374 \text{ lb/hr}$$
$$[= 4304/2377 = 1.81 \text{ kg/s}]$$
$$\text{product} = (24\,379 - 14\,374) = 10\,005 \text{ lb/hr}$$
$$[= (3.07 - 1.81) = 1.26 \text{ kg/s}]$$
$$\text{heat to condenser} = 14\,689\,771 \text{ BTU/hr } [= 4304 \text{ kW}]$$

The weight of product calculated above is sufficiently close to the desired value to justify the assumed mass flow of steam to the first effect and further recalculation is not necessary. Had the quantities failed to check, a new

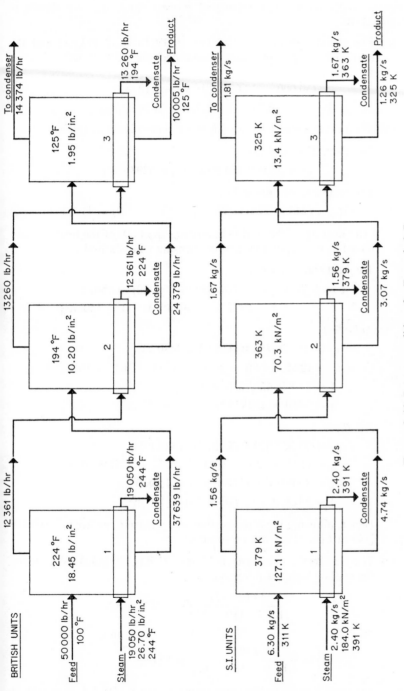

FIGURE 3-10. Process conditions for Example 3-7

value for the steam to the first effect would have been required with consequent recalculation. The calculated process conditions are shown in Figure 3–10.

(a) The steam consumption is thus 19 050 lb/hr [2.40 kg/s] and the economic weight of water evaporated per unit weight of steam supplied is

$$(50\,000 - 10\,005)/19\,050 = 2.10 \text{ lb/lb}$$
$$[(6.30 - 1.26)/2.40 = 2.10 \text{ kg/kg}]$$

(b) The heat transfer area in each effect is calculated as follows:

1st effect. The head load,

$$Q = 18\,078\,450 \text{ BTU/hr } [= 5297 \text{ kW}]$$

The temperature driving force,

$$\Delta t_m = (244 - 224) = 20°\text{F } [= (391 - 379) = 11 \text{ K}]$$

(The heating of the feed is ignored as the bulk of the heat is transferred at the vaporizing and condensing temperatures.)

The heat transfer area,

$$A = Q/U\,\Delta t_m = 18\,078\,450/(600 \times 20) = 1507 \text{ ft}^2$$
$$[= 5297 \times 10^3/(3407 \times 11) = 140 \text{ m}^2]$$

2nd effect

$$Q = 11\,878\,450 \text{ BTU/hr } [= 3480 \text{ kW}]$$
$$\Delta t_m = (224 - 194) = 30°\text{F } [= (379 - 363) = 16 \text{ K}]$$
$$A = 11\,878\,450/(250 \times 30) = 1584 \text{ ft}^2$$
$$[= 3480 \times 10^3/(1420 \times 16) = 147 \text{ m}^2]$$

3rd effect

$$Q = 13\,007\,620 \text{ BTU/hr } [= 3811 \text{ kW}]$$
$$\Delta t_m = (194 - 125) = 69°\text{F } [= (363 - 325) = 38 \text{ K}]$$
$$A = 13\,007\,620/(125 \times 69) = 1508 \text{ ft}^2$$
$$[= 3811 \times 10^3/(710 \times 38) = 140 \text{ m}^2]$$

Hence, allowing for a reasonable safety margin, an area of 1600 ft^2 [149 m^2] per effect will be specified.

(c) The heat to the condenser is 14 689 771 BTU/hr [4304 kW]. Hence, the cooling water requirement is

$$\frac{14\,689\,771}{1.0 \times (110 - 85)} = 979\,318 \text{ lb/hr}$$

or $979\,318/600 = 1632 \text{ gal/min.}$

$$\left[= \frac{4304}{4.18 \times (316 - 303)} = 123 \text{ kg/s} \right]$$

assuming a water outlet temperature of 110°F [316 K], above which scaling becomes a problem.

(ii) *Methods of feeding*

Meisler[2] has worked the above example for a backward feed arrangement, i.e. with the process liquor flowing countercurrent to the vapours and the results are compared in Table 3–9.

TABLE 3–9. COMPARISON OF FEED ARRANGEMENTS[2]

	Forward	Backward
Steam consumption (lb/hr)	19 050	16 850
[kg/s]	2.40	2.12
Economy (lb/lb) [kg/kg]	2.10	2.37
Cooling water (gal/min)	1632	1293
[kg/s]	123	97
Surface area (ft²)	4800	4500
[m²]	446	418

Although backward flow is thermally more efficient, extra capital and maintenance are involved in the backward feed pumps between each effect and problems of air leakage and flow control are greater. In addition to forward and backward flow, parallel flow where dilute feed is introduced into each effect is employed, and also mixed flow which includes all other arrangements. Forward flow incorporates ease of control and the simplest arrangement of equipment, and only one pump is required. The main disadvantage of forward flow is that where the feed is cold, an undue load is put on the first effect and the economy is reduced. Parallel feed is rarely used except in salt evaporators and the only advantage is one of convenience. Mixed feed, for example 2-3-4-1, is most useful where a solution has to be concentrated through a wide range of density and where the process liquor is very viscous.

(iii) *Elevation of Boiling Point*

In Example 3–7, any elevation in boiling point of the liquor due to the presence of dissolved solids was ignored. This is rarely the case in practice except where very dilute liquors are involved and, in carrying out the type of calculation illustrated in Example 3–7, the boiling point elevation must be known in computing the temperature in each effect. The effect of boiling point elevation for brine solutions is shown in Figure 3–11, where the effective temperature driving force is reduced from 120°F [67 K] to 99°F [55 K] in going from 1 to 3 effects. The important point is that the boiling point elevation is lost from the total driving force once for each effect.

(iv) *Steam Pressure*

In most process plants, the evaporation unit is the prime consumer of process steam and hence a most important factor in balancing the steam flow sheet. The steam pressure is usually decided by the amount and pressure of the steam leaving the generators producing power for the process and the designer may be faced with a compromise situation. In the past evaporators have been fabricated from cast iron, which limits the steam pressure to 25–35 lb/in.²*a* [170–240 kN/m²] and this still applies to some extent with horizontal and very large vertical tube evaporators. Where evaporators are fabricated from steel plate or other rolled metal, the unit may be built to

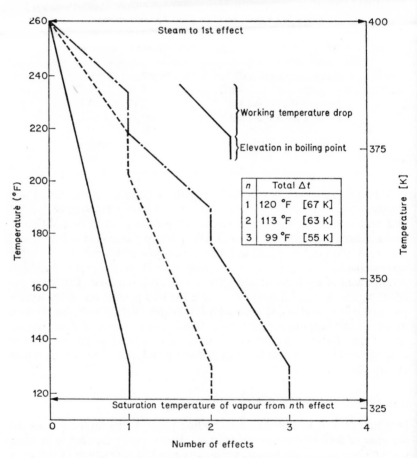

FIGURE 3–11. Effect of boiling point elevation[2]

withstand any pressure up to 250 lb/in.² [1700 kN/m²] or even higher, although the savings in heat transfer area must be offset against increased fabrication costs.

(v) *Vacuum Operation*

Use of a vacuum in evaporators is firstly to increase the available temperature difference and secondly to avoid degradation of heat sensitive materials. Traditionally, steam has been supplied at 220–250°F [377–394 K] and vacuum must be employed to provide a satisfactory driving force especially in multiple effect units. Where steam is supplied at higher temperatures, as in the sugar beet industry, the final effect may be operated at atmospheric pressure and relatively modern installations supply steam at up to 100 lb/in.^2a [690 kN/m^2] from the final effect for other process uses. There is a tendency to operate with the highest vacuum available as the consequent reduction in boiling point increases the temperature driving force. The overall coefficient decreases very rapidly with decreased boiling point, however, and this may outstrip any advantage from increases in Δt_m. The optimum vacuum depends on the type of evaporator, the cost of vacuum and the process liquor involved, though in general, a value of 26 in. Hg, 2.0 lb/in.^2a [13.8 kN/m^2] is usually employed. Where heat sensitive materials such as fruit juices are processed, the highest possible vacuum may be used to preserve the quality of the product even at the expense of decreased capacity.

(vi) *Continuous and Batch Operation*

In common with most process operations, continuous operation of evaporators is preferable wherever possible. Where the final product is very viscous, however, the coefficient in the last effect will be low and there may be advantages in batch operation as the low coefficient will prevail for only a part of the operating cycle. In this situation, the increase in mean coefficient and hence in throughput may offset the down time of the unit. As an alternative, the bulk of the evaporation may be carried out in a continuous multi-effect unit from which liquor with an acceptable viscosity is withdrawn. This is then evaporated to the desired product concentration in a batch single-effect evaporator, sometimes termed a 'finishing pan'.

3.6.3 Evaporator Accessories

(i) *Condensers*

Two important types of condenser are used on evaporator installations:

(a) surface condensers, in which the vapour and the cooling water are separated by a metal wall. These are expensive and only used where the vapour must be recovered separately from the cooling water.
(b) jet condensers, in which the vapour and the cooling water are in direct contact. They are cheaper than surface condensers and use less water.

These two types are further subdivided according to whether the same pump removes the air and hot water or a separate pump is used for the air. The former is known as a wet condenser and the latter, a dry condenser. Similarly, a barometric condenser is one positioned so that water drains from it by a barometric hot leg into a sealed tank, whilst a unit in which the hot water is removed by a pump is a low level condenser. The latter is confined to small installations, whilst barometric dry condensers are the usual practice for large evaporators.

The amount of water required to operate a condenser is obtained by a simple heat balance as shown in the following example.

Example 3–8

3000 lb/hr [3.78 kg/s] steam at 26 in. Hg vacuum [13.5 kN/m²] is to be condensed in a jet condenser using cooling water at 60°F [288 K]. If the cooling water exit temperature is 110°F [316 K] and the steam is not super-heated, what is the cooling water requirement?

At 26 in. Hg [13.5 kN/m²], steam temperature = 125°F [325 K] and latent heat = 1023 BTU/lb [2379 kJ/kg]. Therefore,

latent heat to be removed = 3000 × 1023 = 3 069 000 BTU/hr

[= 3.78 × 2379 = 8993 kW]

sensible heat to be removed = 3000 × 1.0 (125 − 110) = 45 000 BTU/hr

[= 3.78 × 4.18 (325 − 316) = 13 kW]

Hence the total heat transferred to the water is 3 114 000 BTU/hr [9006 kW] and the water required is

$$\frac{3\ 114\ 000}{1.0 \times (110 - 60)} = 62\ 280 \text{ lb/hr}$$

or   62 280/600 = 103.8 gal/min

$$\left[\frac{9006}{4.18\ (316 - 288)} = 7.85 \text{ kg/s} \right]$$

One of the problems in condenser design is the estimation of the air which enters the unit. This arises from several sources including the cooling water, the feed liquor, the steam to the first effect, leaks and any non-condensables liberated by reaction during evaporation. It may be assumed that cooling water contains about 2 per cent by volume of dissolved air, and air in the steam may be eliminated when the first effect (normally at greater than atmospheric pressure) is vented to atmosphere. The other items are very variable though an estimate of 0.5 per cent air in the vapours going to the condenser is not unreasonable.

In sizing the vacuum pump, it is important to remember that the air leaves the condenser saturated with water vapour and adequate connections should be provided between the evaporator and the condenser, bearing in mind the low density of steam under vacuum and hence the high linear velocity.

(ii) *Pumps*

There are relatively few problems associated with evaporator feed pumps and it is the removal of product and condensate which merit closer attention. Centrifugal pumps are rarely suitable for removing viscous materials under vacuum unless they are self-priming, and normally reciprocating or positive displacement pumps are employed. Similar considerations apply to condensate removal, though here the problem of air removal is also involved. This is achieved by oversizing reciprocating pumps or installing steam traps in the case of self-priming centrifugal pumps. Any steam space which is above atmospheric pressure may be vented directly to the air, otherwise vent lines are connected directly to the condenser.

The condensate from the first effect is normally returned as boiler feed and this is made up with some of the condensate from the second effect, providing this is free from contamination due to entrainment. The rest of the condensate from the second and subsequent effects is normally wasted unless it is of value as process water.

In providing vacuum pumps, steam jets have many advantages over pumps, not least being cheapness and absence of moving parts. At a vacuum of 24 in. Hg [20.0 kN/m²] and less, the ejector, which is usually a single stage unit, uses more steam than a reciprocating pump, though above 28 in. Hg [6.9 kN/m²] a three stage ejector is much more economical.

(iii) *Thermocompression*

In a single stage evaporator the vapours produced contain almost as much heat as the steam supplied, although they are at a lower temperature necessary to provide a driving force. These vapours are then condensed with

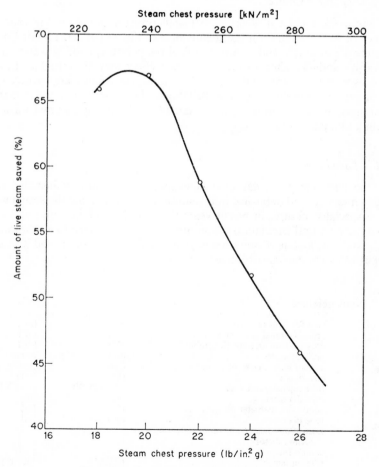

FIGURE 3–12. Thermocompressor economy[2]

water as a convenient means of removal—a waste of both heat and water. If, however, the vapours are recompressed to the saturation temperature of the original steam, they can be re-introduced to the steam chest with considerable savings—the work of recompression being low compared with the latent heat. Such an operation is known as thermocompression and this may be achieved in a centrifugal compressor or, where steam is available at high pressure, in a steam jet booster. The latter works on the principle of an ejector and is used to compress vapours instead of non-condensable gases. The pressure at which steam leaves a thermocompressor depends on the pressure and the proportion of steam to vapours. Figure 3–12 shows the savings in live steam, when steam is available at 150 lb/in.^{2} g [1135 kN/m^2] and vapours leave the evaporator at 15 lb/in.^{2} g [205 kN/m^2].[2]

Apart from large savings in steam consumption, thermocompression vaporators are particularly suited to the processing of heat sensitive materials.

(iv) *Evaporator Controls*

In multi-effect units it is usual to install pressure gauges in the vapour space of each unit and on the steam chest of the first effect. Liquor flow-meters are desirable and a record of feed temperature, cooling water temperature and the temperature of the last effect may be valuable. Level glasses are important, and where dark or viscous fluids are involved a hydrostatic gauge should be specified. Sight glasses are important in many instances and these should be illuminated from inside the unit and washed with a thin jet of liquor.

3.6.4 Conclusion

In the past one particular type of evaporator was always installed for a given service, and published information was so scarce that the design was of a specialist nature. In recent years the forced circulation unit and the long-tube natural circulation evaporator have been applied to a vast range of problems, and it is quite likely that standard horizontal and vertical types will soon become obsolete.

3.7 Nomenclature

A	area for heat transfer	(ft^2)	[m^2]
a	cross-sectional area for flow	(ft^2)	[m^2]
a'	cross-section area for flow/tube	(ft^2)	[m^2]
B	baffle spacing	(in.)	[mm]
C_F	annual fixed charge of exchanger/unit area	(£/ft^2)	[£/m^2]
C_W	cost of water	(£/lb)	[£/kg]
C_p	heat capacity/unit mass	(BTU/lb °F)	[J/kg K]
D	tube diameter	(ft)	[m]
D_e	equivalent diameter on shell side	(ft)	[m]
d	tube diameter	(in.)	[m]
F	logarithmic mean temperature difference correction factor	(—)	[—]
F_c	caloric fraction	(—)	[—]
f	friction factor	(ft^2/in.^{2})	[m^2/m^2]
G	mass velocity of fluid	(lb/hr ft^2)	[kg/s m^2]

g	acceleration due to gravity	(ft/hr²)	[m/s²]
h	film coefficient of heat transfer	(BTU/hr ft² °F)	[W/m² K]
j	$\sqrt{}$ (total number of tubes)	(—)	[—]
k	thermal conductivity	(BTU/hr ft² °F/ft)	[W/mK]
L	tube length between tube sheets	(ft)	[m]
m	mass flow of fluid	(lb/hr)	[kg/s]
N	total number of tubes	(—)	[—]
n	number of tube-side passes	(—)	[—]
ΔP	pressure drop	(lb$_f$/in.²)	[kN/m²]
P_T	tube pitch	(in.)	[mm]
p	temperature ratio defined in equation (3–12)	(—)	[—]
Q	heat load	(BTU/hr)	[W], [kW]
R	total resistance to heat transfer	(hr ft² °F/BTU)	[m² K/W]
Re	Reynolds number	(—)	[—]
r	individual restistance to heat transfer	(hr ft² °F/BTU)	[m² K/W]
r	temperature ratio defined in equation (3–12)	(—)	[—]
r_d	resistance to heat transfer due to scale	(hr ft² °F/BTU)	[m² K/W]
r_m	resistance to heat transfer due to tube wall	(hr ft² °F/BTU)	[m² K/W]
s	specific gravity	(—)	[—]
ΔT_f	temperature term defined in equation (3–14)	(°F)	[K]
T_1, T_2	temperature of the hot fluid in and out respectively	(°F)	[K]
Δt_m	logarithmic mean temperature difference	(°F)	[K]
t_1, t_2	temperature of the cold fluid in and out respectively	(°F)	[K]
U	overall coefficient of heat transfer	(BTU/hr ft² °F)	[W/m² K]
u	linear velocity	(ft/s)	[m/s]
x	wall thickness	(ft)	[m]
y	clearance between tubes	(in.)	[mm]
ρ	density of fluid	(lb/ft³)	[kg/m³]
λ	latent heat of condensation	(BTU/lb)	[J/kg]
ϕ	viscosity ratio $(\mu/\mu_w)^{0.14}$	(—)	[—]
θ	operating time	(hr/yr)	[s/yr]
μ	viscosity of fluid	(lb/ft hr)	[N s/m²]

Subscripts

A	average value
C	clean conditions
D	design value
i	inside
io	inside value based on the outside area
o	outside
s	shell
t	tube
w	wall

REFERENCES

1. Coulson, J. M. and Richardson, J. F. *Chemical Engineering*, Vol. 1, 1st ed. London: Pergamon Press, 1957.
2. Kern, D. Q. *Process Heat Transfer*. New York: McGraw-Hill Book Co., 1950.
3. McAdams, W. H. *Heat Transmission*. New York: McGraw-Hill Book Co., 1951.
4. Jacob, M. *Heat Transfer*. Volume 1. New York: Wiley, 1956.
5. Holland, F. A., Moores, R. M., Watson, F. A. and Wilkinson, J. K. *Heat Transfer*. London: Heinemann Educational Books Ltd, 1970.
6. *Standards of Tubular Exchanger Manufacturers Association*. New York, 1959.
7. *British Standard 2041*, 'Tubular Heat Exchangers for use in the Petroleum Industry', 1953.
8. *British Standard 1500*. 'Fusion Welded Pressure Vessels for use in the Chemical, Petroleum and Allied Industries', 1958.
9. *Boiler and Pressure Vessel Code, section viii*, 'Unfired Pressure Vessels', ASME, 1959.
10. Perry, J. H. *Chemical Engineers Handbook*, 3rd ed. New York: McGraw-Hill Book Co. 1950.
11. Olin, E., et al., *Chem. and Met. Eng.* 1925, **32**, 370.
12. Bliss, J. D. *Chem. Eng.* 1947, **54**, 5, 134.

Chapter 4: Towers

4.1 Introduction

The design of columns for distillation, absorption or stripping operations consists essentially of three steps: the choice of the internals and whether packing, plates or a hybrid contacting device are best employed, the number of trays or the height of the packing, and the calculation of the column diameter. Each of these steps will be considered in detail in this and in subsequent chapters.

The actual shell of the column may be regarded as a simple pressure vessel—it is the internals which affect the performance and determine the final overall dimensions. Whilst change from one type of internal to another may only affect the overall size by a relatively small amount, the total operating and capital costs may be altered substantially. Thus the choice of internal device is extremely important, and in this chapter an attempt will be made first of all to classify briefly the available contacting devices and then to consider in greater depth which type should best be employed for any particular application.

4.2 Contacting Devices

A method of classification described by Fair[1] is shown in Table 4–1.

TABLE 4–1. CLASSIFICATION OF CONTACTING DEVICES[1]

Type of Device	Examples
Crossflow trays	Bubble caps, sieve, valve, others
Random packings	Rings, saddles, others
Regular packings	Grids, mesh, others
Counterflow trays	Rectangular (slotted) openings, round openings, others
Special devices	Splash decks, sprays, moving internals, others

The order in which the type of devices are listed probably represents the order of their current popularity, though this does not apply to the order in which the examples are presented. Crossflow trays permit passage of the frothy liquid across a plate where it discharges over a weir into a downcomer to the tray below. The vapour is dispersed into the liquid by means of caps, valves or simple perforations. In the counterflow devices, the liquid flows directly downwards over packings or specially designed grids or meshes, including Spraypak and Glitchgrid packings and Kittle trays. These special devices form a category in which many proprietary designs are included. They are described in the literature from time to time but sufficient information is rarely available to enable the chemical engineer to make detailed comparisons between individual designs.

The following sections will confine the discussion to the more widely available type of device and will include sufficient information for a reliable selection to be made.

4.3 The Choice between Packed and Plate Columns

Liquid-vapour mass transfer operations may be carried out in either packed or plate columns, the former providing continuous contact between the two phases whilst a plate column operates on a stagewise basis. Thus the two types of operation are quite different and the design engineer has to make a choice between them.

Perry,[2] while giving information on design factors, makes only passing reference to the relative merits of each type of column. In a comprehensive study of the problem of choice, Thibodeaux and Murrill[3] outline a selection scheme considering the factors under four main headings:

(i) factors that depend on the system, i.e. the components,
(ii) factors that depend on the fluid flow movement,
(iii) factors that depend upon the physical characteristics of the column and its internals,
(iv) factors that depend upon the mode of operation.

These factors have been discussed by a number of authors.[4]-[8] Although the method of grouping provides a neat comparison of factors, in practice they are not normally independent.

4.3.1 System Factors

(i) *Scale.* For column diameters of less than approximately 3 ft [1 m], it is more usual to employ packed towers because of the high fabrication costs of the small trays. For a very large column, liquid distribution problems may arise and this fact, coupled with the weight considerations of a large volume of packing, may lead to the choice of a plate tower. A preliminary estimate of tower diameter may be obtained from Figure 4–1.[9]

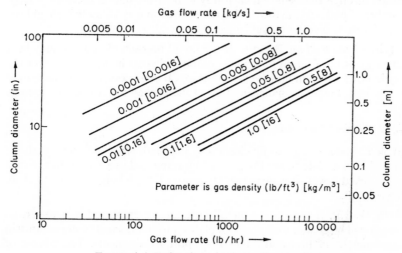

FIGURE 4–1. Estimation of column diameter[9]

(ii) *Foaming*. This is discussed in some detail in Chapter 6, where its effect on tray columns for distillation is discussed. Froth is produced by the bubbling action of the vapour through the liquid, and the use of a packed tower minimizes the effect as contact is then between the vapour and the liquid film flowing over the packing. Examples of packed tower operation with foaming systems include the absorption of hydrogen sulphide and sulphur dioxide from natural gas with sodium carbonate, and the scrubbing of effluent gases.

(iii) *Fouling systems*. If a system contains suspended solids or sludges, a plate column is to be preferred since, in a packed column, solids can accumulate in the voids, coating the packing material and rendering it ineffective.

(iv) *Corrosive systems*. Because ceramics are used for many packings, it is usually more economic to use a packed tower than to employ a corrosion resistant metal for a tray column, with perhaps associated problems in fabrication.

(v) *Heat evolution*. In a process where heat has to be removed, the use of cooling coils on the plates of a tray column provides a method of carrying out this operation, which is almost impossible in a packed tower. A high degree of turbulence is available within the confines of the tray, and the liquid can be removed from the plate and cooled externally before passage to the next plate. Operation in this way with a packed column raises difficulties in the collection and withdrawal of the liquid. The absorption of oxides of nitrogen in water in the nitric acid process provides a good example of this type of operation.

(vi) *Pressure drop*. The difference in pressure between the bottom and the top of a column necessary to achieve the upward flow of vapour is generally less in a packed column. In this case the pressure drop is due to skin friction and the losses due to vapour flowing through the liquid coated voids. In a plate column there is the additional friction generated as the vapour passes through the liquid on each tray to be overcome. If there are a large number of plates in the tower, this pressure drop may be quite high and the use of a packed column could effect considerable saving.

(vii) *Liquid hold-up*. Because of the liquid on each plate there may be a large quantity of liquid in a plate column, whereas in a packed tower the liquid flows as a thin film over the packing. If close control over the system is required, then the smaller hold-up of the packed tower is an advantage.

4.3.2 Physical Considerations

(i) *Maintenance*. If periodic cleaning is required, manholes may easily be provided in a plate tower for internal cleaning and inspection. In the case of a packed tower, it is necessary to remove the packing in order to clean the column.

(ii) *Weight*. If ceramic or metal packings are employed in a tower, its weight will be greater than that of a plate equivalent, a fact of importance in assessing foundation and packing support requirements. The advent of lightweight plastic packings has largely overcome this problem.

(iii) *Sidestreams.* The withdrawal or insertion of sidestreams presents no difficulty with a plate column, though modification at the particular tray may be necessary. With a packed tower such operation is virtually impossible, as a device would be necessary for the collection or distribution of the liquid or vapour without interfering with the smooth operation of the tower.

(iv) *Size and cost.* For diameters of less than 3 ft [1 m], packed towers entail lower fabrication and material costs than their plate equivalents. Above this diameter, generalization is impossible. With regard to height, a packed column is usually shorter than the equivalent plate column, though the latter may require a smaller diameter for the same vapour flowrate.

4.3.3 Mode of operation

(i) *Batch distillation.* Where it is desirable to recover a very high percentage of the more volatile component of a binary mixture, the smaller liquid hold-up of a packed column will perform the duty better than a plate column.

(ii) *Intermittent distillation.* Where it is convenient to retain the maximum hold-up of liquid in a column during periods between operation, the conventional bubble cap plate provides a positive seal for the liquid.

(iii) *Continuous distillation.* There is no one major consideration for this most common operation. All the factors described in this section should be considered.

(iv) *Turndown.* The ratio between the maximum and minimum loadings at which conditions of flooding, and the lowest level of acceptable efficiency, respectively, exist, is known as the turndown ratio. For a feed rate turndown of greater than 2.5:1, a plate column is the only practicable solution.

These then are the main factors to be considered, although it may be that one factor is of overriding importance. In most cases, however, a degree of compromise in the selection of the tower type is usually required.

Having made the choice between a plate or a packed column it is now necessary to consider the type of packing or tray to be used.

4.4 Tower Packings

Packings may be divided into three classes; broken solids, shaped packings and grids. Whilst broken solids are the cheapest packing and may range in size from ½ in. to 6 in. [12–150 mm], they are rarely used because their performance is unpredictable. They cannot be used for a duty where there is possibility of solids depositing on the packing, and although they possess a high specific area the pressure drop is also high.

Grid packings may be made in a variety of materials, the most common of which is wood. A column is easily packed and may be cleaned by dismantling without too much difficulty. Free space is large so that dirty duties, where blocking would occur with other packings, may be handled. For the same reason pressure drop is low, though problems may arise in obtaining good liquid distribution.

It is the shaped packings which will receive detailed attention in this section, as much experience has been gained in the design and operation of columns employing a wide variety of shapes in sizes ranging from $\frac{1}{8}$ in. to 6 in. [3–150 mm].

The proper selection of a packing involves an economic balance wherein allowance is given to cost per unit volume of packing, corrosion, cost of shell and internals, and the packing performance. Packing size should be selected to lie within the limits indicated in Chapter 5 when considering each type in turn. Materials of construction provide a further choice which should be matched to the application under consideration.

Stoneware packing is the most widely used and is the cheapest of materials. It is heavy and must be handled with some care to avoid breakage, though its strength and resistance to thermal shock are adequate for most purposes.

Packing made of carbon is expensive and is normally used with hydrofluoric acid, fluorides, phosphoric acid, and in other conditions of corrosive service. It must not be used in contact with nitric acid or other oxidizing media.

The use of metals for tower packings means that a wide range of materials is available to cover a large number of process conditions. An additional advantage lies in the fact that thin-walled packings may be constructed, providing a higher free gas space than is available with the above materials. The high cost of stainless steel is offset by virtually unlimited life for many applications and, in addition, packings made of aluminium, copper, nickel, monel, iron and titanium are also available. Resistance to thermal shock is good, the packings are mechanically strong and may be cleaned either *in situ* or by removal from the tower.

Plastics, fused silica and glass, have been listed previously under miscellaneous materials, though recent advances in plastics have opened up new possibilities for tower packings.[10] Plastic packings are available in a range of materials and corrosion resistance to a wide range of chemicals is an important advantage, especially at temperatures in excess of 200°F [370 K]. Glass packings[11] are available in several forms and sizes for corrosive duties, although they suffer from lack of mechanical strength and are susceptible to thermal shock.

4.4.1 Types of Packing

Figure 4–2 shows the most commonly used types of packing.

(i) *Raschig Rings*

These are the oldest, cheapest and most widely used type of packing. The height of a ring is always equal to its diameter and the wall thickness is determined by the material of construction. They are less efficient than some other packings and may not necessarily be the most economic in practice. Rings up to 2 in. [50 mm] diameter are normally located in the tower at random, and those above 3 in. [80 mm] are stacked. Random packing provides a higher efficiency though a higher pressure drop and a lower throughput when compared with stacked rings.

Physical characteristics are shown in Table 4–2.

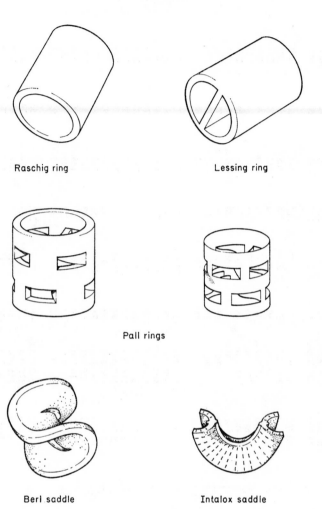

Raschig ring   Lessing ring

Pall rings

Berl saddle   Intalox saddle

FIGURE 4–2. Packing types

(ii) *Lessing Rings*

The addition of a partition across the centre of a Raschig ring forms a Lessing ring.[12] The surface area is increased by 20 per cent, though its effectiveness is less marked than might be expected since it is the internal area which has been increased. The increased strength of the ring in the larger sizes makes it useful for a stacked distribution region below a random packing. The addition of a further portion dividing the internal volume into four parts gives additional strength for distribution duty. Lessing rings are available in stoneware, porcelain and a variety of metals. Their physical characteristics are shown in Table 4–3.

TABLE 4-2. PHYSICAL CHARACTERISTICS OF RASCHIG RINGS

Material	Size in.	Nominal [mm]	Wall thickness in.	Number pieces/ft³	[pieces/m³]	Weight lb/ft³	[kg/m³]	Surface area ft²/ft³	[m²/m³]	Free space %	Packing factor F
Porcelain	1/4	6	1/16	86 500	3 050 000	60	961	217	710	62	1600
	3/8	10	1/16	24 700	870 000	61	978	147	481	67	1000
Stoneware and porcelain	1/2	12	3/32	10 700	377 000	55	882	112	368	64	640
	5/8	16	3/32	5 670	200 000	56	898	94	308	67	380
	3/4	18	3/32	3 090	109 000	50	802	74	242	72	255
	1	25	1/8	1 350	47 500	42	674	58	190	74	160
	1¼	32	3/16	670	23 600	46	737	45	147	71	125
	1½	40	1/4	381	13 400	46	737	36	118	68	130
	2	50	1/4	164	5 770	41	658	28	91.6	74	65
	3	80	5/16	50	1 760	35	561	19	62.2	78	36
	4	100	3/8	20	705	36	578	14	45.8	80	30
	6	150	0.70	8	282	48	770	13	42.6	68	
Metal carbon steel	1/4	6	1/32	88 000	3 100 000	133	2130	224	734	72	700
	3/8	10	1/32	27 000	950 000	94	1510	161	527	81	390
	1/2	12	1/32	11 400	402 000	75	1200	122	400	85	300
	1/2	12	1/16	10 900	384 000	132	2120	111	364	73	410
	5/8	16	1/32	6 130	216 000	62	994	103	337	87	258
	3/4	18	1/32	3 340	118 000	52	834	81	265	89	185
	3/4	18	1/16	3 140	110 000	94	1510	75	246	80	230
	1	25	1/32	1 430	50 400	39	625	62	203	92	115
	1	25	1/16	1 310	46 100	71	1040	56	183	86	137
	1¼	32	1/16	725	25 500	62	994	48	157	87	110
	1½	40	1/16	400	14 100	49	785	39	128	90	83
	2	50	1/16	168	5 910	37	593	29	95.0	92	57
	3	80	1/16	51	1 800	25	401	20	65.5	95	32
Carbon	1/4	6	1/16	85 000	3 000 000	46	737	212	694	55	1600
	3/8	12	1/16	10 600	372 000	27	433	114	374	74	410
	1/2	18	1/8	3 140	111 000	34	545	75	246	67	280
	3/4	25	3/16	1 325	46 700	27	433	57	187	74	160
	1	32	1/4	678	23 900	31	497	45	147	69	125
	1¼	40	2¼	392	13 800	34	545	38	124	67	130
	1½	50	1/4	166	5 850	27	433	28	91.5	74	65
	2	80	5/16	49	1 730	23	369	19	62.2	78	36

TABLE 4–3. PHYSICAL CHARACTERISTICS OF LESSING RINGS

Material	Size in.	Nominal [mm]	Wall thickness in.	Number pieces/ft³	[pieces/m³]	Weight lb/ft³	[kg/m³]	Surface area ft²/ft³	[m²/m³]	Free space %
Porcelain	$\frac{1}{2}$	12	0.07	10 300	363 000	52	834	130	426	64
	$\frac{3}{4}$	18	0.09	2 850	101 000	46	738	85	278	72
	1	25	0.13	1 300	45 800	49	786	68	222	68
	$1\frac{1}{2}$	40	0.18	390	13 700	43	690	47	154	68
	2	50	0.24	160	5 640	43	690	34	101	69
	3	75	0.31	50	1 760	43	690	24	78.5	71
	4	100	0.44	21	740	45	722	18	59.0	70
Stoneware	6	150	0.70	8	282	62	995	16	52.4	60
Metal*	$\frac{1}{2}$	12	$\frac{1}{32}$	11 000	387 000	100	1602	148	485	77
	$\frac{3}{4}$	18	$\frac{1}{32}$	3 170	112 000	71	1140	99	324	80
	1	25	$\frac{1}{16}$	1 250	44 000	95	1520	68	222	78
	$1\frac{1}{4}$	32	$\frac{1}{16}$	674	23 500	81	1300	58	190	79
	$1\frac{1}{2}$	40	$\frac{1}{16}$	391	13 700	65	1040	48	157	81
	2	50	$\frac{1}{16}$	167	5 880	49	786	35	114	83
Cross Partition Stoneware	3	75	0.30	72	2 540	73	1170	41	134	51
	4	100	0.44	30	1 060	74	1190	31	101	51
	6	150	0.70	9	316	75	1200	20	65.5	50

* See footnote to Table 4–2.

TABLE 4-4. PHYSICAL CHARACTERISTICS OF BERL SADDLES

Material	Size in.	Nominal [mm]	Wall thickness in.	Number pieces/ft³	[pieces/m³]	Weight lb/ft³	[kg/m³]	Surface area ft²/ft³	[m²/m³]	Free space %	Packing factor F
Stoneware	$\frac{1}{4}$	6		107 000	3 770 000	56	898	274	896	64	900
	$\frac{1}{2}$	12		16 700	588 000	54	865	142	465	66	240
	$\frac{3}{4}$	18		4 950	174 000	49	785	87	285	71	170
	1	25		2 180	76 800	45	721	76	249	73	110
	$1\frac{1}{2}$	40		645	22 700	40	641	46	151	74	65
	2	50		250	8 800	39	625	32	105	75	45

TABLE 4-5. PHYSICAL CHARACTERISTICS OF INTALOX SADDLES

Material	Nominal size in.	[mm]	Number pieces/ft³	[pieces/m³]	Weight lb/ft³	[kg/m³]	Surface area ft²/ft³	[m²/m³]	Free space %	Packing factor F
Porcelain	$\frac{1}{4}$	6	117 500	4 140 000	54	866	300	981	75	725
	$\frac{3}{8}$	10	49 800	1 760 000	50	802	240	785	76	330
Ceramic	$\frac{1}{2}$	12	20 700	730 000	45	723	190	621	78	200
	$\frac{3}{4}$	18	6 500	229 000	44	706	102	334	77	145
	1	25	2 385	84 000	44	706	78	256	77	98
	$1\frac{1}{2}$	40	709	25 000	42	674	59	193	80	52
	2	50	265	9 350	42	674	36	118	79	40
	3	80	53	1 870	37	593	28	91.5	80	22
Plastic Note 2	1	25	1 650	58 000	5.75	92.3	63	206	91	33
	2	50	190	6 700	4.25	68.2	33	108	93	21
	3	80	41	1 440	3.50	56.2	27	88.4	94	16

Weights are given for Polypropylene. For other materials multiply by the following factors: polyethylene 1.03, plastile 0.92, penton 1.54, kynar 1.95, PVC 1.54.

TABLE 4–6. PHYSICAL CHARACTERISTICS OF PALL RINGS

Materials	Nominal size in.	[mm]	Wall thickness in.	Number pieces/ft³	[pieces/m³]	Weight lb/ft³	[kg/m³]	Surface area ft²/ft³	[m²/m³]	Free space %	Packing factor F
Stoneware	1	25	0.12	1400	49 300	40	642	67	220	73	70
	1½	40	0.18	420	14 800	36	578	47	154	76	48
	2	50	0.20	170	5 990	34	546	38	124	78	28
	3⅜	80	0.30	43	1 510	30	482	23	75.4	80	20
	4	100	0.38	16	564	26	417	17	55.7	82	
Carbon	1	25	0.16	1400	49 300	36	578	69	226	73	
	1⅜	35	0.18	510	18 000	33	530	51	167	76	
	2	50	0.25	164	5 780	27	434	37	121	79	
Metal*	⅝	16	0.016	5950	209 000	37	594	104	341	93	
	1	25	0.024	1400	49 300	30	482	63	207	94	
	1½	40	0.032	375	13 200	24	385	39	128	95	
	2	50	0.040	170	6 000	22	353	31	101	96	
Plastic†	⅝	16	0.03	6050	213 000	7.0	112	104	341	87	97
	1	25	0.04	1440	50 800	5.5	88.4	63	207	90	52
	1½	40	0.04	390	13 700	4.75	76.2	39	128	91	32
	2	50	0.06	180	6 340	4.25	68.2	31	101	92	25
	3½	90	0.06	33	1 160	4.00	64.2	26	85.2	92	16

* See footnote to Table 4–2.
† See footnote to Table 4–5

TABLE 4-7. COST DATA—TOWER PACKINGS[21] COSTS ARE PRESENTED AS INSTALLED COSTS OF PACKINGS. $/ft³ [$/m³]

Packing material	¾ in. [19 mm]		1 in. [25 mm]		1½ in. [38 mm]		2 in. [50 mm]		3 in. [76 mm]		Unspecified size	
	$/ft³	[$/m³]	$/ft³	[$/m³]	$/ft³	[$/m³]	$/ft³	[$/m³]	$/ft³	[$/m³]	$/ft³	[$/m³]
Raschig rings												
Stoneware	—	—	5.8	205	4.8	170	3.9	138	3.2	113	—	—
Porcelain	—	—	7.8	276	6.5	230	5.3	187	4.3	152	—	—
Stainless steel	—	—	78.4	2770	51.2	1820	36.3	1280	25.4	898	—	—
Berl saddles												
Stoneware	21.0	740	16.2	572	8.7	307	—	—	—	—	—	—
Porcelain	24.3	860	17.8	628	9.7	343	—	—	—	—	—	—
Coke	—	—	—	—	—	—	—	—	—	—	3.9	138
Alumina	—	—	—	—	—	—	—	—	—	—	14.1	498
Activated carbon	—	—	—	—	—	—	—	—	—	—	15.9	562
Silica gel	—	—	—	—	—	—	—	—	—	—	30.4	1070

Pall rings[22]. Installed costs of 50 mm pall rings in a mild steel shell is given by cost of packed tower,

$$\$/ft = 23.3X^{1.66} \qquad [\$/m = 556X^{1.66}]$$

(See equation (4.1) where X is the diameter of the shell (ft) [m].)

(iii) *Berl Saddles*

Berl saddles were the original type of saddle packing. The free gas space is less than for rings, but with the better aerodynamic form, the pressure drop is lower and the capacity larger.[13] Recent improvements in Intalox saddles and Pall rings mean that Berl saddles have largely been superseded, though they still find application in vacuum distillation and in cases where high contact efficiency is required. They are manufactured in both stoneware and porcelain.

(iv) *Intalox Saddles*

Intalox saddles, together with Pall rings, now form the two most popular choices of packing material. The shape of the Berl saddle has been modified in the Intalox so that adjacent elements do not blank off any significant portion of the wetting liquid. No stagnant pools of liquid form and no gas bubbles are trapped. These factors, combined with the smooth shape, mean that liquid flows freely with the minimum of hold-up, and the gas is not subjected to violent changes of direction. The result is that the capacity is increased, and a low pressure drop observed.[14] Intalox saddles are available in ceramics or plastics; the latter in a range designed to offer good corrosion resistance to a large number of systems.[10]

(v) *Pall Rings*

The use of Pall rings may be expected to supersede the use of other ring packings in the future. They are economical to produce and highly efficient in use. Their superior characteristics derive from the slotted wall and internal projecting tongues, which lead to better wetting and distribution; this overcomes the problem of liquid tending to migrate to the column wall. Pall rings are available in stoneware, carbon, metals and plastics, and their performance is some 50–100 per cent greater in throughput at a pressure drop 50–70 per cent smaller than with Raschig rings under the same conditions.

4.4.2 Cost Data

Cost data for tower packings is more readily obtainable direct from manufacturers. As a guide, Table 4–7 presents recently published costs and correlations for various tower packings, which may be used to give a first estimate of packing costs. With the exception of Pall rings, all data excludes the cost of the shell, which may be obtained from Chapter 7.

4.5 Choice of Plate Types

There are four main tray types—the bubble cap, sieve tray, ballast or valve tray, and the counterflow trays. The bubble cap tray has enjoyed wide application in industry for many years and performance characteristics have been widely published.[2],[15] Counterflow trays suffer from lack of design information on their performance, though they have the merit of being the cheapest tray device. Against this cost consideration must be balanced the facts that design reliability and flexibility in operation are poor. In recent years the use of bubble cap trays has been largely superseded by the sieve tray and valve tray, and the former are now generally used only for special applications where leakage from one tray to another must be minimized. The primary reasons why sieve and valve trays are to be preferred are as follows:

(i) They are lighter in weight, less expensive and are easier and cheaper to install than bubble cap plates.
(ii) They have higher vapour and liquid handling capacity.
(iii) The peak efficiency is generally higher.
(iv) The pressure drop is lower than in a bubble cap plate due to the usually negligible liquid gradient on a sieve and valve tray.
(v) Maintenance costs are reduced due to the ease of cleaning the simpler construction of a sieve or valve tray.

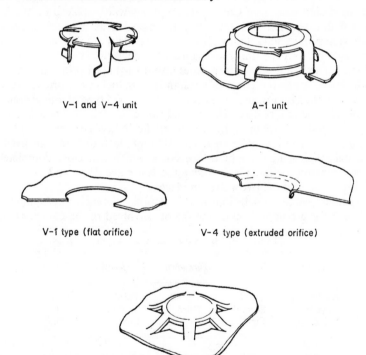

V–1 and V–4 unit   A–1 unit

V–1 type (flat orifice)   V–4 type (extruded orifice)

V–0 unit

FIGURE 4–3. Valve units

Sieve or perforated trays are normally employed if a wide range of flexibility is not required (normally a turndown of 5:1 is considered to be the maximum), and if the lowest tray cost is desired. They should not be used for low liquid rate applications, and if extensive blanking of the tray is likely to be required, their cost may exceed that of an equivalent valve tray. A discussion of design parameters and a design procedure for sieve trays is included in Chapter 6.

Valve trays offer the largest operating range and a greater capacity at a cost which is reasonable compared to sieve trays.[16] A turndown ratio of 10:1 has been quoted as the maximum, but it is important to realize that this figure can only be achieved under the following limitations:

 (i) There must be no pressure drop limitation.
 (ii) Tray spacing must be at least 24 in [0.61 m].
 (iii) Liquid loading must be fairly high.
 (iv) Single-pass tray design is necessary.
 (v) Equal loadings should apply to all trays within a given section in the tower.
 (vi) The system must be completely non-foaming.
 (vii) The vapour densities in the column should be low.

There are four main types of valve unit and these are shown in Figure 4–3.

In the design section in Chapter 6, emphasis will be placed on the V-1 and V-4 devices. The V-1 unit may be regarded as a general-purpose standard size unit, suitable for all services. The legs are formed integrally with the valve and the unit may be used for deck thicknesses up to $\frac{3}{8}$ in. [9.5 mm]. The V-4 design incorporates a venturi-shaped orifice opening in the deck which has the effect of substantially reducing the pressure drop across the unit. This low pressure drop makes the V-4 type of unit especially suitable for vacuum service. The normal range of deck thickness is 10–14 s.w.g. material [3.4–1.9 mm]. In cases where flexibility is not important, the V-0 non-moving unit may be used. The A-1 unit illustrated was the original valve unit with a lightweight orifice cover which can close completely. It has a separate ballast plate to give the two stage effect and a cage, or travel stop, to hold the two valves in place.

To summarize the conclusions of this section, reference may be made to Table 4–8 for a comparison of the factors involved in the choice of tray

TABLE 4–8. COMPARISON OF TRAY TYPES

Tray type	Capacity	Turndown	Efficiency	Cost
Bubble cap	Third	Second	Third	Third
Sieve tray	Second	Third	Second	First
Valve tray	First	First	First	Second

type. These would normally be considered in the order of capacity, turndown, and cost, but it should be emphasized that each case should be reviewed on its own merits before a decision is made on the most suitable type of tray to employ.

4.5.1 Cost Data

The wide variety of tray geometries, as opposed to tray types, make meaningful correlations of cost data difficult to present. Data for bubble cap trays,[17],[18] sieve trays[19] and valve trays[20] have appeared in the literature, but agreement in the comparison of costs is poor. A recent correlation for all these tray types has been produced by Guthrie[21] where multiplying factors are used in conjunction with a base cost to allow for different tray types. The data of Guthrie has been updated to 1970 and is presented below.

The base cost includes the cost of the trays, supports, fittings and fabrication and is given by:

$$\text{base cost} = CX^{1.0} \quad (\$) \qquad (4\text{-}1)$$

where $X = $ height of tray stack, (ft) [m]

$C = $ constant given by Table 4-9.

TABLE 4-9. VALUES OF CONSTANT C (EQUATION (4-1))

Tray diameter ft		2	3	4	5	6	7	8	9	10
[m]		0.61	0.91	1.22	1.52	1.83	2.14	2.44	2.74	3.05
Value of C	(A.A.)	20.1	29.1	40.3	55.9	78.3	101	134	168	222
	[SI]	65.9	95.4	132	183	257	331	440	551	728

The values of C are based on a tray spacing of 24 in. [0.61 m] using mild steel sieve trays. To take account of different tray spacings, tray types and materials of construction, the adjustment factors of Table 4-10 may be used in conjunction with equation (4-2).

$$\text{Cost} = \text{base cost}(F_s + F_t + F_m) \quad (\$) \qquad (4\text{-}2)$$

TABLE 4-10. VALUES OF ADJUSTMENT FACTORS

Tray spacing		F_s	Tray type	F_t	Tray material	F_m
in.	[m]					
24	0.61	1.0	Grid plate, sieve	0.0	Mild steel	0.0
18	0.46	1.4	Valve	0.4	Stainless steel	1.7
12	0.30	2.2	Bubble cap	1.8	Monel	8.9

N.B. If values of F_t and F_m are used individually, 1.0 should be added to the above values.

An example should serve to illustrate the method of cost estimation.

Example 4-1

Compare the costs of sieve trays, valve trays and bubble cap trays in a distillation column of 6 ft [1.83 m] diameter employing a tray spacing of 18 in. [0.46 m] and containing a total of 20 trays. The material of construction is stainless steel.

$$\text{Base cost} = CX^{1.0} \qquad (4\text{-}1)$$

$$X = 18/12 \times 19 = 28.5 \text{ ft}$$

$$[= 0.46 \times 19 = 8.73 \text{ m}]$$

For 6 ft [1.83 m] diameter, $C = 78.3$ [257], and

$$\text{base cost} = 78.3 \times 28.5$$

$$[= 257 \times 8.73] = \$2240$$

(a) Cost of stainless steel sieve trays, 18 in. [0.46 m] spacing:

$$F_s = 1.4 \qquad F_t = 0.0 \qquad F_m = 1.7$$
$$\text{Cost} = 2240(1.4 + 0.0 + 1.7) = \$6950 \qquad (4\text{–}3)$$

(b) Cost of stainless steel valve trays, 18 in. [0.46 m] spacing:

$$F_s = 1.4, \qquad F_t = 0.4, \qquad F_m = 1.7$$
$$\text{Cost} = 2240(1.4 + 0.4 + 1.7) = \$7850$$

(c) Cost of stainless steel bubble cap trays, 18 in. [0.46 m] spacing:

$$F_s = 1.4, \qquad F_t = 1.8, \qquad F_m = 1.7$$
$$\text{Cost} = 2240(1.4 + 1.8 + 1.7) = \$11\ 000$$

These costs are for the trays only; the costs of column shells are presented in Chapter 7.

Having presented a guide to the selection of internal contacting devices, the following section will be devoted to a summary of the methods available for calculating the number of trays in the column or the necessary height of packing needed to effect the desired separation.

4.6 Plate Calculations

The method of calculating the number of stages in a distillation problem will depend upon the system itself and the degree of accuracy required in the final answer. A discussion of the factors involved has recently been presented[23] and a classification of continuous distillation problems by considering methods of calculation has been produced. This classification is presented in Table 4–11.

TABLE 4–11. CLASSIFICATION OF CONTINUOUS DISTILLATION PROBLEMS BY METHODS OF CALCULATION[23]

System	Vapour-liquid equilibrium	Stages	Calculation	Remarks
Binary	Ideal	Few	Graphical	
Binary	Ideal	Many	Graphical/ analytical	Computer useful
Binary	Non-ideal	Few	Graphical	
Multicomponent	Ideal	Few	Analytical	Hand, if not repetitive
Binary	Non-ideal	Many	Graphical/ analytical	Computer useful
Multicomponent	Ideal	Many	Analytical	Computer essential
Multicomponent	Non-ideal	Few	Analytical	Computer essential
Multicomponent	Non-ideal	Many	Analytical	Computer essential

Some of these systems will be considered briefly here in order to outline the most suitable method of approach. References to detailed examples of particular methods are included for further evaluation.

4.6.1 Binary Calculations

The well-known method of McCabe and Thiele[24] depends upon the simplifying assumptions of constant heat of vaporization, no heat losses, no heat of mixing and constant molal flows of vapour and reflux in any section of the column. The method is applicable to both ideal and non-ideal systems and, if curved operating lines are used, for systems with unequal heats of vaporization.

Vapour–liquid equilibrium data is required for these and for all other separation processes. The vapour–liquid equilibrium ratio, K, is a function of temperature, pressure and composition, and for a component a,

$$K_a = y_a/x_a \qquad (4\text{-}4)$$

A wide range of published vapour–liquid equilibrium data has been reviewed by Hala et al.[25] while estimation methods have been fully discussed by Holland.[26] These data may be converted to an equilibrium x–y relationship by:

$$y_a = \alpha_{ab}x_a/1 + (\alpha_{ab} - 1)x_a \qquad (4\text{-}5)$$

where α_{ab} = relative volatility between a and b

$$= K_a/K_b \qquad (4\text{-}6)$$

The operating lines may be determined by a series of mass balances and the resulting graphical analysis is illustrated by the following example.

Example 4-2

A binary mixture of n-heptane and n-octane at its boiling point, containing 70 mol per cent of heptane, is to be continuously distilled to give a top product of 98 mol per cent and a bottom product of 1 mol per cent heptane. The reflux ratio is 3.0 and the relative volatility of n-heptane to octane is 2.0. Using the McCabe–Thiele construction, obtain the number of theoretical plates required for this separation.

(i) Calculate the x-y equilibrium diagram. From equation (4–5),

$$y_a = \alpha x_a/1 + (\alpha - 1)x_a = 2.0 \qquad (4\text{-}5)$$

Putting values of x_a into equation (4–5) over the range $x_a = 0$ to $x_a = 1.0$, values of y_a are obtained as follows:

x_a	0	0.1	0.2	0.3	0.4	0.5	0.6	0.7	0.8	0.9	1.0
y_a	0	0.182	0.333	0.461	0.571	0.667	0.750	0.823	0.889	0.947	1.0

The equilibrium relationship is plotted in Figure 4–4.

(ii) The McCabe–Thiele construction may now be completed.

(a) The q-line is vertical through $x_F = 0.70$.

(b) The operating line in the rectifying section may be drawn through the points (x_D, x_D) and $(0, x_D/R + 1)$, i.e. points (0.98, 0.98) and (0, 0.245).

The rectifying operating line is therefore the line AB and the operating line in the stripping section is drawn between B and the point C, (x_S, x_S) = (0.01, 0.01).

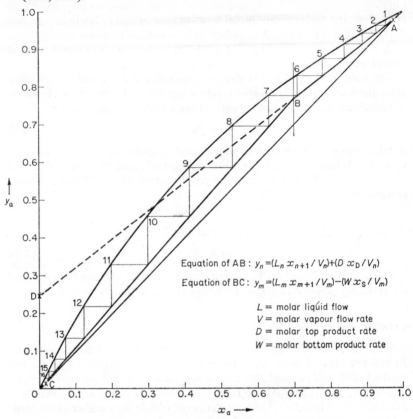

Equation of AB : $y_n = (L_n \, x_{n+1} / V_n) + (D \, x_D / V_n)$

Equation of BC : $y_m = (L_m \, x_{m+1} / V_m) - (W \, x_S / V_m)$

L = molar liquid flow
V = molar vapour flow rate
D = molar top product rate
W = molar bottom product rate

FIGURE 4–4. McCabe–Thiele construction for Example 4–2

By stepping off the various stages as shown in Figure 4–4,

number of ideal stages = 17 ± 1.

Hence   number of theoretical plates = 16 ± 1.

Another graphical method for binary systems is that of Ponchon–Savarit.[27],[28] The particular merit of this method is its use in applications where large heat effects prevail and H-x-y data are available. A detailed discussion of the method with many examples is given by Henley and Staffin[29] to which further reference should be made.

A four-step procedure for the rapid estimation of the actual number of stages may be carried out as follows:

(i) Calculate the minimum number of stages, N_{min}, at total reflux.

(ii) Calculate the minimum reflux, R_{min}, at infinite stages.

(iii) Obtain the number of theoretical stages by an empirical method dependent on N_{min} and R_{min}.

(iv) Estimate the tray efficiency to calculate the actual number of plates.

Each of these steps will now be considered in turn.

4.6.2 Minimum Stages at Total Reflux

The Fenske equation[30] relates the minimum number of stages, N_{min}, to the compositions of the components in the distillate and bottoms using an average value of the relative volatility, α_{av},

$$N_{min} = \frac{\log\{(x_a/x_b)_D(x_b/x_a)_S\}}{\log \alpha_{av}} \quad (4\text{-}7)$$

$$N_{min} = (n + 1)$$

where n = number of theoretical plates.

If α_{ab} is not constant, one of the following averages may be used:

$$\alpha_{ab} = \{(\alpha_{ab})_D(\alpha_{ab})_S\}^{0.5} \quad (4\text{-}8)$$

or

$$\alpha_{ab} = \{(\alpha_{ab})_D(\alpha_{ab})_F^2(\alpha_{ab})_S\}^{0.25} \quad (4\text{-}9)$$

If the volatility varies very widely, the method described by Winn[31] may be more applicable, though this is usually confined to multicomponent systems.

4.6.3 Minimum Reflux at Infinite Stages

Underwood[32] has presented an analytical solution in general terms for the calculation of R_{min}. For the particular case of a binary mixture, the equations reduce to

$$\{\alpha_{ab} \cdot x_{aF}/(\alpha_{ab} - \theta)\} + \{(1 - x_{aF})/(1 - \theta)\} = 1 - q \quad (4\text{-}10)$$

$$R_{min} = \{\alpha_{ab} \cdot x_{aD}/(\alpha_{ab} - \theta)\} + \{(1 - x_{aD})/(1 - \theta)\} - 1 \quad (4\text{-}11)$$

where

$$\alpha_{ab} > \theta > 1.0$$

q is the ratio of moles of saturated liquid in the feed to the total moles of feed, i.e. heat to vaporize 1 mol of feed/molal latent heat of feed.

N.B. $q = 1.0$ and 0 for feed at its boiling point and for a saturated vapour feed respectively. θ is the root of equation (4–10).

Equation (4–10) is first solved for θ and then equation (4–11) is solved for R_{min}. A graphical method for R_{min} may be employed where convenient.[28]

4.6.4 Theoretical Stages

Having calculated the minimum number of plates at the minimum reflux ratio, the number of plates corresponding to finite reflux ratios may be determined. Gilliland's method[33] has been used for many years though a more recent method of Erbar and Maddox[34] is an improved technique. They produced the relationship shown in Figure 4–5. The way in which Figure 4–5, is used is best illustrated by means of an example.

Example 4–3

Using the 'short cut' method of estimating the number of ideal stages, find the number of theoretical plates for the system described in Example 4–2.

(i) Using the Fenske relationship, equation (4–7), to find N_{min},

$$N_{min} = \frac{\log \{(x_a/x_b)_D (x_b/x_a)_S\}}{\log \alpha_{av}} \qquad (4-7)$$

Now $x_{aD} = 0.98$, $x_{bD} = 0.02$, $x_{bS} = 0.99$, $x_{aS} = 0.01$, $\alpha_{av} = 2.0$. Hence

$$N_{min} = \log [(0.98/0.02)(0.99/0.01)]/\log 2 = 12.2$$

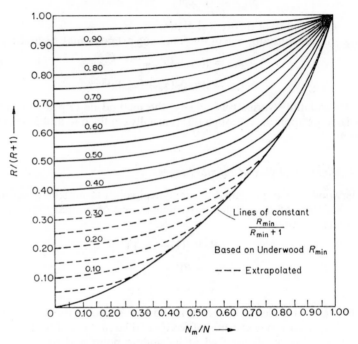

FIGURE 4–5. The correlation of Erbar and Maddox[34]

(ii) Using Underwood's relationship to find R_{min},

$$\{\alpha_{ab} \cdot x_{aF}/(\alpha_{ab} - \theta)\} + \{(1 - x_{aF})/(1 - \theta)\} = 1 - q \qquad (4-10)$$

$$R_{min} = \{\alpha_{ab} \cdot x_{aD}/(\alpha_{ab} - \theta)\} + \{(1 - x_{aD})/(1 - \theta)\} - 1 \qquad (4-11)$$

$\alpha_{ab} = 2.0$, $x_{aF} = 0.70$, $q = 1.0$, and hence from equation (4–10), $\theta = 1.18$. Then

$$R_{min} = \frac{2.0 \times 0.7}{2.0 - 1.18} + \frac{1 - 0.98}{1 - 1.18} - 1 = 1.29$$

(iii) Using the Erbar and Maddox Correlation for N (Figure 4–5),

$$R_{min}/R_{min+1} = 1.29/2.29 = 0.56$$

If $R = 3$, $R/R + 1 = 0.75$. From Figure 4–5, $N_{min}/N = 0.77$. Hence

number of plates $= 12.2/0.77 = 15.9$, i.e. 16

Example 4-4

Using the correlation of Erbar and Maddox (Figure 4–5), obtain the relationship between the number of ideal plates and the reflux ratio if $R_{min} = 1.29$ and $N_{min} = 12.2$.

By selecting values of the reflux ratio R, corresponding values of N may be obtained from Figure 4–5 as shown in Table 4–12. The relationship between

TABLE 4–12. THE CALCULATION OF THE NUMBER
OF PLATES FOR EXAMPLE 4–4.

	R	$R/R + 1$	N_{min}/N	N
	1.5	0.60	0.61	20.0
	2	0.67	0.67	18.8
$R_{min}/(R_{min}+1)$	3	0.75	0.77	15.9
$= 0.56$	4	0.80	0.82	14.9
	5	0.83	0.86	14.2
	7.5	0.88	0.91	13.4
	10	0.91	0.92	13.3

R and N is shown in Figure 4–6. The variation of the number of plates with the reflux ratio shows the problem of selecting the optimum value of R.

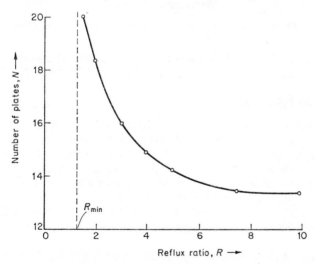

FIGURE 4–6. Relationship between the reflux ratio and the number of plates in Example 4–4

When the value of R is increased near the value of R_{min}, there is a large reduction in the number of plates required for the given separation. Hence the capital cost of the column will fall rapidly with a small increase in R at this point, with a comparatively small increase in steam costs due to the increased boil-up rate. At high values of R there is only a very small reduction in the number of plates, although the vapour rate in the column is high, giving high steam costs together with high capital costs due to the increased diameter of the column. The optimum value of the reflux ratio may be estimated by calculating the capital charges and the total running costs as indicated in Figure 4–7.

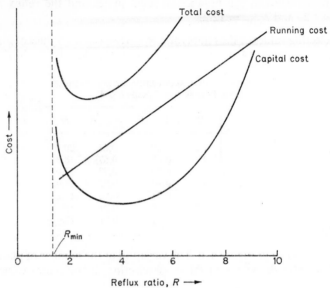

FIGURE 4–7. The relationship between reflux ratios and costs

There is no simple relationship between R_{min} and the optimum value of R, but in practice values range between 1.20 and 1.30 times R_{min}.[23]

4.6.5 Tray Efficiencies

The calculations discussed previously determined the number of theoretical plates or stages. By definition, the exit vapour stream leaving a tray is in thermodynamic equilibrium with the exit liquid stream. In practice it is normally impossible to produce this performance on any plant equipment, though it can be approached on small trays where the liquid is almost completely mixed. In this case, the liquid composition is virtually constant and equal to the exit concentration, and the vapour entering the liquid should also be of constant composition as it rises from a completely mixed tray below.

The approach to equilibrium conditions on a tray has been defined by Murphree[35] as the ratio of the actual change in a vapour concentration through the tray to the change which would have occurred if the vapour had actually reached a state of equilibrium with the liquid, i.e.

$$E_{MV} = (y_n - y_{n-1})/(y_n^* - y_{n-1}) \qquad (4\text{--}12)$$

where the subscripts n and $n - 1$ refer to the outlet and inlet vapour streams respectively. The Murphree tray efficiency can also be expressed in liquid terms as

$$E_{ML} = (x_{n+1} - x_n)/(x_{n+1} - x_n^*). \qquad (4\text{--}13)$$

In these equations, y_n^* and x_n^* refer to the vapour and liquid concentrations which would exist if the exit vapour and liquid were in equilibrium with the exit liquid and vapour compositions x and y respectively.

In practice, the Murphree tray efficiency can only represent reality in the special case of completely mixed tray liquids where only one value of y_n and x_n apply. In the more usual case, that is with incomplete mixing, the above equations only apply at a point in a pool of liquid and point Murphree efficiencies are defined as

$$E_{OG} = (y - y_{n-1})/(y^* - y_{n-1}) \qquad (4\text{-}14)$$

$$E_{OL} = (x_{n+1} - x)/(x_{n+1} - x^*) \qquad (4\text{-}15)$$

where the lack of a subscript denotes the actual concentrations at a given point in the pool. y^* is the vapour concentration in equilibrium with x and x^* is the liquid concentration in equilibrium with y ($y \neq y^*$ and $x \neq x^*$ since the concentrations change over the tray).

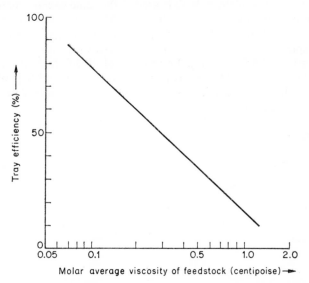

FIGURE 4–8. The tray efficiency correlation of Drickamer and Bradford[37]

The difficulty in the prediction of tray efficiency lies in the determination of the relationship between point efficiencies, mass transfer characteristics, tray layout, and hydraulic parameters. This topic has been the subject of extensive research with a major contribution in the Tray Efficiency Research Programme of the A.I.Ch.E.[36]

The most satisfactory method of obtaining tray efficiency is from experience gained from plant equipment using a similar system. Product streams may be analysed to determine the actual separation which is achieved over N_t stages; the number of equilibrium stages, N, to effect the same separation may then be found by an appropriate method and the overall column efficiency E_o is given by

$$E_o = N/N_t \qquad (4\text{-}16)$$

Designers and manufacturers of separation equipment are able in this way to build up an 'experience file' and are able to make a reasonable estimate of efficiency for physically similar systems. Data on efficiencies obtained in this way are tabulated in Perry,[2] though care should be exercised to ensure that the tray type and geometry and the operating conditions are similar in both applications.

Empirical correlations relating the overall column efficiency to the physical properties of the feed are useful providing their limitations are fully understood. The correlation of Drickamer and Bradford[37] relates molar feed viscosity to efficiency based on experience from fifty-four petroleum refinery systems. This correlation is shown in Figure 4–8. The viscosity may be calculated from the components of the feed thus:

$$\mu_L = x_1\mu_1 + x_2\mu_2 + x_3\mu_3 + \ldots + x_n\mu_n$$

with the temperature being taken as the arithmetic mean of the top and bottom temperatures of the column. The data may also be used for absorbers where the viscosity of the rich oil is used at the same average temperature. The data was originally obtained from bubble cap trays and sieve plates but may be applied to valve trays without modification.[38] The range of tower diameters was between 4 and 7.5 ft [1.2–2.3 m] with an average flow path length of 2.5 ft [0.76 m]. For towers of 10.5 ft [3.2 m] and 13.5 ft [4.1 m] diameter with flow paths of 4.0 ft [1.2 m] and 5.5 ft [1.67 m], the predicted values of the efficiency were found to be too low by 8 per cent and 18 per cent respectively.

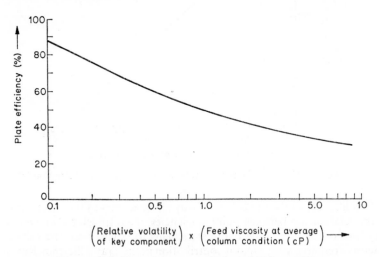

FIGURE 4–9. The efficiency correlation of O'Connell[39]

O'Connell[39] extended the scope of the above correlation by incorporating the relative volatility of the light key component to the heavy key with the average liquid viscosity. Both parameters are estimated at the average of the top and the bottom tower temperatures. The correlation, which is reproduced in Figure 4–9, covers commercial hydrocarbon and chlorinated hydrocarbon fractionation systems and commercial alcohol fractionation.

A more recent correlation for the Murphree plate efficiency of bubble cap and sieve trays has been proposed by English and Van Winkle[40] which was later simplified by Eduljee[41] to the equations presented below.

$$E_{MV} = 12.62(F_a)^{-\frac{1}{4}}(h_w)^{\frac{1}{4}}(Sc_L)^{1/7} \qquad (4-16)$$

$$[= 5.65(F_a)^{-\frac{1}{4}}(h_w)^{\frac{1}{4}}(Sc_L)^{1/7}]$$

where E_{MV} = Murphree plate efficiency (%)
F_a = fractional area of tray available (%),
   for vapour flow, i.e. free area fraction (dimensionless),
h_w = weir height (in.) [mm],
Sc_L = liquid Schmidt Number ($\mu/\rho D$) (dimensionless).

In situations where the above correlations are not applicable and no previous experience of the system is available, the efficiency may be estimated by the scale up of laboratory efficiency data. Detail of this type of method remains largely confidential, although a few reports of the procedure to be adopted have been published.[42],[43],[44] It is important to ensure that the actual system conditions should be employed, using continuous operation with the same tray device as will be used in the full size plant.

The A.I.Ch.E. *Bubble Cap Manual*[36] lists four main factors which affect the tray efficiency: the rate of mass transfer in both vapour and liquid phases, the degree of mixing on the tray, and the amount of liquid entrainment between the trays. These factors are discussed in detail by Smith[15] and the conclusions will be summarized here.

The point efficiency, E_{OG} or E_{OL}, may be determined from equations (4-17), (4-18), (4-19).

$$E_{OG} = 1 - \exp(-N_{OG}) \qquad (4-17)$$

$$E_{OL} = 1 - \exp(-N_{OL}) \qquad (4-18)$$

$$E_{OG} = E_{OL}/\{E_{OL} + \lambda(1 - E_{OL})\} \qquad (4-19)$$

where N_{OG} and N_{OL} are the numbers of overall transfer units (which have been discussed elsewhere, together with methods for their estimation) and

$$\lambda = mG_m/L_m \qquad (4-20)$$

The Murphree tray efficiency, E_{MV}, may then be calculated from

$$E_{MV} = 1/\lambda\{\exp(\lambda E_{OG} - 1)\} \qquad (4-21)$$

The effect of entrainment is included in the relationship between E_{MV} and the overall column efficiency,

$$E_o = E_{MV}/\{1 + (e' \cdot E_{MV}/L_m)\} \qquad (4-22)$$

where e' = entrainment estimated from Figure 6-4 and equation 6-2

L_m = liquid flow rate (mol/hr ft^2) [kmol/s m^2]

Equation (4-22) is subject to the following limitations:

(i) There is no back mixing on the tray.
(ii) No vertical concentration gradient exists in the liquid.
(iii) A linear equilibrium relationship ($y^* = mx + c$) applies over the concentration range on the tray.

(iv) The value of E_{OG} is constant on the tray.

(v) Vapour and liquid rates are constant across the tray.

(vi) There is constant composition across the column for the entering vapour.

4.7 Multicomponent Calculations

The design of most multicomponent distillation systems are now handled by computers and calculations by hand are rare. The basic methods of Lewis and Matheson[45] and Thiele and Geddes[46] are used and the application of computing to distillation problems has been covered by Hanson *et al.*[47] and Holland.[48]

The Lewis–Matheson model starts with assumed terminal compositions and works from each end of the column towards the feed tray, considering at each stage the constraints of heat and mass balances and equilibrium conditions. At the feed tray, there must be agreement between the compositions calculated from each end of the column and the feed composition. If there is no agreement, the iterative procedure must be repeated. In the Thiele–Geddes model, the number of stages and a temperature profile is assumed and calculations from one end of the column arrive at a composition at the other end.

Material balance requirements must be met, and in order to achieve this criterion the profile may be altered until agreement is reached.

The earlier method outlined for a rapid estimation of the number of stages for a binary system may be used for multicomponent separations by considering the light and heavy key components alone. Although there are obvious limitations on the use of a standard method of this type, it is widely used for first order approximations.

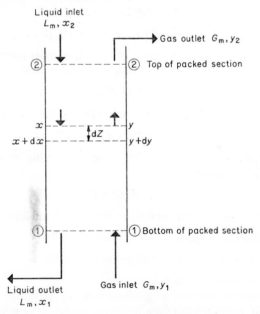

Liquid inlet
L_m, x_2

Gas outlet G_m, y_2

② Top of packed section

① Bottom of packed section

Liquid outlet
L_m, x_1

Gas inlet G_m, y_1

FIGURE 4–10. Diagrammatic section of a countercurrent packed tower

4.8 Transfer Unit Calculations

The height of packing in a countercurrent contactor, as opposed to a stagewise contactor, may most conveniently be calculated from the expression

$$Z = N_{OG}H_{OG} \qquad (4\text{-}23)$$

where Z = total height of contacting zone (ft) [m],
N_{OG} = number of overall transfer units (dimensionless),
H_{OG} = height of an overall transfer unit (ft) [m]

A diagrammatic representation of a countercurrent contactor is shown in Figure 4–10.

4.8.1 Number of Transfer Units

The number of transfer units depends upon the required separation, with more units being required for a difficult separation than a simple one. Detailed methods of determining the value of N_{OG} are available in a number of excellent texts[4],[5],[28],[49] and are all based upon the original definition

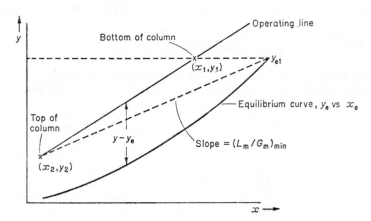

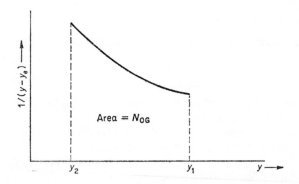

FIGURE 4–11. Evaluation of the number of overall transfer units, N_{OG}

of Chilton and Colburn,[50],[51] who defined the overall number of transfer units N_{OG} as

$$N_{OG} = \int_{y_1}^{y_2} \left(\frac{dy}{y - y_e} \right) + \frac{1}{2} \ln \left(\frac{1 - y_1}{1 - y_2} \right) \tag{4-24}$$

where $(y - y_e)$ is the driving force in the gas phase for a particular value of y. For the case of a dilute gas the second term is usually neglected.

In an analogous manner, the number of overall liquid phase transfer units N_{OL} may be defined as

$$N_{OL} = \int \frac{dx}{x_e - x} \tag{4-25}$$

Equation (4-24) may be conveniently solved by graphical integration as shown in Figure 4-11. An example of the graphical integration is included in Chapter 5.

For the special case of dilute solutions where both the operating and equilibrium lines are straight, i.e. $y_e = mx$, and the inlet liquor is solute free, it can be shown that the number of overall transfer units is given by[6]

$$N_{OG} = \frac{1}{1 - \lambda} \ln \left\{ (1 - \lambda) \frac{y_1}{y_2} + \lambda \right\} \tag{4-26}$$

where
$$\lambda = mG_m/L_m = y_{e_1}/(y_1 - y_2) \tag{4-20}$$

and y_{e_1} is the value of y in equilibrium with x_1.

Colburn[51] plotted the relationship between N_{OG}, y_1/y_2 and λ as shown in Figure 4-12.

For problems involving stripping or desorption operations, the use of N_{OL} is generally more convenient. The equation for N_{OL} is given by

$$N_{OL} = \frac{1}{1 - 1/\lambda} \ln \left\{ \left(1 - \frac{1}{\lambda} \right) \left(\frac{x_1 - x_{e_2}}{x_2 - x_{e_2}} \right) + \frac{1}{\lambda} \right\} \tag{4-27}$$

Figure 4-12 may be used to solve this equation if the abscissa and ordinate are considered to represent $(x_1 - x_{e_2})/(x_2 - x_{e_2})$ and N_{OL} respectively, the parameter being $1/\lambda$. Figure 4-12 may also be used to examine, fairly quickly, the effect of the required degree of separation (y_1/y_2) on the number of transfer units. Colburn has suggested that the economic range of λ is 0.7-0.8.

4.8.2 Height of a Transfer Unit

The height of an overall transfer unit, H_{OG}, is defined by the equations

$$H_{OG} = G_m/K_G a \cdot P \tag{4-28}$$

$$H_{OL} = L_m/K_L a \cdot \rho_L \tag{4-29}$$

where L_m, G_m = molar liquid and gas flow rates (mol/hr ft^2) [mol/s m^2],
  P = average column pressure (atm) [N/m^2],
  $K_G a$ = overall coefficient (mol/hr ft atm) [mol/s m(N/m^2)],
  $K_L a$ = overall coefficient (mol/hr ft (unit Δc))
  [mol/s m(unit Δc)],
  ρ_L = liquid density (lb/ft^3) [kg/m^3].

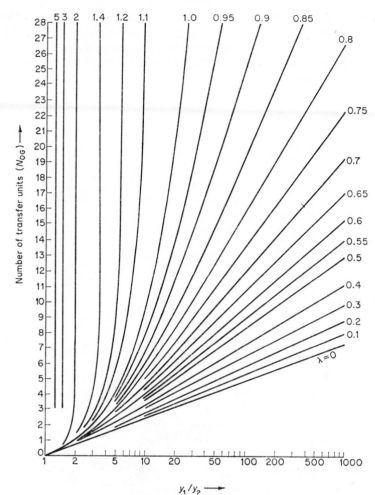

FIGURE 4–12. Colburn's relationship between the number of transfer units and the operating conditions when $x_2 = 0$[51]

If the driving force is taken over the gas film only, the height of a gas film transfer unit H_G is obtained:

$$H_G = G_m/k_g \, a \cdot P \qquad (4\text{--}30)$$

Similarly for the liquid film,

$$H_L = L_m/k_L \, a \cdot \rho_L \qquad (4\text{--}31)$$

where $k_G a$ and $k_L a$ now refer to the film coefficients.

It can be shown[7],[28] that the individual film coefficients and the height of an overall transfer unit are related by the equations

$$H_{OG} = H_G + \lambda H_L \qquad (4\text{--}32)$$

and

$$H_{OL} = H_L + H_G/\lambda \qquad (4\text{--}33)$$

where $\lambda = mG_m/L_m$ = slope of the equilibrium line/slope of the operating line.

Values of the heights of transfer units abound in the literature[5],[49] though, where possible, experimental data should be used for the required values. In the absence of such information, the nomographs of Figures 4–13 and 4–14 may be used to estimate the heights of the individual film transfer units H_G and H_L.

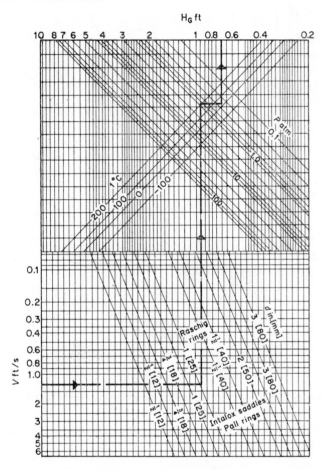

FIGURE 4–13. Nomograph for the estimation of the height of a gas phase transfer unit (British Units); for SI use the following factors:

V_f (ft/s) × 0.35 = [m/s] P(atm) × 101 = [kN/m²]
t(°C) + 273 = [K] H_G (ft) × 0.305 = [m]

Equations (4–32) and (4–33) may be used to provide the value of H_{OG} or H_{OL}, from which the packed height may be calculated. Figure 4–14 should not be used for organic liquids or for distillation applications. In these cases, Table 4–13 should be consulted for the value of H_L for the packing under consideration.

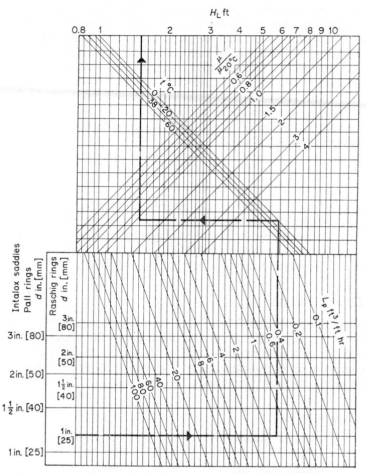

FIGURE 4–14. Nomograph for the estimation of the height of a liquid phase transfer unit (British Units); for SI use the following factors:

$$L_P \text{ (ft}^3/\text{ft hr)} \times 2.58 \times 10^{-4} = [\text{m}^3/\text{m s}]$$

$$t \text{ (°C)} + 273 = [\text{K}] \qquad\qquad H_L \text{ (ft)} \times 0.305 = [\text{m}]$$

TABLE 4–13. VALUES OF H_L FOR DISTILLATION

Packing size (in.) [mm]		$\frac{1}{2}$[12]	$\frac{3}{4}$[18]	1[25]	1$\frac{1}{2}$[40]	2[50]
Raschig Type	ft	0.24	0.30	0.34	0.47	0.58
	[m]	0.073	0.092	0.104	0.143	0.177
Intalox	ft	0.20	0.26	0.29	0.40	—
	[m]	0.061	0.079	0.089	0.122	—
Pall rings	ft	—	—	—	0.40	0.49
	[m]	—	—	—	0.122	0.150

The above analysis is most useful where there is essentially only one component in the solute. Multicomponent absorption systems are best handled by calculating the number of equivalent plates and then converting to transfer units. The height equivalent to a theoretical plate (HETP) is usually obtained from test data on the particular system and is completely empirical. Both HETP and HTU have been the subject of extensive research, and many correlations and data have been published. Representative values of HETP for a number of commercial operations have been presented by Eckert[52] with the majority of values in the range 1.5–3.0 ft [0.46–0.9 m]. The relationship between HETP and H_{OG} is given by equation (4–34) and that between the number of transfer units, N_{OG}, and the number of theoretical plates, N, by Equation 4–35:

$$H_{OG} = \text{HETP}[(\lambda - 1)\ln \lambda] \qquad (4\text{–}34)$$

$$N_{OG} = N(\log_e \lambda)/(\lambda - 1) \qquad (4\text{–}35)$$

These equations strictly apply only when both operating and equilibrium lines are straight, though systems with curved operating and equilibrium lines may be treated by dividing the column into sections such that each section approximates to the straight line condition. In the special case where both lines are straight and parallel, $H_{OG} = \text{HETP}$.

4.9 Column Diameter

Having selected the column internals and calculated the number of actual plates or the height of packing, the diameter of the column may now be determined. It is most convenient to discuss the methods available in the context of either packed columns or plate towers, and this part of the design will be considered in the next two chapters.

4.10 Nomenclature

C	constant in cost equations	(—)	[—]
D	product rate	(mol/hr, lb/hr)	[kmol/s, kg/s]
D	diffusion coefficient	(ft²/hr)	[m²/s]
e	entrainment rate	(mol/hr ft²)	[kmol/s m²]
E_{MV}, E_{ML}	Murphree tray efficiency	(—)	[—]
E_o	overall Column efficiency	(—)	[—]
E_{OL}, E_{OG}	Murphree point efficiency	(—)	[—]
F_a	fractional free area, equation (4–16)	(—)	[—]
F_m	adjustment factor, material, equation (4–2)	(—)	[—]
F_s	adjustment factor, tray spacing, equation (4–2)	(—)	[—]
F_t	adjustment factor, tray type, equation (4–2)	(—)	[—]
G_m	molar gas flow rate	(mol/hr ft²)	[kmol/s m²]
h_w	weir height	(in.)	[mm]
HETP	height equivalent to a theoretical plate	(ft)	[m]
H_g	height of a gas phase transfer unit	(ft)	[m]
H_e	height of a liquid phase transfer unit	(ft)	[m]
H_{OG}, H_{OL}	heights of overall transfer units	(ft)	[m]
K	equilibrium ratio	(—)	[—]
$K_G a, K_L a$	overall coefficients	(various units)	
L_m	molar liquid flow rate	(mol/hr ft²)	[kmol/s m²]
L_p	wetting rate	(ft³/ft hr)	[m³/s m]
m	slope of equilibrium line	(—)	[—]
n	number of theoretical plates	(—)	[—]
N	number of theoretical stages	(—)	[—]
N_t	number of actual stages	(—)	[—]
N_{OG}, N_{OL}	number of overall transfer units	(—)	[—]

P	total pressure	(atm)	[N/m²]
q	ratio of mols of saturated liquid in feed to the total mols of feed	(—)	[—]
R	reflux ratio	(—)	[—]
Sc	Schmidt Number	(—)	[—]
V	vapour flow rate	(mol/hr, lb/hr)	[kmol/s, kg/s]
W	bottoms product rate	(mol/hr, lb/hr)	[kmol/s, kg/s]
x	liquid composition	(—)	[—]
X	parameter in cost equations	(—)	[—]
y	vapour composition	(—)	[—]
Z	contacting height	(ft)	[m]
α_{ab}	relative volatility between components a and b	(—)	[—]
ρ	density	(lb/ft³)	[kg/m³]
μ	viscosity	(cP, lb/ft hr)	[N s/m²]
λ	ratio $= mG_m/L_m$	(—)	[—]
θ	factor in equations (4–10) and (4–11)	(—)	[—]

Subscripts

a	component a
b	component b
L	liquid
min	minimum
$n-1, n, n+1$	stages.

REFERENCES

1. Fair, J. R. *Chem. Eng.* 5 July, 1965, **72**, 107.
2. Perry, J. H. ed. *Chemical Engineering Handbook*, 4th ed. New York: McGraw-Hill, 1963.
3. Thibodeaux, L. J. and Murrill, P. W. *Chem. Eng.* 18 July, 1966, **73**, 155.
4. Treybal, R. E. *Mass Transfer Operations*, 2nd ed. New York: McGraw-Hill, 1968.
5. Leva, M. *Tower Packings and Packed Tower Design*. Akron Ohio: U.S. Stoneware Co., 1953.
6. Hengstebeck, R. J. *Distillation Principles and Design Procedures*. New York: Reinhold, 1961.
7. Sherwood, T. K. and Pigford, R. L. *Absorption and Extraction*. New York: McGraw-Hill, 1952.
8. Billet, R., Conrad, S., Grubb, C. M. *Inst. Chem. E. Symp.* Ser. No. 32., 1969.
9. Wall, K. J. *Chem. Proc. Eng.* 1967, **48**, 7, 56.
10. Bulletin S. 42, Norton Chemical Process Products Division, Akron, Ohio, U.S.A.
11. Q.V.F. Ltd, Stoke on Trent. Staffs. England.
12. Lessing. *J. Soc. Chem. Ind.* 1921, **40**, 115.
13. Molstad, McKinney, Abbey. *Tr. Am.I.Ch.E.* 1943, **39'** 605.
14. Bulletin No. TP 33 1965. Hydronyl Ltd, Stoke on Trent. Staffs., U.K.
15. Smith, B. D. *The Design of Equilibrium Stage Processes*. New York: McGraw-Hill, 1963.
16. Kitterman, L. and Ross, M. C. *H.carb. Proc.* May 1967.
17. Zimmerman, D. T. *Cost. Eng.* 1968, **13**, 4, 9.
18. Anon. *Ind. Chem.* 1964, **40**, 1, 34.
19. Anon. *Ind. Chem.* 1964 **40**, 2, 91.
20. *Capital Cost Estimation*. London: Inst. Chem. Eng. 1969.
21. Guthrie, K. M. *Chem. Eng.* 27 March 1969, **76**, 114.
22. Billet, R. and Raichle, L. *Chem. Eng.* 1967, **74**, 4, 148.
23. Fair, J. R. and Bolles, W. L. *Chem. Eng.* 22 Apr. 1968, **75**, 156.
24. McCabe, W. L. and Thiele, E. W. *Ind. Eng. Chem.* 1925, **17**, 605.
25. Hala., E. *et al.* *Vapour-Liquid Equilibrium*, 2nd ed. Oxford: Pergamon Press, 1967.
26. Holland, F. A., Moores, R. M., Watson, F. A., Wilkinson, J. K. *Heat Transfer*. London: Heinemann Educational Books, 1970.

27. Ponchon, M. *Tech. Moderne*, 1921, **13**, 20, 55.
28. Coulson, J. M. and Richardson, J. F. *Chemical Engineering*, Vol. II. 2nd ed. Oxford: Pergamon Press, 1968.
29. Henley, E. J. and Staffin, H. K. *Stagewise Process Design*. New York: John Wiley and Sons, 1963.
30. Fenske, M. R. *Ind. Eng. Chem.* 1932, **24**, 482.
31. Winn, F. W. *Pet. Ref.* 1958, **37**, No. 5, 216.
32. Underwood, A. J. V. *Chem. Eng. Prog.* 1948, **44**, 603.
33. Gilliland, E. R. *Ind. Eng. Chem.* 1940, **32**, 1220.
34. Erbar, J. H. and Maddox, R. N. *Pet. Ref.* 1961, **40**, No. 5, 183.
35. Murphree, E. V. *Ind. Eng. Chem.* 1925, **17**, 747.
36. A.I.Ch.E. *Bubble-Tray Design Manual*. New York.
37. Drickamer, H. G. and Bradford, J. R. *Tr.A.I.Ch.E.* 1943, **39**, 319.
38. *Ballast Tray Design Manual*, Bulletin No. 4900, F. W. Glitsch and Sons, Inc., Dallas, Texas.
39. O'Connell, H. E. *Tr.A.I.Ch.E.* 1946, **42**, 741.
40. English, G. E. and Van Winkle, M. *Chem. Eng.*, 1963, **11**, 241.
41. Eduljee, H. E. *Chem. Eng.*, 17 Feb. 1964, **6**.
42. Weissberger, A. ed. *Techniques of Organic Chemistry*, Vol. IV—*Distillation*, 2nd ed. New York: Interscience/Wiley, 1965.
43. Swanson, R. W. and Gerster, J. A. *J. Chem. Eng. Data*, 1962, **7**, 132.
44. Veatch, F., Callahan, J. L., Idol, J. D. and Milberger, E. C. *Chem. Eng. Prog.* 1960, **56**, 10, 65.
45. Lewis, W. K. and Matheson, G. L. *Ind. Eng. Chem.* 1932, **24**, 494.
46. Thiele, E. W. and Geddes, R. L. *Ind. Eng. Chem.* 1933, **25**, 289.
47. Hanson, D. N., Duffin, J. H., Somerville, G. F. *Computation of Multistage Separation Processes*. New York: Reinhold, 1962.
48. Holland, C. D. 'Multicomponent Distillation'. Englewood Cliffs N.J.: Prentice-Hall, 1963.
49. Morris, G. A. and Jackson, J. *Absorption Towers*. London: Butterworths Scientific Publications, 1953.
50. Chilton, T. H. and Colburn, A. P. *Ind. Eng. Chem.* 1935, **27**, 255.
51. Colburn, A. P. *Tr.A.I.Ch.E.* 1939, **35**, 211.
52. Eckert, J. S. *Chem. Eng. Prog.* 1963, **59**, 5, 76.

Chapter 5: Packed Towers

5.1 Introduction

Consideration of the factors affecting the selection of packed columns has been discussed in Chapter 4. Packed towers are most commonly used in absorption and stripping systems and for distillation. A typical absorption system is shown in Figure 5–1, where lean solvent enters the top of the absorber and rich gas enters at the bottom. The streams flow countercurrently

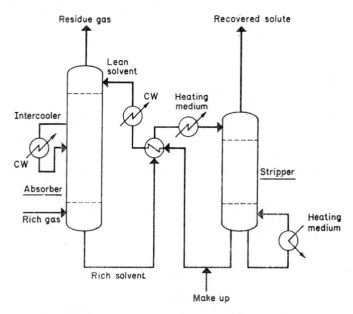

FIGURE 5–1. A typical absorber-stripper system

with the residue gas leaving the top of the column which has lost its solute, and the rich solvent is removed at the bottom. A regenerator or stripper then recovers the solvent for re-cycle and the solute as product.

This chapter will deal with the selection and sizing of packing and the determination of column height and diameter for both absorption and distillation applications.

5.2 Type and Size of Packing

Figure 5–2 shows the relative efficiency of different types of packings as a function of the nominal packing size.[1] Although the efficiency is higher for small packings, it is generally accepted that it is uneconomical to use these

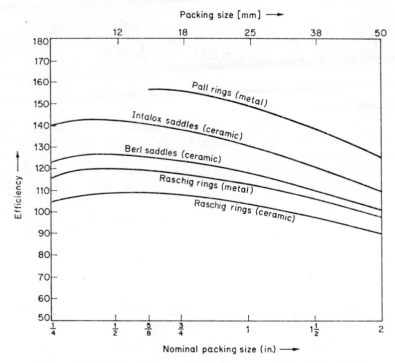

FIGURE 5–2. Relative efficiency of different types and sizes of packing[1]

small sizes in an attempt to improve the performance of a column. It is preferable to use the largest recommended size of a particular type of packing and to increase the packed height to compensate for the small loss of efficiency.[1],[2].

For a preliminary estimate of tower diameter, Figure 4–1 may be consulted or, in the first instance, a gas velocity of 3 ft/s [1 m/s] selected. From a knowledge of the volumetric inlet gas flow, the diameter may then be estimated. With the estimated tower diameter, the following recommendations regarding packing size and the associated problems of liquid distribution may be of value in selecting the size of the packing.

(a) *Raschig rings.* The size of packing should not exceed $\frac{1}{30}$ of the column diameter if liquid distribution problems are to be minimized. Redistribution should be considered every 20 ft [6 m] or each $2\frac{1}{2}$–3 column diameters, whichever is the lowest.

(b) *Berl or Intalox saddles.* The maximum packing size should not exceed $\frac{1}{15}$ of the column diameter with redistribution every 5–8 column diameters, or 20 ft [6 m], whichever is smaller.

(c) *Pall rings.* The maximum packing size should not exceed the range of $\frac{1}{10}$–$\frac{1}{15}$ of the column diameter. Redistribution should be considered every 20 ft [6 m] or every 5–10 column diameters, whichever is smaller.

The practical aspects of gas and liquid distribution and liquid redistribution will be considered later in Chapter 7.

Example 5-1

8000 lb/hr [1 kg/s] of a vapour enters a packed column. If the density is 0.1 lb/ft³ [1.6 kg/m³] make a first estimate of the tower diameter and select suitable packing sizes for each type of packing.

From Figure 4–1, the approximate column diameter is 38 in. [1 m]. If a velocity of 3 ft/s [1 m/s] is used, the cross sectional area is given by

$$8000/(0.1 \times 3.0 \times 3600) = 7.4 \text{ ft}^2$$

$$[1/(1.6 \times 1.0) = 0.625 \text{ m}^2]$$

Hence the approximate diameter is 3.1 ft [0.9 m].
From the recommendations made in section 5.2, the maximum packing sizes are:

Raschig rings: $38 \times \frac{1}{30} = 1.25$ in.; use 1 in. [25 mm]
                                              or $1\frac{1}{2}$ in. [38 mm]

Intalox saddles: $38 \times \frac{1}{15} = 2.5$ in.; use 2 in. [50 mm]

Pall rings: $38 \times \frac{1}{10} = 3.8$ in.; use 3 in. [75 mm]

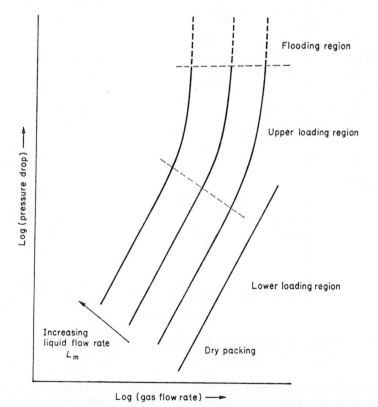

FIGURE 5–3. Typical pressure drop characteristics in a packed tower

5.3 Flooding

Figure 5–3 shows the traditional approach to the performance of packed beds.[3]

The parameter of the curves is the liquid rate which may be fixed by external process conditions. If this is not the case, a value of the molar liquid rate, L_m, may be selected by joining the points y_{e1} and y_2, as shown in Figure 4–11, to obtain the ratio $(L_m/G_m)_{min}$. The slope of this line is then increased by a factor of 1.3 to obtain a reasonable liquid flow.

Three regions are indicated in Figure 5–3; a lower loading zone, an upper loading zone and a flooding region. It is only the flooding point which can be visually observed. This occurs when liquid is seen to be held up to an extent in the bed such that standing waves appear on the liquid surface. Although flooding is a complex phenomenon, it may be defined[1] as that point in gas–liquid loading where the liquid phase becomes continuous in the voids and the gas phase becomes discontinuous in the same voids of the bed. It is fairly obvious that liquid carried upwards with the gas at flooding causes a drop in the efficiency of the column.

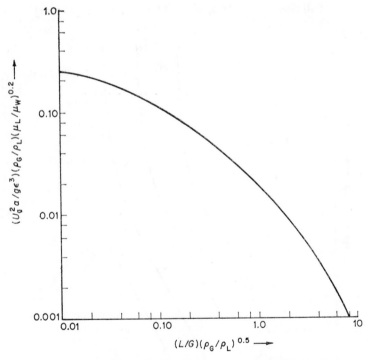

FIGURE 5–4. The correlation of Lobo for flooding velocity in packed towers

Sherwood[4] attempted the first correlation of flooding for various packings and this was subsequently improved by Lobo[5] whose correlation is presented in Figure 5–4. In the figure,

L = liquid flow rate	(lb/hr)	[kg/s]
G = mass vapour flow	(lb/hr)	[kg/s]
ρ_G, ρ_L = densities of gas and liquid respectively	(lb/ft³)	[kg/m³]
U_g = gas velocity at flooding	(ft/s)	[m/s]
a = gas surface area per unit volume of packing	(ft²/ft³)	[m²/m³]
ε = free space of packing	(–)	[–]
μ_L, μ_W = viscosity of liquid under tower conditions and that of water at 20°C [293 K] respectively	(cP)	[N s/m²]
g = acceleration due to gravity	(ft/s²)	[m/s²]

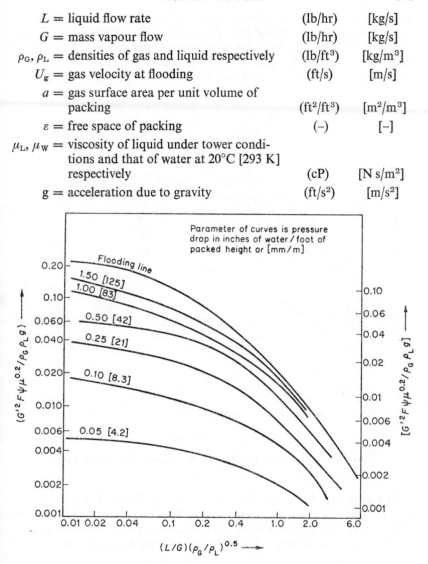

FIGURE 5–5. Generalized pressure drop correlation for packed beds[1]

These correlations have been further improved and the best representation of the data is shown in Figure 5–5[1], where G' = gas flow rate (lb/ft² s) [kg/m² s]. The packing factor, F, has replaced the term a/ε^3 and its values have been tabulated, where available, in the tables of the physical characteristics of packings in Chapter 4. In order to use Figure 5–5 in determining the gas rate G' and hence the column diameter, it is necessary to select a value for the acceptable pressure drop per unit height of packing. This selection is normally made so that the operation occurs at the maximum economical pressure drop. The choice must be made between high capital cost and low operation costs for a low pressure drop tower and low capital investment, and higher operating costs for a column with a higher pressure drop. The normal maximum of 90 per cent flooding is only achieved with careful control of the pressure drop across the column, and a more usual range of 0.2–0.4 in. water/ft [16.7–33.3 mm/m] is employed in scrubbing operations, while vacuum applications may cover the whole range of pressure drop.

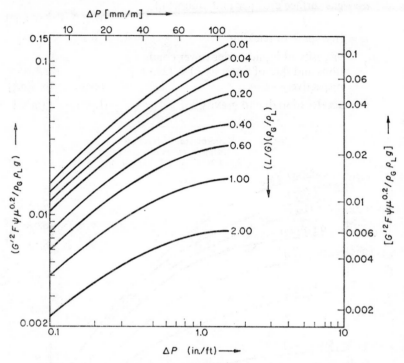

FIGURE 5–6. Alternative presentation of the generalized pressure drop correlation[5a]

The generalized pressure drop correlation of Figure 5–5 has been re-plotted in a form which may be more convenient for certain applications.[5a] This plot is presented in Figure 5–6 and a nomograph to enable a quick estimation of the flooding velocity to be made is presented in Figure 5–7.

5.4 Pressure Drop

In calculating the pressure drop through a given packing, a variety of methods may be used.

(i) For conditions below or near the loading point, the following equation is applicable:

$$\Delta P = a \cdot 10^{bl} \cdot g_F^2/\rho_G \qquad (5\text{-}1)$$

where ΔP = pressure drop (in. water/ft of packing),

l, g_F = liquid and gas flow rates (lb/s ft²),

ρ_G = gas density (lb/ft³),

a, b = constants obtained from Table 5–1.

Equation (5–1) may be converted to SI units by multiplying the value of ΔP obtained from that equation by the factor 83.3 to obtain units of [mm/m] and for liquids other than water, ΔP should be corrected by the factor $(\rho_w/\rho_L)^{(6)}$. For conditions near to flooding, the pressure drop is higher

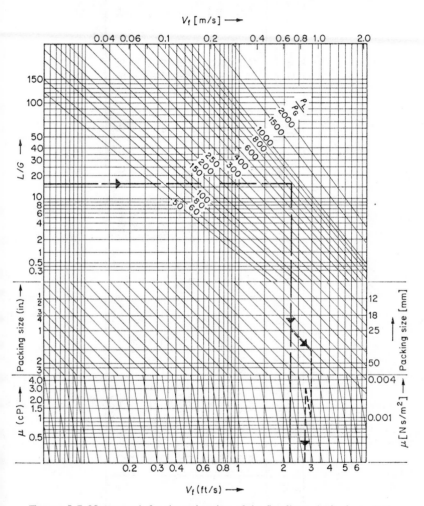

FIGURE 5–7. Nomograph for the estimation of the flooding velocity in packed towers[10]

than the value predicted by equation (5–1) and account should be taken of this fact where appropriate.

(ii) Data for ceramic Berl saddles, plastic Intalox saddles, and plastic Pall rings are presented in Figures 5–8, 5–9 and 5–10 for a wide range of gas and liquid flow rates.

(iii) For other packings at unspecified loading conditions, the generalized pressure drop correlations of Figures 5–5 and 5–6 may be used.

TABLE 5-1. TABLE OF CONSTANTS FOR USE WITH EQUATION 5-1[10]

		$\frac{1}{2}$ in. [12.5 mm]	$\frac{5}{8}$ in. [16 mm]	$\frac{3}{4}$ in. [19 mm]	1 in. [25 mm]	$1\frac{3}{8}$ in. [35 mm]	$1\frac{1}{2}$ in. [38 mm]	2 in. [50 mm]	3 in. [80 mm]
Ceramic Raschig rings	a	1.96	—	0.82	0.53	—	0.30	0.24	0.18
	b	0.56	—	0.38	0.22	—	0.20	0.14	0.15
Metal Raschig rings	a	—	—	—	0.42	—	0.29	0.23	—
	b	—	—	—	0.21	—	0.20	0.14	—
Ceramic Intalox saddles	a	1.04	—	0.52	0.52	—	0.13	0.12	—
	b	0.37	—	0.25	0.16	—	0.15	0.10	—
Ceramic Pall rings	a	—	—	—	0.35	—	0.20	0.13	—
	b	—	—	—	0.16	—	0.17	0.12	—
Metal Pall rings	a	—	0.43	—	0.15	0.09	—	0.06	—
	b	—	0.17	—	0.15	0.16	—	0.12	—

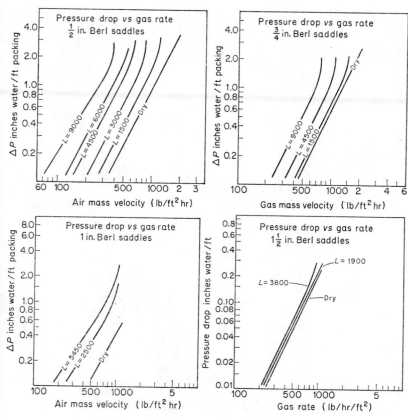

FIGURE 5–8. Pressure drop characteristics for ceramic Berl saddles[10] (British Units; for SI units use the following factors:

$$\Delta P \text{ (in./ft)} \times 83.3 = [\text{mm/m}]$$
$$L_1 G (\text{lb/hr ft}^2) \times 1.35 \times 10^{-3} = [\text{kg/s m}^2])$$

Example 5–2

A column is packed with 2 in. [50 mm] ceramic Raschig rings. The tower is to operate with 15 000 lb/hr ft² [20.3 kg/s m²] of water and 1000 lb/hr ft² [1.35 kg/s m²] of air. Compare the values of the pressure drop through the packing obtained by the above methods.

(i) Using Table 5–1,

$$a = 0.24$$
$$b = 0.14$$
$$\rho_g = 0.075 \text{ lb/ft}^3$$
$$l = 15\,000/3600 = 4.16 \text{ lb/s ft}^2$$
$$g_F = 1000/3600 = 0.278 \text{ lb/s ft}^2$$
$$P = 0.24 \times 10^{0.14 \times 41.6} (0.278^2/0.075) \qquad (5\text{–}1)$$
$$= 0.95 \text{ in./ft } [79 \text{ mm/m}]$$

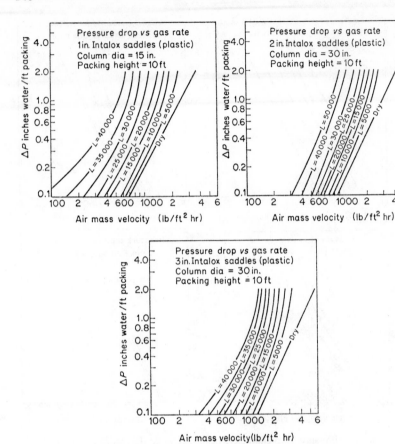

FIGURE 5–9. Pressure drop characteristics for plastic Intalox saddles[10] (British Units; for SI units use the following factors:

$$\Delta P \text{ (in./ft)} \times 83.3 = [\text{mm/m}]$$
$$L, G \text{ (lb/hr ft}^2) \times 1.35 \times 10^{-3} = [\text{kg/s m}^2])$$

(ii) Using the generalized pressure drop coefficient, the packing factor is taken from Table 4–2, $F = 65$. Hence

$$(L/G)(\rho_G/\rho_L)^{\frac{1}{2}} = (15\ 000/1000)(0.075/62.4)^{\frac{1}{2}}$$
$$= 0.52$$

$$(G'^2 F \psi \mu^{0.2})(\rho_G \cdot \rho_L g) = \frac{(1000/3600)^2 \times 65.1 \times (1)^{0.2}}{0.075\ 62.4\ 32.2}$$
$$= 0.034$$

From Figure 5–5, the pressure drop is approximately 1.0 in./ft [83.3 mm/m]. From Figure 5–6, the pressure drop is approximately 1.1 in./ft [91.5 mm/m].

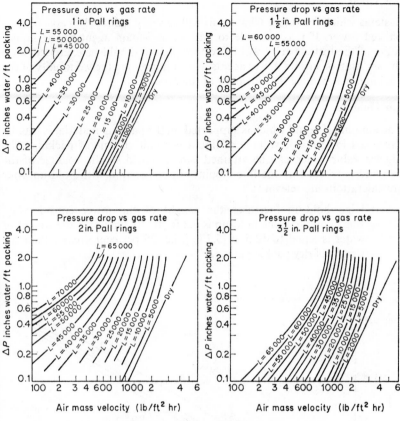

FIGURE 5–10. Pressure drop characteristics for plastic Pall rings[10] (British
Units; for SI units use the following factors;

$$\Delta P \text{ (in./ft)} \times 83.3 = [\text{mm/m}]$$
$$L, G \text{ (lb/hr ft}^2) \times 1.35 \times 10^{-3} = [\text{kg/s m}^2])$$

Example 5–3

If the column of Example 5–2 were packed with 2 in. [50 mm] plastic
Intalox saddles, what would the pressure drop be, assuming the flow rates
remain unchanged?

The packing factor for the plastic Intalox saddles is obtained for Table
4–5 as $F = 21$. Thus

$$(L/G)(\rho_G/\rho_L)^{\frac{1}{2}} = 0.52 \qquad \text{as before}$$

and $\quad (G'^2 F \psi \mu^{0.2})/(\rho_G \cdot \rho_L g) = 0.011$

From Figure 5–5 the pressure drop is approximately 0.2 in./ft [16.7
mm/m]. From Figure 5–6 the pressure drop is approximately 0.18 in./ft
[15.0 mm/m].

The most accurate estimate of pressure drop for this packing is obtained
directly from Figure 5–9, giving a value of 0.23 in./ft [19.2 mm/m].

5.5 Foam

Systems which foam (see Chapter 6) will severely reduce the capacity of a packed tower. If it is possible to add an anti-foam agent to the liquid, some of the lost capacity will be restored. Care should be exercised since the addition of surface active agents, whilst improving the wettability of the packing, may reduce the efficiency of the process.

5.6 Hold-up

The amount of liquid which is supported on the packing material is usually expressed in terms of volume of liquid per unit volume of packing. The highest value which can be attained is determined by the voidage of the bed and is reached at the flood point. In connection with hold-up, four considerations are relevant:

(i) High hold-up increases the weight of the tower considerably. For example, if the hold-up is 0.2 ft^3/ft^3 [0.2 m^3/m^3], the weight of water is equal to 12.5 lb [5.7 kg], i.e. 25 per cent greater than the weight of dry packing alone.

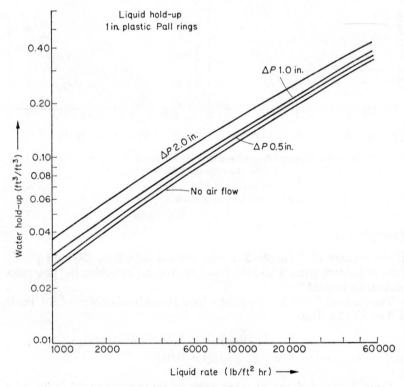

FIGURE 5–11. Typical hold-up characteristics for tower packings[10] (British Units; for SI units use the following factors:

$$L \text{ (lb/hr ft}^2) \times 1.35 \times 10^{-3} = [\text{kg/s m}^2]$$
$$\Delta P \text{ (in./ft)} \times 83.3 = [\text{mm/m}]$$
$$H \text{ (ft}^3/\text{ft}^3) \times 1.0 = [\text{kg/kg}])$$

(ii) A high hold-up will considerably increase the pressure drop as shown by the typical values in Figure 5–11. An increase of 50 per cent in the hold-up can increase the pressure drop by 100 per cent—an important factor is vacuum distillation.

(iii) A packing with a high hold-up will require a larger drainage time on shut-down.

(iv) In the case of batch distillation hold-up should be as low as possible, since the volume of liquid involved cannot be driven through the column and will remain on the packing undistilled.

5.7 Degree of Wetting

When considering the flow of liquid over packings, it is convenient to express the liquid rate as a wetting rate, defined as

$$\text{wetting rate, } L_p = \frac{\text{liquid rate (ft}^3/\text{hr ft}^2)}{\text{specific area of packing (ft}^2/\text{ft}^3)} \quad (\text{ft}^3/\text{hr ft}) \ [\text{m}^3/\text{s m}]$$

Morris and Jackson[7] recommend that the minimum wetting rate for all packings except rings greater than 3 in. [76 mm] diameter and grids of pitch greater than 2 in. [50 mm] should be 0.85 ft^3/hr ft [2.2 × 10^{-5} m^3/s m], and for all remaining packings the value is 1.3 ft^3/hr ft [3.4 × 10^{-5} m^3/s m]. A convenient nomograph produced more recently is shown in Figure 5–12.

A larger packing size may be chosen to improve the wetting rate by reducing the specific area of packing, although the recommendations of size laid down in section 5.2 should be adhered to. If the wetting rate is unavoidably low, allowance must be made for lower performance by increasing the height of packing.

5.8 Column Diameter

The diameter of a packed tower may be determined in three ways:

(i) A preliminary estimate of column diameter may be obtained from Figure 4–1[8] which will indicate whether or not the first estimate is within the normal size range for a packed tower. This graph should only be used to make this initial determination or to estimate the packing size.

(ii) Use may be made of the nomograph in Figure 5–7 to determine the flooding velocity V_F (ft/s) [m/s] for a selected packing size. To obtain a satisfactory value for the operating gas velocity, V_F should be multiplied by the following factors for the type of packing chosen.

Packing	Factor
All Raschig rings	0.60
1 in. [25 mm] and 1½ [38 mm] Intalox saddles	0.80
2 in. [50 mm] and 3 in. [76 mm] Pall rings	0.90

The cross-sectional area may then be calculated and the diameter obtained. It will be necessary to check that the resulting pressure drop is acceptable.

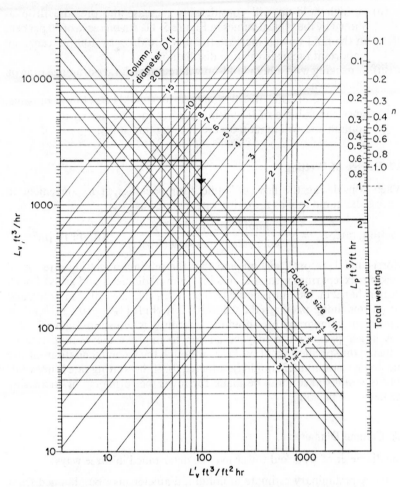

FIGURE 5-12. Nomograph for the estimation of the degree of wetting in a packed column[10] (British Units; for SI units use the following factors:

$$D \text{ (ft)} \times 0.305 = \text{[mm]}$$
$$d \text{ (in.)} \times 25.4 = \text{[mm]}$$
$$L_p(\text{ft}^3/\text{ft hr}) \times 2.58 \times 10^{-4} = \text{[m}^3/\text{m s]}$$
$$L'_v(\text{ft}^3/\text{ft}^2 \text{ hr}) \times 0.85 \times 10^{-4} = \text{[m}^3/\text{m}^2 \text{ s]}$$
$$L_v(\text{ft}^3/\text{hr}) \times 7.87 \times 10^{-6} = \text{[m}^3/\text{s]})$$

(iii) If the pressure drop per unit height of packing can be specified either by prevailing conditions or from the recommendations in section 5-4, use may be made of Figures 5-5 and 5-6. The correlations provide the gas rate in lb/s ft², from which the diameter may be calculated.

Example 5-4

Compare the diameters obtained by each of the three methods given in section 5.8, for a system with a water rate of 80 000 lb/hr [10.1 kg/s] and an air rate of 8000 lb/hr [1.01 kg/s], using packings of $1\frac{1}{2}$ in. [38 mm] Raschig rings and 2 in. [50 mm] plastic Intalox saddles.

(i) For both packings, Figure 4–1 gives a diameter = 35 in. [0.89 m].

(ii) From the nomograph of Figure 5–7, when $L/G = 10$, $\rho_L/\rho_G = 830$:

For $1\frac{1}{2}$ in. Raschig rings, $V_F = 3.6$ ft/s [1.1 m/s]

For 2 in. Intalox saddles, $V_F = 4.4$ ft/s [1.34 m/s]

$$\text{Gas rate} = 80\ 000\ \text{ft}^3/\text{hr}\ [10.63\ \text{m}^3/\text{s}]$$

$$\text{Actual velocities for rings} = 3.6 \times 0.6$$
$$= 2.16\ \text{ft/s}\ [0.66\ \text{m/s}]$$

$$\text{Actual velocities for saddles} = 4.4 \times 0.80$$
$$= 3.52\ \text{ft/s}\ [1.07\ \text{m/s}]$$

$$\text{Column areas for rings} = 80\ 000/3600 \times 2.16$$
$$= 10.3\ \text{ft}^2\ [0.956\ \text{m}^2]$$

Hence column diameter for rings = 3.52 ft [1.08 m]

$$\text{Column areas for saddles} = 80\ 000/3600 \times 3.52$$
$$= 6.31\ \text{ft}^2\ [0.586\ \text{m}^2]$$

Hence column diameter for saddles = 2.84 ft [0.866 m]

(iii) Assume that a pressure drop = 0.8 in./ft [66.6 mm/m] is acceptable. From Figure 5–5, at $(L/G)(\rho_v/\rho_L)^{\frac{1}{2}} = 0.35$, the ordinate is 0.032, i.e.

$$(G'^2 F\psi\mu^{0.2})/\rho_G \cdot \rho_L g = 0.032$$
$$G'^2 F . 1 \times 1(0.075 \times 62.4 \times 32.2) = 0.032$$

Hence $G'^2 F = 4.80$

For $1\frac{1}{2}$ in. [38 mm] Raschig rings,

$$F = 130 \quad \text{(Table 4–2)}$$
$$G = (4.80/130)^{\frac{1}{2}}$$
$$= 0.192\ \text{lb/s ft}^2\ [0.935\ \text{kg/s m}^2]$$
$$\text{Area} = 8000/3600 \times 0.192$$
$$= 11.6\ \text{ft}^2\ [1.08\ \text{m}^2]$$

Hence   diameter = 3.84 ft [1.17 m]

For 2 in. [50 mm] Intalox saddles,

$$F = 21 \quad \text{(Table 4–5)}$$

Hence $G' = 0.478$ lb/s ft^2 [2.33 kg/s m^2]

and   diameter = 2.16 ft [0.659 m]

It can be seen that the values obtained show wide variation depending upon which method is used. The explanation lies in the selection of the acceptable pressure drop to use with the generalized pressure drop correlation. If a lower pressure drop is demanded, the tower diameter will be increased.

5.9 Height of Packing

The methods of obtaining the packed height in a column have been discussed earlier in Chapter 4. The method will be illustrated by an example on the design of a packed column for distillation. The height is calculated from equation (5–2) or (5–3):

$$\text{height} = N_{OG}(H_G + \lambda H_L)/n \qquad (5\text{–}2)$$

or

$$\text{height} = N_{OL}(H_L + H_G/\lambda)/n \qquad (5\text{–}3)$$

where n is the degree of wetting (from Figure 5–12).

If $n > 1$, $n = 1$ is used in equations (5–2) and (5–3). For distillation applications, the height obtained should be increased by 43 per cent[10] or be multiplied by $1/0.70$. Values of H_G and H_L, the heights of the individual transfer units, may be obtained from Figure 4–13 or 4–14, or in the case of distillation H_L may be obtained from Table 4–13.

5.10 Design of a Packed Tower for Distillation

Example 5–5

The separation of a mixture of 60 mol per cent of A(MW = 78) and 40 mol per cent of B (MW = 92) is to be effected by distillation in a packed tower at atmospheric pressure, the feed being at its boiling point. The annual throughput of 50 000 tons per annum corresponds to 13 100 lb/hr [1.65 kg/s] after allowing for an annual shutdown. The rate of top product removal is to be 8000 lb/hr [1.01 kg/s] and the column is operated with a reflux of A ratio of 4 : 1. The top and bottom products should contain 95 per cent and 5 per cent benzene respectively. The average column conditions give rise to the following data:

Average liquid density $\rho_1 = 43.3$ lb/ft^3 [695 kg/m^3]

Vapour density at top and bottom of column respectively = 0.168 and 0.182 lb/ft^3 [=2.69 and 2.92 kg/m^3]

Average liquid viscosity $\mu = 0.32$ cP [3.2 $\times$ 10^{-4} N s/m^2]

Vapour liquid equilibrium data is presented in Table 5–2 and plotted in Figure 5–13.

TABLE 5–2. EQUILIBRIUM DATA FOR BENZENE–TOLUENE AT ATMOSPHERIC PRESSURE[10]

Molar fraction of A in liquid, x_a	0	0.1	0.2	0.3	0.4	0.6	0.7	0.8	0.9	1.0
Molar fraction of A in vapour in equilibrium y_a	0	0.23	0.38	0.51	0.71	0.78	0.86	0.92	0.97	1.0

The internal flow rates are calculated from a series of mass balances:
In the rectifying section,

$$L_n/D = R \qquad (5\text{–}4)$$

$$V_n = L_n + D \qquad (5\text{–}5)$$

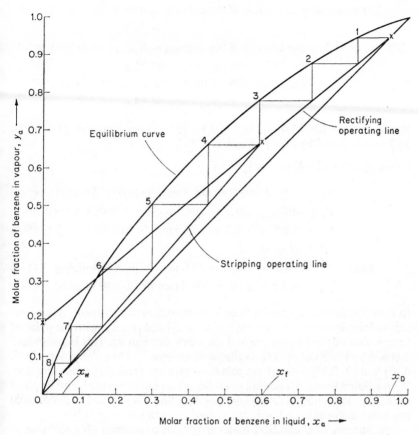

FIGURE 5-13. Equilibrium data and McCabe–Thiele construction for Example 5-5

In the stripping section, as the feed is a liquid at its boiling point,

$$L_m = L_n + F \qquad (5\text{-}6)$$

and $$V_m = L_m - W \qquad (5\text{-}7)$$

The overall balance is given by

$$F = D + W \qquad (5\text{-}8)$$

where L_n, L_m = liquid flows in the rectifying and stripping sections respectively (mol/hr) [kmol/s],

V_n, V_m = vapour flows in the rectifying and stripping sections respectively (mol/hr) [kmol/s],

F = feed rate (mol/hr) [kmol/s],

D = top product removal rate (mol/hr) [kmol/s],

W = bottom product removal rate (mol/hr) [kmol/s].

The mean MW at the top of the column $= 0.95 \times 78 + 0.05 \times 92$
$$= 78.7$$
The mean MW at the bottom of the column $= 0.05 \times 78 + 0.95 \times 92$
$$= 91.3$$
The mean MW of the feed $= 0.6 \times 78 + 0.4 \times 92$
$$= 83.6$$

Hence $F = 13\,100/83.6 = 156.7$ mol/hr [0.0197 kmol/s] and $D = 8000/78.7 = 101.7$ mol/hr [0.0128 kmol/s].

From equations (5–4) to (5–8),

$$L_n = 40.0 \times 101.7 = 406.8 \text{ mol/hr [0.0512 kmol/s]} \quad (5\text{–}4)$$
$$V_n = 406.8 + 101.7 = 508.5 \text{ mol/hr [0.0640 kmol/s]} \quad (5\text{–}5)$$
$$L_m = 406.8 + 156.7 = 563.5 \text{ mol/hr [0.0709 kmol/s]} \quad (5\text{–}6)$$
$$F = D + W$$
Hence, $W = 156.7 - 101.7 = 55.0 \text{ mol/hr [0.0069 kmol/s]} \quad (5\text{–}8)$
$$V_m = 563.5 - 55.0 = 508.5 \text{ mol/hr [0.0640 kmol/s]} \quad (5\text{–}7)$$

In constructing the operating lines in each section of the tower, it is necessary to join the points (x_d, x_d), $(0, x_d/(R + 1))$ and (x_w, x_w) with the point of intersection of the former line and the q-line through x_p. This is a standard method and full details are available elsewhere.[3] Thus the points (0.95, 0.95) and $(0, 0.95/(4 + 1))$ are joined to give the rectifying operating line. With a liquid feed at its boiling point the q-line will be vertical, and its point of intersection with the rectifying operating line when joined to (0.05, 0.05) produces the operating line in the stripping section.

The number of theoretical plates may then be stepped off according to the McCabe–Thiele method.[9] This shows that eight theoretical stages are required for the separation, i.e. seven theoretical plates are required.

As the equilibrium curve is not a straight line, the number of overall transfer units N_{OG} must be obtained by graphical integration as shown in Table 5–3 and Figure 5–14.

TABLE 5–3. ESTIMATION OF N_{OG}

Vapour in operation (y)	Vapour in equilibrium (y_e)	$(y_e - y)$	$1/(y_e - y)$
0.95	0.985	0.035	28.6
0.93	0.98	0.05	20.0
0.90	0.97	0.07	14.3
0.80	0.90	0.10	10.0
0.70	0.815	0.115	8.7
0.60	0.735	0.135	7.4
0.50	0.66	0.16	6.3
0.40	0.575	0.175	5.7
0.30	0.47	0.17	5.9
0.20	0.36	0.16	6.35
0.10	0.21	0.11	9.1
0.05	0.125	0.075	13.3

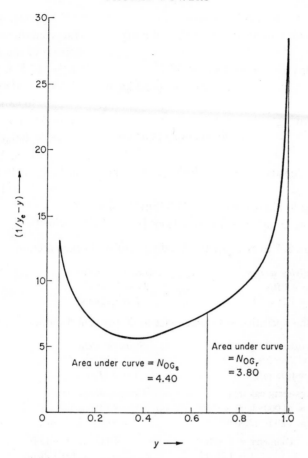

FIGURE 5-14. Graphical integration for Example 5-5

The area under the curve from $y = 0.95$ to $y = 0.67$ is equal to the number of overall transfer units in the rectifying section, i.e.

$$\text{area} = N_{OGr} = 3.80$$

The area under the curve from $y = 0.05$ to 0.07 is equal to the number of overall transfer units in the stripping section, i.e.

$$\text{area} = N_{OGs} = 4.40$$

The next step in the procedure is to estimate the flooding velocity for a selected packing in each section of the column. The separation in the example is an easy one and a tower is likely to result in which the pressure drop will be of little importance. Therefore, to keep the cost as low as possible, a preliminary selection of $1\frac{1}{2}$ in. [38 mm] Raschig rings will be made. In subsequent calculations the rectifying and stripping sections of the tower will be considered separately.

Rectifying section	*Stripping section*
(for conditions at the top of the column)	(for conditions at the bottom of the column)

Liquid flow rate, $L = 406.8 \times 78.7$ $\qquad$ $L = 563.5 \times 91.3$
$\qquad\qquad\qquad = 32\,000$ lb/hr [4.03 kg/s] $\qquad = 51\,400$ lb/hr
$\qquad\qquad\qquad\qquad\qquad\qquad\qquad\qquad$ [6.48/kg/s]

Vapour flow rate, $V = 508.5 \times 78.7$ $\qquad$ $V = 508.5 \times 91.3$
$\qquad\qquad\qquad = 40\,000$ lb/hr [5.04 kg/s] $\qquad = 46\,400$ lb/hr
$\qquad\qquad\qquad\qquad\qquad\qquad\qquad\qquad$ [5.85 kg/s]

Vapour density $\rho_V = 0.168$ lb/ft^3 [2.69 kg/m^3] $\quad$ $\rho_V = 0.182$ lb/ft^3
$\qquad\qquad\qquad\qquad\qquad\qquad\qquad\qquad\qquad\qquad$ [2.92 kg/m^3]

Average liquid density $\rho_L = 43.3$ lb/ft^3 [694 kg/m^3]
Average liquid viscosity $= 0.32$ cP [3.2×10^{-4} N s/m^2]

Use Figure 5–7 to estimate the flooding velocity in each section.

Rectifying section	*Stripping section*
$V_f = 4.1$ ft/s	$V_f = 4.0$ ft/s
[$= 1.25$ m/s]	[$= 1.22$ m/s]

Assume the operation is to take place at 60 per cent flooding,

Rectifying section	*Stripping section*
$V = 2.46$ ft/s	$V = 2.40$ ft/s
[$= 0.75$ m/s]	[$= 0.73$ m/s]
Cross-sectional area	Cross-sectional area
$= 40\,000/(3600 \times 0.168 \times 2.46)$	$= 46\,400/(3600 \times 0.182 \times 2.40)$
$= 26.9$ ft^2 [2.50 m^2]	$= 37.6$ ft^2 [3.49 m^2]
Diameter $= 5.84$ ft	Diameter $= 6.12$ ft
[$= 1.78$ m]	[$= 1.86$ m]

A uniform column diameter of 6 ft [1.83 m] would be chosen as most convenient. The cross-sectional area is then 28.3 ft^2 [2.63 m^2]. The next step is to check that the packing is sufficiently wetted by the liquid flow.

Rectifying section	*Stripping section*
$L_v = 32\,000/43.3$	$L_v = 51\,400/43.3$
$= 739$ ft^3/hr	$= 1187$ ft^3/hr
$= [5.83 \times 10^{-3}$ m^3/s]	$= [9.34 \times 10^{-3}$ m^3/s]
$D = 6.0$ ft [1.83 m]	$D = 6.0$ ft [1.83 m]
$d_p = 1\frac{1}{2}$ in. [38 mm]	$d_p = 1\frac{1}{2}$ in. [38 mm]

From Figure 5–12,

$L_p = 0.7$ ft^3/ft hr	$L_p = 1.05$ ft^3/ft hr
$= [1.8 \times 10^{-5}$ m^3/m s]	$= [2.71 \times 10^{-5}$ m^3/m s]
$n = 1$	$n > 1$

i.e. wetting is satisfactory for both sections of the tower. The height of an overall transfer unit may be calculated estimating the individual film transfer unit heights. In the case of distillation, Figure 4–14 does not apply and Table 4–13 is used, to give

Rectifying section	Stripping section
$H_L = 0.47$ ft	$H_L = 0.47$ ft
[= 0.143 m]	[= 0.143 m]

Figure 4–13 may be used to obtain H_G at a mean temperature of 95°C [368 K],

Rectifying section	Stripping section
$H_G = 1.4$ ft	$H_G = 1.4$ ft
[= 0.427 m]	[= 0.427 m]

Slope of operating line

Rectifying section:
$$= (0.95 - 0.67)/(0.95 - 0.60)$$
$$= 0.80$$

Stripping section:
$$= (0.67 - 0.11)/(0.60 - 0.10)$$
$$= 1.12$$

The slope of the equilibrium line, which is curved, may be approximated to the slope of the tangent at the mid-point of each operation.

Rectifying section	Stripping section
Mean $x_r = (0.95 + 0.60)/2$	Mean $x_s = (0.60 + 0.05)/2$
$= 0.795$	$= 0.325$

Slope of tangent at 0.775	Slope of tangent at 0.325
$= (0.992 - 0.725)/(0.9 - 0.5)$	$= (0.74 - 0.15)/(0.5 - 0)$
$= 0.67$	$= 1.18$

The ratio of the slope of the equilibrium line to the slope of the operating line λ is given by:

Rectifying section	Stripping section
$\lambda_r = 0.67/0.80$	$\lambda_s = 1.18/1.12$
$= 0.838$	$= 1.05$

The height of each packed section may be calculated from equation (5–2) using the recommended factor of $1/0.70$:

Height $= N_{OG}(H_G + \lambda H_L)/0.70n$ (5–2)

Rectifying section	Stripping section
Height $= 3.80 (1.4 + 0.838 \times 0.47)/$ 0.7 × 1	Height $= 4.40 (1.4 + 1.05 \times 0.47)/$ 0.7 × 1
$= 9.74$ ft [2.95 m]	$= 12.0$ ft [3.65 m]

Thus the total packed height is calculated to be 21.74 ft [6.61 m] and, although safety factors have been incorporated in the design method employed, this height would probably be further increased to 24 ft [7.3 m]. If in practice, it is found that the column is providing a separation equivalent to more theoretical plates than are necessary, the product take-off rate may be increased, lowering the reflux ratio and effecting a saving in heat consumption per unit weight of product.

It remains to check that the pressure drop over the column is acceptable.

In the *rectifying section*, packed height $= 11.0$ ft [3.35 m], and so

$$\text{liquid flow rate } L_r = 32\,000/3600 \times 28.3$$
$$= 0.314 \text{ lb/ft}^2 \text{ s } [1.53 \text{ kg/s m}^2]$$
$$\text{vapour flow rate } G_r = 40\,000/3600 \times 28.3$$
$$= 0.393 \text{ lb/ft}^2 \text{ s } [1.92 \text{ kg/s m}^2]$$

For $1\frac{1}{2}$ in. [38 mm] ceramic Raschig rings, $a = 0.30$ and $b = 0.20$. Hence

$$\Delta P_r = 0.30 \cdot 10^{0.20 \times 0.314} \times (0.393)^2/0.168 \qquad (5\text{-}1)$$
$$= 0.32 \text{ in./ft of packing } [26.6 \text{ mm/m}]$$

In the *stripping section*, packed height $= 13.0$ ft [3.95 m], and

$$L_s = 51\,400/28.3 \times 3600$$
$$= 0.505 \text{ lb/ft}^2 \text{ s } [2.46 \text{ kg/s m}^2]$$
$$G_s = 0.456 \text{ lb/ft}^2 \text{ s } [2.22 \text{ kg/s m}^2]$$
$$\Delta P_s = 0.30 \times 10^{0.20 \times 0.505} (0.456)^2/0.182 \quad (5\text{-}1)$$
$$= 0.43 \text{ in./ft of packing } [28.3 \text{ mm/m}]$$
$$\text{Total pressure drop} = (0.32 \times 11) + (0.43 \times 13)$$
$$= 3.52 + 5.59$$
$$= 9.11 \text{ in. water } [202 \text{ mm H}_2\text{O}]$$

5.11 Optimum Design

It will be apparent that the choice of parameters in the design of packed towers is wide and to attempt to optimize a particular design involves a considerable amount of effort. The considerations involved in determining the optimum design of an absorber have been considered by Kim and Molstadt.[11] The total costs of the absorption and subsequent recovery of the solute in a stripping column include:

(i) the fixed cost of the absorber plus the energy costs of the gas pumping system. A figure of half the fixed cost has been suggested as a realistic factor to add to the fixed costs to cover energy requirements.
(ii) the value of the unrecovered solute, taken as the market cost of its replacement.
(iii) the cost of recovery of the solute in the stripper separation.

The cost, which is to be minimized, is the total annual cost of recovery of one mol/hr [1 kmol/s] of solute vapour fed to the absorber in the gas stream and can be shown to be given by:

$$\frac{C_T}{G_m A y_1}\left(1 - \frac{1}{Y}\right) = C_a \frac{Z}{G_m} y_1 \left(1 - \frac{1}{Y}\right) + \frac{C_s N_h}{Y - 1} +$$
$$C_{st} N_h \left[\frac{r\{1 - ((m_s G_m/L_m)y_1(1 - 1/Y))\}}{(m_s - 1)(G_m/L_m)y_1(1 - 1/Y)} + 1\right] \qquad (5\text{-}10)$$
$$Y = y_1/y_2 \qquad (5\text{-}11)$$

In equations (5–10) and (5–11),

C_T	total annual costs	(£/yr)	[£/yr]
C_a	annual cost of power and fixed charges for unit volume of the absorption column	(£/ft³ yr)	[£/m³ yr]
C_{St}	total cost of stripping operation expressed as £/mol of vapour produced at the base of the stripper including fixed charges, cost of water and steam	(£/mol)	[£/kmol]
C_s	cost of solute	(£/mol)	[£/kmol]
A	cross-sectional area of absorber	(ft²)	[m²]
y_1, y_2	inlet and exit molar fraction of solute in gas stream in the absorber	(—)	[—]
Z	$H_{OG} N_{OG}$ (see section 4)	(ft)	[m]
H_{OG}	height of an overall transfer unit	(ft)	[m]
N_h	number of hours of operation per year	(—)	[—]
m	slope of operating line in the absorber	(—)	[—]
m_s	slope of operating line in the stripper	(—)	[—]
r	ratio of actual to minimum reflux ratio in the stripping column	(—)	[—]
G_m, L_m	gas and liquid flow rates in the absorber	(mol/hr ft²)	[kmol/s m²]

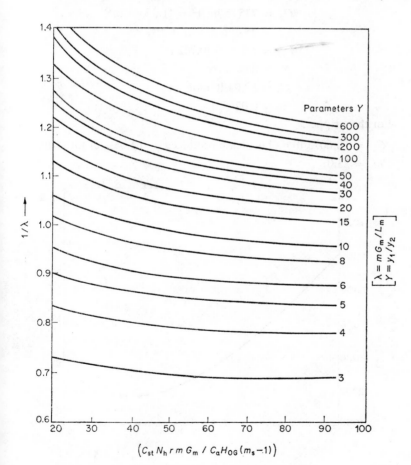

$$\left(C_{st}\, N_h\, r\, m\, G_m\, /\, C_a H_{OG}\, (m_s - 1) \right)$$

FIGURE 5–15. Cost optimization of packed towers[11]

Equation (5–10) may be partially differentiated with respect to the absorption factor and the resultant equation made equal to zero. The final equation has been presented in graphical form[11] and is reproduced in Figure 5–15. The method of application of Figure 5–15 is to evaluate the abscissa to obtain a series of values of the absorption factor, λ and the ratio of inlet to exit solute concentrations in the gas stream, Y. These values may then be substituted simply into equation (5–10) and the minimum value of $C_T/G_m A(y_1 - y_2)$ obtained. The method is best illustrated by an example.

Example 5–6[11]

Acetone is to be removed from an air stream in a packed absorption column and recovered in a stripper. The following data is available from which the optimum value of Y is to be found.

For the absorption column,

$$C_a = £1.8/\text{yr}$$
$$m = 2.7$$
$$G_m = 735 \text{ lb/hr ft}^2 = [1.0 \text{ kg/s m}^2]$$
$$= 25.4 \text{ lb m s/hr ft}^2$$
$$H_{OG} = 2.5 \text{ ft} = [0.762 \text{ m}]$$
$$N_h = 8400 \text{ hr/yr}$$
$$C_s = £2.04/\text{lb mol}$$
$$y_1 = 0.02$$

For the stripper,

$$C_{st} = £0.0075/\text{lb mol of steam produced at the base of the stripper}$$
$$m_s = 23.0$$

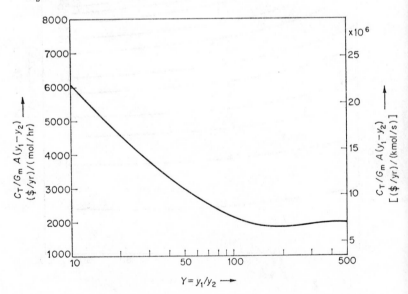

FIGURE 5–16. Graphical optimization for the problem of Example 5–6[11]

Substitution of the relevant values into the ordinate of Figure 5–15 gives a value of 54.1. From Figure 5–15, a series of values of $1/\lambda$ and Y are obtained. These values inserted into equation (5–10) enable Figure 5–16 to be produced, which shows a minimum value at $Y = 200$. The total cost is then £777/yr for a recovery rate of 1 lb mol/hr.

Thus the cost of recovery of acetone is

$$\frac{777 \text{ £/yr/(mol/hr)}}{8400 \text{ (hr/yr)} \times 58 \text{ (lb/lb mol)}} = \text{£0.0016/lb acetone}$$

The optimum value of $1/\lambda$ is 1.22 and the resultant liquid rate should be checked for satisfactory performance, especially with respect to pressure drop and wetting of the packing.

5.12 Nomenclature

a	constant in equation (5–1)	(—)	[—]
a	specific area of packing	(ft²/ft³)	[m²/m³]
A	cross-sectional area	(ft²)	[m²]
b	constant in equation (5–1)	(—)	[—]
C_a	cost of power and fixed charges per unit volume	(£/ft³ yr)	[£/m³ yr]
C_s	cost of solute	(£/mol)	[£/kmol]
C_{st}	total cost of stripping operation	(£/mol)	[£/kmol]
C_T	total annual costs	(£/yr)	[£/yr]
d_p	packing size	(in.)	[mm]
D	product rate	(mol/hr)	[kmol/s]
F	packing factor	(—)	[—]
F	feed rate	(mol/hr)	[kmol/s]
g	gas rate	(lb/s ft²)	[kg/s m²]
g	acceleration due to gravity	(ft/s²)	[m/s²]
G	gas rate	(lb/hr)	[kg/s]
G'	gas rate	(lb/ft² s)	[kg/m² s]
G_m	molar gas rate	(mol/hr)	[kmol/s]
H_G, H_L	heights of individual transfer units	(ft)	[m]
l	liquid rate	(lb/s ft²)	[kg/s m²]
L	liquid rate	(lb/hr, lb/s ft²)	[kg/s, kg/s m²]
L_m	molar liquid rate	(mol/hr)	[kmol/s]
L_p	wetting rate	(ft³/hr ft²)	[m³/s m²]
L_v	volumetric liquid flow	(ft³/hr)	[m³/s]
L'_v	volumetric liquid flow (Figure 5–12)	(ft³/hr ft²)	[m³/s m²]
m	slope of equilibrium line	(—)	[—]
n	degree of wetting (Figure 5–12)	(—)	[—]
N_h	hours of operation/year	(hr/yr)	[hr/yr]
N_{OG}, N_{OL}	number of overall transfer units	(—)	[—]
ΔP	pressure drop	(in./ft)	[mm/m]
r	ratio of actual to minimum reflux ratio	(—)	[—]
R	reflux ratio	(—)	[—]
U_g	gas velocity at flooding (Figure 5–4)	(ft/s)	[m/s]
V	vapour rate	(lb/hr)	[kg/s]
V_f	flooding velocity (Figure 5–7)	(ft/s)	[m/s]
W	bottoms product rate	(mol/hr)	[kmol/s]
x	liquid composition	(—)	[—]
y	vapour composition	(—)	[—]
Y	ratio of inlet to exit concentrations (y_1/y_2)	(—)	[—]

Z	packed height	(ft)	[m]
ρ	density	(lb/ft^3)	[kg/m^3]
λ	ratio $= mG_m/L_m$	(—)	[—]
ε	voidage	(—)	[—]
μ	viscosity	(cP, lb/ft hr)	[N s/m^2]
ψ	ratio $= \rho_w/\rho_L$	(—)	[—]

Subscripts

g, G	gas
l, L	liquid
m	stripper in equation (5–10)
r	rectifying section
s	stripping section
v	vapour
W	water

REFERENCES

1. Eckert, J. S. *Chem. Eng. Prog.* 1963, **59**, 5, 76.
2. Eckert, J. S. *Chem. Eng. Prog.* 1961, **57**, 9, 54.
3. Coulson, J. M. and Richardson, J. F. *Chemical Engineering*, Vol. II, 2nd ed. Oxford: Pergamon Press, 1968.
4. Sherwood, T. K., Shipley, G. H. and Holloway, F. A. L. *Ind. Eng. Chem.* 1938, **30**, 765.
5. Lobo, W. E., Freud, L., Hashmall, F., and Zenz, F. A. *T.Am.I.Ch.E.* 1945, **41**, 693.
5a. Prahl, W. H. *Chem. Eng.* 11 Aug. 1969, **76**, 89.
6. Leva, M. *Tower Packings and Packed Tower Design*, 2nd ed. Akron, Ohio: U.S. Stoneware Co., 1953.
7. Morris, G. A. and Jackson, J. *Absorption Towers*. London: Butterworths Scientific Publications, 1953.
8. Wall, K. J. *Chem. Proc. Eng.*, 1967, **48**, 7, 56.
9. McCabe, W. L. and Thiele, E. W. *Ind. Eng. Chem.* 1925, **17**, 605.
10. *Tower Packings*, Hydronyl Ltd, Stoke-on-Trent, Staffs, 1965.
11. Kim, J. C. and Molstad, M. C. *H.C. Proc.* 1966, **45**, 12, 107.

Chapter 6: Sieve and Valve Tray Design

6.1 Introduction

The popularity of bubble cap trays in the past has been due to operating experience obtained over years of service, resulting in a well established design procedure,[1] and indeed chemical plant manufacturers have been reluctant to use simpler types of trays whose hydraulic characteristics were, until recently, relatively unknown.

However, since the early 1950s the literature relating to research and operating experience has grown enormously, due primarily to the economics of plant construction which has emphasized the need for simple, efficient and cheaper contacting devices. In this chapter an attempt will be made to present the recent developments and design information together with acceptable and accurate design procedures for a simple cross-flow sieve tray and valve trays with downcomers.

In normal operation of a tray, the vapour flows through the perforations into the liquid on the tray and expands either directly, or via the contacting device, into the liquid to form a foaming, turbulent mixture. As this foam moves across the tray, a high interfacial area for efficient vapour–liquid mass transfer is produced. A schematic diagram of conditions on a tray are shown in Figure 6–1. Liquid descends from a tray through a downcomer and

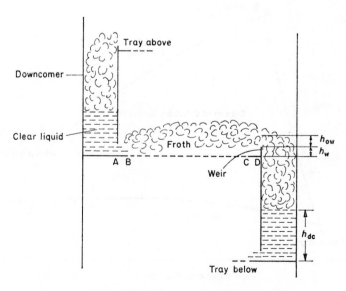

FIGURE 6–1. Schematic diagram of conditions on a sieve tray

on to the next tray to point A, and then starts to flow across the unperforated distribution area to B, where the holes or valves start. The active portion of the tray lies approximately between B and C over the perforated area, though aeration due to turbulence occurs in the areas bounded by AB and CD. These distribution and redistribution areas should not exceed 7 per cent of the tower area.[1] The area CD, which is optional, provides a partial calming area for foam collapse before the foam passes over the outlet weir and into the downcomer leading to the next tray. Secondary foam formation usually occurs in the downcomer as a result of turbulence and splashing, and the downcomer design must provide for this consideration and allow for foam collapse.

There are two major differences between sieve trays and bubble cap trays. Firstly, the vapour passes through the large number of perforations on a sieve tray and emerges through the liquid in a vertical direction. Secondly, there is no built-in liquid seal and only the passage of vapour effectively prevents loss of liquid through the holes. With a valve tray, the mode of operation is dependent upon the type of unit which is chosen. Suggestions for the most suitable type of valve for a particular operation have been included in Chapter 4.

Design methods for trays follow a simple procedure whereby a tray diameter, spacing and layout are estimated and the capacity, pressure drop, and flexibility of the suggested layout are then calculated.[1],[2],[3] A change in the original estimates may then be necessitated by process or cost considerations and the procedure may be repeated until the desired specifications are met. In this chapter design factors for sieve trays and valve trays are considered separately, though there is similarity between both the method employed and the design considerations.

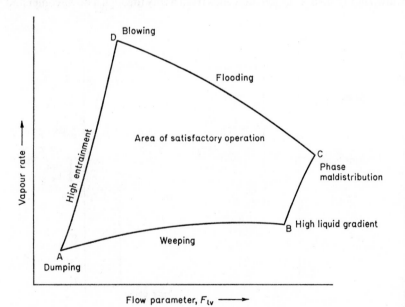

FIGURE 6–2. Sieve tray performance diagram

6.2 Sieve Trays

Consideration of the operating characteristics of a sieve tray will assist in the formation of a clear picture of the satisfactory limits of operation of the tray. In Figure 6–2, the area ABCD encloses the region of satisfactory operation. At point A, dumping or raining occurs at very low vapour rates. The line AB represents weeping, where the vapour flow rate is insufficient to keep all the liquid on the plate. Zenz[4] has recently discussed the determination of weepage rate from sieve trays, but in general a small amount of weeping can be tolerated and is usually acceptable.

The line AD, corresponding to high vapour velocities at low liquid flow rates, represents the limits of tolerable entrainment; at point D blowing or jetting occurs. The high vapour rate at point D carries large drops of liquid up to the next tray without sufficient time for good vapour-liquid contact on the tray, and this point represents the flooding condition at high vapour rates. Flooding, the limits of which are shown as DC, may be caused by one or more of several factors and is detected by a sudden increase in pressure drop and a sharp decrease in efficiency. Factors causing flooding may be listed as:[5]

(i) jetting of liquid to the tray above,
(ii) foam expansion into the tray above,
(iii) the complete lifting off of the liquid on the plate by high vapour flows,
(iv) excessive entrainment of the liquid by the vapour,
(v) excess downcomer back-up causing the downcomers to be flooded often by excessive foaming.

At point B, where high liquid flows are prevailing, the liquid gradient on the tray will become too high to be tolerated, while at similar liquid flows but high vapour rates at C phase maldistribution will cause a sharp drop in efficiency.

The design of a sieve tray is an exercise in keeping the area of operation within the bounds of ABCD. The factors involved will now be considered in turn, with a summary of the design method and a worked example included at the end of this section.

6.2.1 Tower Diameter

The diameter of the tray is determined by the point of incipient flooding. This may be estimated from Fair's correlation[6] which is presented in Figure 6–3. The flow parameter F_{lv} is equal to $(L/V)(\rho_v/\rho_L)^{0.5}$ and accounts for liquid flow effects on the tray, while C_{sb} the capacity parameter, has been adapted from the work of Souders and Brown[7] and is equal to $U_n[\rho_v/(\rho_L - \rho_v)]^{0.5}$, where U_n is the vapour velocity based on the net area of the tray, i.e. (tower area − downcomer area). The use of the correlation is limited by certain restrictions:

(i) The curves presented are for a liquid surface tension equal to 20 dynes/cm [0.02 J/m²]. For other liquids the following correction should be applied:

$$C_{sb}/(C_{sb})_{\sigma=20} = (\sigma/20)^{0.2} \; [= (\sigma/0.02)^{0.2}] \qquad (6\text{--}1)$$

where σ = liquid surface tension (dynes/cm) [J/m^2].

(ii) The curves apply for a hole/active area ratio = 0.10. For values of hole/active area ratio equal to 0.08 and 0.06, the value of C_{sb} from the chart should be multiplied by 0.90 and 0.80 respectively.

(iii) Values of C_{sb} obtained from the chart apply to non- or low-foaming systems. The capacity will be reduced considerably for severe foaming systems. (See section 6.3.3 under valve trays.)

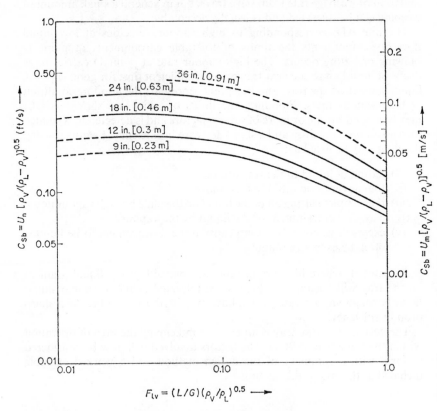

FIGURE 6–3. Sieve tray flooding capacity[6]

6.2.2 Plate Spacing

The selection of the tray spacing depends upon the size and duty of the column. Large columns usually employ a spacing of 2–3 ft [0.6–0.9 m] to allow high vapour velocities and ample room for manways to permit maintenance of the column. These columns are usually erected in the open, where headroom is no problem. For small columns of less than 4 ft diameter [1.2 m], the plate spacing is reduced to avoid the problem of supporting a tall, slender column. In this case, a spacing of 18 in. [0.46 m] is common, though spacings as low as 6 in. [0.15 m] have been used.[1]

For cryogenic columns, for example those used for the distillation of liquid air, spacings of 2 in.–4 in. [50–100 mm] have been used.[8] This, however, is a special case, as the heat leakage through the insulation is of such importance that the column size must be an absolute minimum.

6.2.3 Entrainment

Some entrainment always occurs in the normal operation of a sieve tray column, and since entrainment affects efficiency and can cause flooding it is necessary to be able to estimate its magnitude under both design and flooding conditions. Fair[6] has defined the concept of fractional entrainment as

$$\psi = e/(L + e) \qquad (6\text{-}2)$$

where e = liquid entrainment (mol/hr) [mol/s],
L = liquid flow rate (mol/hr) [mol/s].

In addition, he has presented a chart which predicts the fractional entrainment under various percentage flooding conditions. This is reproduced in Figure 6–4.

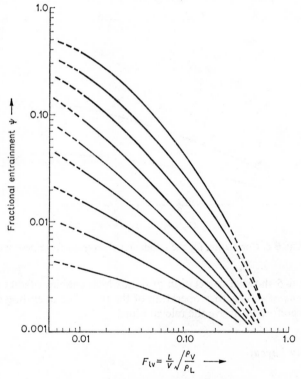

FIGURE 6–4. Sieve tray fractional entrainment[6]

6.2.4 Weepage

The study of weeping from sieve trays has attracted the attention of many workers,[4],[9],[10] though theoretical prediction has proved difficult and experimental data has been obtained from visual studies on trays. The liquid will not drain through the holes of the tray if surface tension effects and vapour pressure drop through the perforations are present to prevent it.

For no weeping,

$$\Delta P_{dry} + h_\sigma \geqslant h_w + h_{ow} \qquad (6\text{–}3)$$

or

$$[\Delta P_{dry} + h_\sigma \geqslant 0.1 h_w + h_{ow}] \qquad (6\text{–}4)$$

where ΔP_{dry} = dry tray pressure drop (in.) [cm],
h_w = weir height (in.) [mm],
h_{ow} = height of liquid crest over weir from equation (6–14) or (6–15) (in.) [cm],
h_σ = head loss due to bubble formation (in.) [cm],
d_h = hole diameter (in.) [mm],
σ = liquid surface tension (dyne/cm^2) [J/m^2],
$h_\sigma = 0.04\, \sigma/\rho_L d_h \qquad (6\text{–}5)$
$[= 4.14 \times 10^4\, \sigma/\rho_L d_h] \qquad (6\text{–}6)$

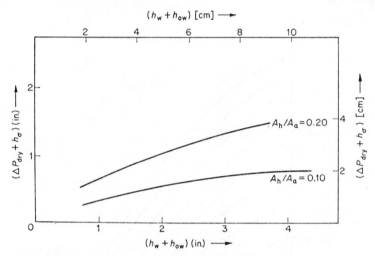

FIGURE 6–5. Correlation for the estimation of weepage from sieve trays[1]

Figure 6–5 shows a correlation of data which enables prediction of the lower limits of satisfactory operation of the tray. The operating condition should therefore lie above the relevant line.

6.2.5 Tray Layout

(i) *Tray Types*

Generally speaking, three types of tray are used for sieve plates. These are shown in Figure 6–6. The reverse flow tray is employed for low liquid flow rates of 0–50 g.p.m. [0–0.003 m^3/s] and has all the downcomers located on one side of the column, thus forcing the liquid to flow around a centre baffle. The tray has a high active area at the expense of downcomer area, and because of the length of the flow path, the liquid gradient may become a problem.

The crossflow tray is most commonly used and is simple and economic in construction, and is used for liquid flow rates of approximately 50–500 g.p.m. [0.003–0.03 m³/s] and gives high plate efficiencies.

In the double pass tray, the liquid flow, which usually exceeds 500 g.p.m. [0.03 m³/s], is divided into two and each half flows across half the tray.

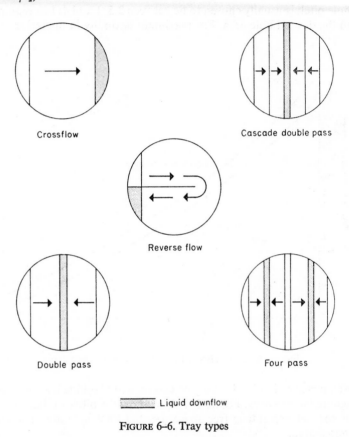

Crossflow   Cascade double pass

Reverse flow

Double pass   Four pass

▓▓▓ Liquid downflow

FIGURE 6–6. Tray types

This type of tray is usually required in large diameter towers. It is efficient and has a large capacity and less liquid gradient than the crossflow type, but is more expensive to construct.

(ii) *Tray Areas*

The tower area (A_t) is the total superficial cross-sectional area of the tower. The downcomer area (A_d) is the cross-sectional area of the segment formed by the overflow weir and the tower wall. A high value of A_d permits low liquid velocities in the downcomer and allows the froth to collapse, but only at the expense of reduced tower area for vapour contact. For a cross-flow tray, A_d is usually taken as 12 per cent of A_t.

The hole area is usually taken as 10 per cent of the tower area; a higher value than this can lead to excessive weeping and a lower value can lead to high pressure drops. Hole sizes usually lie in the range $\frac{1}{8}$–$\frac{1}{2}$ in. [3.2– 12.7 mm]—most commonly $\frac{3}{16}$ in. [4.75 mm].

(iii) *Hole Pitch*

The hole pitch is usually chosen to lie between 2.5 : 1 and 4 : 1 as required to obtain the desired hole area. For maximum flexibility in operation, a hole

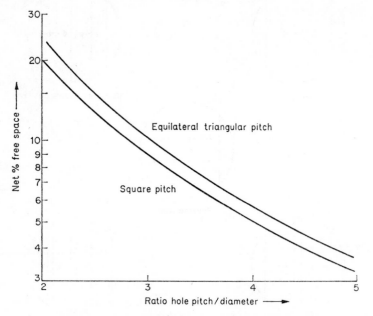

FIGURE 6–7. The effect of hole layout on the free area

pitch/diameter ratio of 2.5 : 1 may be chosen and blanking used to produce the required active area. Figure 6–7 shows how the pitch-to-diameter ratio affects the net percentage free space, i.e. the total hole area per unit of perforated area.

The hole pattern may be chosen as either equilateral triangular or square and the effect this has on the net free space is shown in Figure 6–7, from which the ratio of the hole area to the total perforated area (open area) may be estimated. The perforated area is not the same as the active area since the latter includes the calming zones adjacent to the inlet and outlet weirs where aeration still occurs. The open area is a term which refers to the identi- fication of perforated metal sheets.

(iv) *Blanking*

Owing to varying liquid and vapour loadings through the tower, the hole area requirements may vary considerably. Rather than manufacture dif- ferent trays to meet these needs, it is usual to use the same tray layout and vary the active area by means of blanking strips. However, unless the need is temporary, blanking should not exceed 25 per cent of the tray area. To avoid formation of 'dead areas' the width of blanking strips should not exceed 7 per cent of the tray diameter for small columns of 3–10 ft diameter [1–3 m] or 5 per cent for larger towers.

Blanking strips should be bolted or screwed into position to provide a good seal without distortion or buckling, and means for their fixing and removal should be made as simple as possible.

(v) Inlet Calming Zone

This zone should have a minimum depth of 2 in. [50 mm] which would normally be provided by the plate support ring. Khamdi[11] has shown that for any tray the depth may be estimated from the following equation:

$$\text{depth of calming zone} = 1.4\{h_{\text{ow}}(h_{\text{w}} + h_{\text{ow}}/3)\}^{0.5} \qquad (6\text{–}7)$$

$$[= 0.551\{h_{\text{ow}}(0.1h_{\text{w}} + h_{\text{ow}}/3)\}^{0.5}] \qquad (6\text{–}8)$$

where h_{ow} = height of liquid crest over weir (in.) [cm],
                 h_{w} = weir height (in.) [mm].

(vi) Outlet Calming Zone

Hydraulically none is required[11] although, as for the inlet calming zone, a 2 in. [50 mm] zone is usually provided by the tray supporting ring.

(vii) Baffles

Sargent[12] has shown that for a given vapour rate, an outlet splash baffle gives greater foam heights, increases entrainment and slightly increases downcomer hold-up. Together with other reported work[13],[17] the indications are that baffles are not to be recommended, though some manufacturers do include them in standard proprietary trays.

6.2.6 Hydraulic Parameters

(i) Dry Tray Pressure Drop (ΔP_{dry})

The calculation of the pressure drop due to the passage of vapour through the perforations may be estimated from the simple orifice equation which can be re-arranged to give a direct value of the dry tray drop:

$$\Delta P_{\text{dry}} = 0.186 \, (\rho_{\text{V}}/\rho_{\text{L}})U_{\text{h}}^2(1/C_{\text{vo}})^2 \qquad (6\text{–}9)$$

$$[= 5.08(\rho_{\text{V}}/\rho_{\text{L}})U_{\text{h}}^2(1/C_{\text{vo}})^2] \qquad (6\text{–}10)$$

where U_{h} = vapour velocity through holes (ft/s) [m/s],
          C_{vo} = dry orifice coefficient (—) [—].

C_{vo} has been the subject of many investigations and is found to be a function of vapour velocity, the ratio of hole diameter to tray thickness, Reynolds number and the condition of the hole 'lip'. Chan[14] has made a comparison of the results of several workers and has shown that the simple relation of

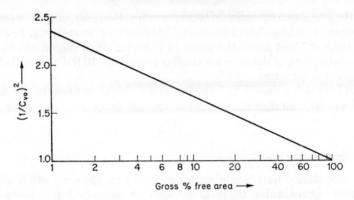

FIGURE 6–8. Orifice coefficient and free area for sieve trays[15]

Prince[15] shown in Figure 6–8 gives the smallest mean error of 11 per cent. The gross percentage free area is the ratio of hole area to tower area.

In cases where the fractional entrainment is greater than 0.10, a correction factor equal to $(15X + 1)$ should be multiplied by the value of ΔP_{dry} obtained in this way to give the corrected value of dry tray pressure drop. The parameter X is defined as

$$X = \psi F_{1v}/(1 - \psi) \qquad (6–11)$$

(ii) *Total Pressure Drop* (ΔP_T)

By considering the head loss due to the aerated liquid in the tray and the dry tray pressure drop, an additive value of the total tray pressure drop can be obtained. This method is considered more accurate than the alternatives of a dimensionless group approach or empirical correlations.[16]

The head loss due to the aerated liquid, h_a, is obtained via Figure 6–9 and Q_p, the aeration factor.

$$n_a = Q_p(h_w + h_{ow}) \qquad (6–12)$$

$$[= Q_p(0.1h_w + h_{ow})] \qquad (6–13)$$

where h_w = weir height (in.) [mm],
   h_{ow} = height of liquid crest over the weir (in.) [cm].

h_{ow} is obtained from

$$h_{ow} = 0.48(q/l_w)^{0.67} \qquad (6–14)$$

$$[= 66.6(q/l_w)^{0.67}] \qquad (6–15)$$

where q = liquid flow rate (US gal/min) [m³/s],
   l_w = weir length (in.) [mm].

(iii) *Liquid Gradient* (Δ)

The head required to produce the crossflow of the aerated mass on the tray is known as the liquid gradient, Δ. If this value is high problems of

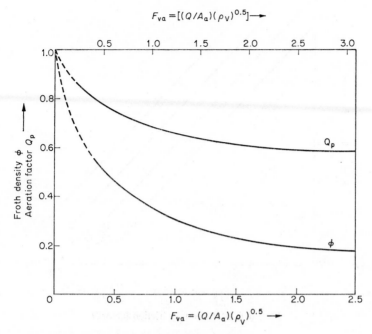

$$F_{va} = [(Q/A_a)(\rho_V)^{0.5}] \longrightarrow$$

FIGURE 6–9. Aeration factor and relative froth density for sieve trays[15a]

liquid maldistribution and weeping can occur, but in general the gradient on sieve trays is small. The criterion for tray stability is that the liquid gradient should be less than half the value of the dry tray pressure drop.

The value of Δ may be determined from the method of Hughmark and O'Connell[17] in which a friction factor, f, is obtained from a modified Reynolds number and calculated from

$$f = R_h g \Delta / 12 U_f^2 L_f \qquad (6\text{--}16)$$

$$[= R_h g \Delta / 10^4 U_f^2 L_f] \qquad (6\text{--}17)$$

where R_h = hydraulic radius of aerated mass (ft) [m],
   U_f = velocity of aerated mass (ft/s) [m/s],
   L_f = distance between weirs (ft) [m],
   g = acceleration due to gravity (ft/s²) [m/s²].
The plot of Reynolds modulus against the friction factor for various weir heights is presented in Figure 6–10. The method of calculating Δ is illustrated in a worked example later in this chapter.

iv) *Pressure drop in the Downcomer* (h_{dc})

If an excessive pressure drop exists in the downcomer, flooding will frequently result. This pressure drop may be determined by a simple pressure

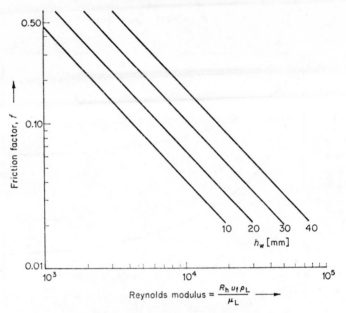

FIGURE 6–10. Sieve tray friction factor[1]

balance and, knowing the aeration factor, the height of froth in the down comer may be calculated.

$$h_{dc} = \Delta P_T + h_w + h_{ow} + \Delta + h_{da} \qquad (6-18)$$

where h_{da} is the downcomer apron pressure drop, given by

$$h_{da} = 0.03(q/100A_{da})^2 \qquad (6-19)$$

$$[= 16.5(q/A_{da})^2] \qquad (6-20)$$

where q = liquid flow rate (US gal/min) [m³/s],
    A_{da} = weir length and apron clearance (in.²) [m²].

As h_{da} is the height of clear liquid in the downcomer, h_{dc}/Q_p will giv the total height of aerated mass. It is suggested that this latter figure shoul not exceed half of the tray spacing for a foaming system or 90 per cent o tray spacing for non-foaming systems.

These then are some of the most important factors affecting sieve tra design. In order to make the procedure clear, a worksheet on the desig of a column is now included where each step is itemized, leading to the fina specifications for the tower. To illustrate the method, Example 5–5 wi be repeated using sieve trays instead of a packed column.

6.2.7 Work Sheet for Sieve Tray Design

(i) Vapour

Mass flow rate, V	(lb/hr)	[kg/s]
Density, ρ_V	(lb/ft³)	[kg/m³]
Volumetric flow rate Q	(ft³/s)	[m³/s]

ii) *Liquid*

Mass flow rate, L	(lb/hr)	[kg/s]
Density, ρ_L	(lb/ft^3)	[kg/m^3]
Volumetric flow rate	(US gal/min)	[m^3/s]
Surface tension, σ	(dynes/cm)	[J/m^2]

iii) *Flow parameter*

$$F_{lv} = (L/V)(\rho_V/\rho_L)^{0.5} \tag{6-21}$$

iv) *Vapour capacity C_{sb}* (Figure 6–3)

Assume plate spacing = . . . ft . . . m
For F_{lv} = . . . C_{sb} = . . .
Correct for surface tension

$$C_{sb} = (C_{sb_{\sigma=20}})(\sigma/20)^{0.2} \tag{6-1}$$

$$C_{sb} = U_{nf}[\rho_V/(\rho_L - \rho_V)]^{0.5} \tag{6-22}$$

$$U_{nf} = C_{sb}[(\rho_L - \rho_V)/\rho_V]^{0.5} \tag{6-23}$$

$$= . . . \text{ft/s} = . . . \text{m/s}.$$

v) *Selection of Tray*

Where possible, a single crossflow tray should be used with segmental downcomers. Use straight weirs of length = 0.77 × tower diameter when downflow area (A_d) = 0.12 × tower area (A_t).

$$\text{Net area} = A_n = A_t - A_d \tag{6-24}$$

Select weir height: $\frac{1}{2}$ in. [12.5 mm] for vacuum columns, 2 in. [50 mm] for atmospheric columns as first approximations.
Select hole size: $\frac{3}{16}$ in. [4.75 mm] initially.

vi) *Tower Diameter*

Choose percentage flooding F^*, e.g. 80 per cent. Hence

$$U_n^* = F^* \times U_{nf} = . . . \text{ft/s} . . . \text{m/s} \tag{6-25}$$

$$A_t^* = \text{tower area} = Q/U_n^* = . . . \text{ft}^2 . . . \text{m}^2 \tag{6-26}$$

Calculate tower diameter D_t = . . . ft . . . m.
Select tower to nearest convenient size.

$$\text{Final tower area } A_t = . . . \text{ft}^2 . . . \text{m}^2$$

vii) *Tabulation of Tower Areas*

$$\text{Area of tower } A_t = . . . \text{ft}^2 . . . \text{m}^2$$

$$\text{Downcomer area } A_d = 0.12\,A_t = . . . \text{ft}^2 . . . \text{m}^2$$

$$\text{Net area } A_n = 0.88\,A_t = . . . \text{ft}^2 . . . \text{m}^2$$

$$\text{Active area } A_a = (A_t - 2A_d) \tag{6-27}$$

$$= 0.76A_t = . . . \text{ft}^2 . . . \text{m}^2$$

$$\text{Hole area } A_h = 0.10\,A_t = . . . \text{ft}^2 . . . \text{m}^2$$

(viii) *Check of Approach to Flooding*

$$U_n = Q/A_n \ldots \text{ft/s} \ldots \text{m/s} \tag{6-28}$$

$$(U_n/U_n^*)F^* = F$$

Actual approach to flooding $= F = \ldots$

(ix) *Calculation of Entrainment* (Figure 6-4)

For a value of $F_{lv} = \ldots$, fractional entrainment $\psi = \ldots$ N.B. ψ should not exceed 0.2.

$$\text{Total entrainment } e = \psi L/(1 - \psi) = \ldots \text{lb/hr} \ldots \text{kg/s} \tag{6-29}$$

(x) *Tray Pressure Drop*

(a) *Dry tray pressure drop* ΔP_{dry}. Calculate velocity through holes,

$$U_h = Q/A_h = \ldots \text{ft/s} \ldots \text{m/s} \tag{6-30}$$

$$\text{Tray thickness/hole diameter} = \ldots$$

$$A_h/A_a = \ldots$$

Determine discharge coefficient C_{vo} from Figure 6-8.

$$C_{vo} = \ldots$$

$$\Delta P_{dry} = 0.186(\rho_V/\rho_L)U_h^2(1/C_{vo})^2 \tag{6-9}$$

$$[= 5.08(\rho_V/\rho_L)U_h^2(1/C_{vo})^2] \tag{6-10}$$

$$= \ldots \text{in} \ldots \text{cm}$$

If $\psi > 0.10$, correct ΔP_{dry} by

$$\Delta P_{dry} \text{ corrected}/\Delta P_{dry} = (15X + 1.0) \tag{6-31}$$

where

$$X = \psi F_{lv}/(1 - \psi) \tag{6-11}$$

(b) *Aerated liquid drop* h_a. Determine aeration factor of liquid on tray (Q_p) from Figure 6-9.

$$F_{va} = (Q/A_a)\rho_V^{0.5} = \ldots \tag{6-32}$$

Hence $Q_p = \ldots$

Determine the height of liquid crest over weir (h_{ow}) from

$$h_{ow} = 0.48 \, (q/l_w)^{0.67} \tag{6-14}$$

or $$h_{ow} = [66.6 \, (q/l_w)^{0.67}] \tag{6-15}$$

where $q =$ liquid flow (US gal/min) [m^3/s],
$l_w =$ weir length (in.) [m].

$$\text{Aerated liquid drop } h_a = Q_p(h_w + h_{ow}) \tag{6-12}$$

$$[= Q_p(0.1h_w + h_{ow}] \tag{6-13}$$

$$h_a = \ldots \text{in} \ldots \text{cm}.$$

(c) *Total tray pressure drop.* Total pressure drop is given by

$$\Delta P_{\mathrm{T}} = \Delta P_{\mathrm{dry}} + h_{\mathrm{a}} \qquad (6\text{-}33)$$

$$= \ldots \text{in.} \ldots \text{cm}$$

Is ΔP_{T} acceptable? If not, can weir height or ΔP_{dry} be reduced?

(xi) *Estimation of Weep Point*

The point of operation must be above the relevant line on Figure 6–5.

$$h_{\sigma} = 0.04\sigma/\rho_{\mathrm{L}}d_{\mathrm{h}} \qquad (6\text{-}5)$$

$$[= 4.14 \times 10^{4}\sigma/\rho_{\mathrm{L}}d_{\mathrm{h}}] \qquad (6\text{-}6)$$

where σ = surface tension (dynes/cm) [J/m²],
ρ_{L} = liquid density (lb/ft³) [kg/m³],
d_{h} = hole diameter (in.) [mm].

$$\Delta P_{\mathrm{dry}} + h_{\sigma} = \ldots \text{in.} \ldots \text{cm}$$

$$h_{\mathrm{w}} + h_{\mathrm{ow}} = \ldots \text{in.} \ldots \text{cm}$$

Is the operating point above the relevant line for hole area/active area $= \ldots$? If so, operation is above the weep point and is satisfactory.

(xii) *Calculation of Downcomer Residence Time*

Downcomer velocity on basis of clear liquid is

$$V_{\mathrm{d}} = L/3600A_{\mathrm{d}}\rho_{\mathrm{L}} = \ldots \text{ft/s} \qquad (6\text{-}34)$$

$$[= L/A_{\mathrm{d}}\rho_{\mathrm{L}} = \ldots \text{m/s}] \qquad (6\text{-}35)$$

$$\text{Residence time} = (\text{tray spacing } /V_{\mathrm{d}} = \ldots \text{s} \qquad (6\text{-}36)$$

This residence time must be greater than 3 s.

(xiii) *Calculation of Liquid Gradient*

For sieve trays the liquid gradient is likely to be small but should be checked by the following procedure for designs where long flow paths and high liquid rates prevail.

(a) Calculate froth height on tray (h_{f}),

$$h_{\mathrm{f}} = h_{\mathrm{a}}/2Q_{\mathrm{p}} - 1 = \ldots \text{in.} \ldots \text{cm} \qquad (6\text{-}37)$$

(b) Calculate hydraulic radius of aerated mass (R_{h}),

$$R_{\mathrm{h}} = \text{cross-section/wetted perimeter}$$

$$= h_{\mathrm{f}}D_{\mathrm{f}}/(2h_{\mathrm{f}} + 12D_{\mathrm{f}}) \qquad (6\text{-}38)$$

$$[= h_{\mathrm{f}}D_{\mathrm{f}}/(2h_{\mathrm{f}} + 100D_{\mathrm{f}})] \qquad (6\text{-}39)$$

where D_{f} = flow width normal to liquid flow

$$= (L_{\mathrm{w}} + D)/2 \text{ (ft) [m]} \qquad (6\text{-}40)$$

$$L_{\mathrm{w}} = \text{weir length (ft) [m]}$$

(c) Calculate velocity of aerated mass (U_f)

$$U_f = 0.0267 \, q/h_f \phi D_f \tag{6-41}$$

$$[= 100q/h_f \phi D_f] \tag{6-42}$$

where ϕ is obtained from Figure 6–9.

(d) Calculate Reynolds modulus (R_{eh})

$$R_{eh} = R_h U_f \rho_L / \mu_L \tag{6-43}$$

where R_h = hydraulic radius (see (b)) (ft) [m],
      U_f = velocity of aerated mass (see (c)) (ft/s) [m/s],
      ρ_L = liquid density (lb/ft³) [kg/m³],
      μ_L = liquid viscosity (lb/ft s) [Ns/m²].

(e) Determine friction factor (f). From Figure 6–10,

$$f = \ldots$$

(f) Calculate liquid gradient, (Δ)

$$\Delta = 12 f U_f^2 L_f / R_h g \tag{6-44}$$

$$[= 10^4 f U_f^2 L_f / R_h g] \tag{6-45}$$

where g = acceleration due to gravity (ft/s²) [m/s²],
      h_f = distance between weirs (ft) [m].

(xiv) *Height of Aerated Mass in Downcomer*

Height of clear liquid = $h_{dc} = \Delta P_T + h_w + h_{ow} + \Delta + h_{da}$ (6-18)

$$[= \Delta P_T + 0.1 h_w + h_{ow} + \Delta + h_{da}] \tag{6-46}$$

where h_{da} = downcomer apron pressure drop (in.) [cm],
      $= 0.03(q/100 A_{da})^2$ (6-19)
      $[= 13.1(q/A_{da})^2]$ (6-20)
      A_{da} = weir length × downcomer clearance (in.²) [m²],
      $h_{dc} = \Delta P_T + h_w + h_{ow} + \Delta + h_{da}$
      $= \ldots$ in. $\ldots$ cm.

This height should be less than half the tray spacing.

(xv) *Summary of Design*

Type of tray		
Tower diameter	(ft)	[m]
Tray spacing	(ft)	[m]
Active area	(ft²)	[m²]
Hole area	(ft²)	[m²]
Downflow area	(ft²)	[m²]
Hole/tower area	(—)	[—]
Hole/active area	(—)	[—]
Hole size	(in.)	[mm]
Weir length	(ft)	[m]
Weir height	(in.)	[mm]
Downcomer clearance	(in.)	[mm]
Tray thickness	(in.)	[mm]

To illustrate the design procedure above, a worked example will now be considered.

Example 6–1

Design a crossflow sieve tray for the system described earlier in Example 5–5, considering conditions on the top trays.

The example will be worked following the procedure outlined in section 6.2.7.

(i) *Vapour*

$$\text{Mass flow rate, } V = 40\ 000\ \text{lb/hr} \ [= 5.04\ \text{kg/s}]$$
$$\text{Density, } \rho_V = 0.168\ \text{lb/ft}^3 \ [= 2.69\ \text{kg/m}^3]$$
$$\text{Volumetric flow rate, } Q = 66.08\ \text{ft}^3/\text{s} \ [= 1.87\ \text{m}^3/\text{s}]$$

(ii) *Liquid*

$$\text{Mass flow rate, } L = 32\ 000\ \text{lb/hr} \ [= 4.03\ \text{kg/s}]$$
$$\text{Density, } \rho_L = 43.3\ \text{lb/ft}^3 \ [= 694\ \text{kg/m}^3]$$
$$\text{Volumetric flow, } q = 739\ \text{ft}^3/\text{hr} = 12.32\ \text{ft}^3/\text{min}$$
$$= 92.16\ \text{US gal/min} \ [= 5.83 \times 10^{-3}\ \text{m}^3/\text{s}]$$
$$\text{Viscosity, } \mu_L = 0.32\ \text{cP} = 2.5 \times 10^{-4}\ \text{lb/ft s}$$
$$[= 3.2 \times 10^{-4}\ \text{N s/m}^2]$$
$$\text{Surface tension, } \sigma = 20\ \text{dyne/cm} \ [= 0.02\ \text{J/m}^2]$$

(iii) *Flow Parameter*

$$F_{lv} = (L/V)(\rho_V/\rho_L)^{0.5} \tag{6–21}$$
$$= (32\ 000/40\ 000)(0.168/43.3)^{0.5}$$
$$= 0.0498$$

(iv) *Vapour Capacity* (C_{sb})

Assume plate spacing $= 18$ in. $[= 0.458$ m$]$. From Figure 6–3, at $F_{lv} = 0.0498$,

$$C_{sb} = 0.28\ [= 0.0855]$$
$$U_{nf} = 0.28\{(\rho_L - \rho_V)/\rho_V\}^{0.5} \tag{6–23}$$
$$= 4.48\ \text{ft/s} \ [= 1.37\ \text{m/s}]$$

(v) *Tray Selection*

For the relatively low flow rates in this example, a single crossflow tray with segmental downcomers should be suitable.

If $A_d = 0.12 A_t$, the weir length is $0.77\ D_t$.

$$\text{Net area, } A_n = A_t - A_d = 0.88\ A_t \tag{6–24}$$

Select a weir height of 2.0 in. [50 mm], a hole size of $\frac{3}{16}$ in. [4.75 mm] and a tray thickness of $14g = 0.074$ in. $= [1.88$ mm].

(vi) *Tower Diameter*

Select a design percentage flooding $F^* = 80$ per cent $= 0.80$.

$$U_n^* = F^* \times U_{nf} = 0.80 \times 4.48$$
$$= 3.58 \text{ ft/s} \qquad (6\text{-}25)$$
$$[= 1.09 \text{ m/s}]$$
$$A_t = Q/0.88U_n^*$$
$$= 66.08/3.58 \times 0.88$$
$$= 21.0 \text{ ft}^2$$
$$[= 1.95 \text{ m}^2]$$

Column diameter $= 5.18$ ft

Select tray diameter $= 5.25$ ft $= [1.60$ m]

Tower area $= 21.6$ ft^2 $= [2.01$ m$^2]$

(vii) *Tabulation of Tower Areas*

Tower area $A_t = 21.6$ ft^2 $[= 2.01$ m$^2]$

Downcomer area $A_d = 0.12A_t = 2.52$ ft^2 $[= 0.236$ m$^2]$

Net area $A_n = 0.88A_t = 18.5$ ft^2 $[= 1.72$ m$^2]$

Active area $= 0.76A_t = 16.0$ ft^2 $[= 1.49$ m$^2]$

Hole area $= 0.10A_t = 2.16$ ft^2 $[= 0.201$ m$^2]$

(viii) *Flooding Check*

$$U_n = Q/A_n \qquad (6\text{-}28)$$
$$= 66.08/18.5 \ [= 1.87/1.72]$$
$$= 3.57 \text{ ft/s} \ [= 1.09 \text{ m/s}]$$
$$F = F^*(U_n/U_n^*)$$
$$= 0.80(3.57/3.58)$$
$$= 0.797 = 79.7\%$$

(ix) *Calculation of Entrainment*

From Figure 6–4, at $F_{1v} = 0.0498$, $F = 79.7$ per cent. $\psi = 0.06$, i.e. < 0.2 and is therefore satisfactory.

$$\text{Total entrainment, } e = \psi L/(1 - \psi) \qquad (6\text{-}29)$$
$$= 0.06 \times 32\,000/(1 - 0.06)$$
$$[= 0.06 \times 4.03/(1 - 0.06)]$$
$$= 2040 \text{ lb/hr} \ [= 0.258 \text{ kg/s}]$$

x) *Tray Pressure Drop*

a) Hole velocity,

$$U_h = Q/A_h \qquad (6\text{-}30)$$
$$= 66.08/2.16 \, [= 1.87/0.201]$$
$$= 30.6 \, \text{ft/s} \, [= 9.31 \, \text{m/s}]$$

Tray thickness/hole diameter $= 0.074/0.187 \, [= 1.88/4.75]$
$$= 0.396$$

Hole area/active area $= 2.16/16.0 \, [= 0.201/1.49]$
$$= 0.135$$

From Figure 6–8, $C_{vo} = 0.75$. Hence

$$\Delta P_{dry} = 0.186(\rho_V/\rho_L)U_h^2(1/C_{vo})^2 \qquad (6\text{-}9)$$
$$= 0.186(0.186/43.3)(30.6)^2(1/0.75)^2 = 1.21 \, \text{in.}$$
$$[= 5.08(\rho_V/\rho_L)U_h^2(1/C_{vo})^2 = 4.07 \, \text{cm}] \qquad (6\text{-}10)$$

As $\psi < 0.10$, there is no correction to be applied to ΔP_{dry}.

(b) Aerated liquid drop h_a.

$$F_{va} = (Q/A_a)\rho_V^{0.5} \qquad (6\text{-}32)$$
$$= (66.08/16.0)(0.168)^{0.5}$$
$$= 1.69 \, [= 2.06]$$

From Figure 6–9,

Aeration factor $Q_p = 0.60$

$$l_w = 0.77 \times 5.25 \times 12$$
$$= 48.5 \, \text{in.} \, [= 1.23 \, \text{m}]$$
$$h_{ow} = 0.48(q/l_w)^{0.67} \qquad (6\text{-}14)$$
$$= 0.48 \times (92.16/48.5)^{0.67} = 0.74 \, \text{in.}$$
$$= [66.6 \times (5.83 \times 10^{-3}/1.23)^{0.67}] = [1.86 \, \text{cm}] \qquad (6\text{-}15)$$

Weir height $h_w = 2.0 \, \text{in.} \, [= 50 \, \text{mm}]$

Aerated liquid drop $h_a = Q_p(h_w + h_{ow}) \qquad (6\text{-}12)$
$$= 0.60 \, (2.0 + 0.74) = 1.64 \, \text{in.}$$
$$[= Q_p(0.1h_w + h_{ow}) \qquad (6\text{-}13)$$
$$= 0.6(5.0 + 1.86 = 4.17 \, \text{cm}]$$

(c) Total tray pressure drop

$$\Delta P_T = \Delta P_{dry} + h_a \qquad (6\text{-}33)$$
$$= 1.21 + 1.64$$
$$= 2.85 \, \text{in.} \, [= 7.24 \, \text{cm}]$$

This is an acceptable pressure drop for atmospheric duty.

(xi) *Weep Point*

$$h_\sigma = 0.04\sigma/\rho_1 d_h \tag{6-5}$$
$$= 0.04 \times 20/43.3 \times 0.187 = 0.099 \text{ in.}$$
$$[= 4.14 \times 10^4 \sigma/\rho_L d_h \tag{6-6}$$
$$= 4.14 \times 10^4 \times 0.02/694 \times 475 = 0.25 \text{ cm}]$$
$$\Delta P_{dry} + h_\sigma = 1.31 \text{ in. } [= 3.32 \text{ cm}]$$
$$h_w + h_{ow} = 2.74 \text{ in. } [= 6.86 \text{ cm}]$$

The point of operation is above the relevant line on Figure 6–5 so that weeping is not a problem.

(xii) *Downcomer Residence Time*

$$V_d = L/3600 \, A_d \rho_L$$
$$= 32\,000/3600 \times 2.52 \times 43.3 \tag{6-34}$$
$$= 0.0815 = [L/A_d \rho_L] \tag{6-35}$$
$$\text{Residence time} = \text{Tray spacing}/V_d = 1/0.0815 = 12.3 \text{ s} \tag{6-36}$$

The residence time is greater than the minimum of 3 s and is therefore satisfactory.

(xiii) *Liquid gradient* (Δ)

(a) Height of froth,

$$h_f = h_a/(2Q_P - 1) \tag{6-37}$$
$$= 1.64/(2 \times 0.60 - 1)$$
$$= 8.2 \text{ in. } [= 20.8 \text{ cm}]$$

(b) Hydraulic radius (R_h),

$$D_f = (L_w + D_T)/2 \tag{6-40}$$
$$= (4.04 + 5.25)/2$$
$$= 4.65 \text{ ft } [= 1.42 \text{ m}]$$
$$R_h = h_f D_f/(2h_f + 12 D_f) \tag{6-38}$$
$$= 8.2 \times 4.65/(2 \times 8.2 + 12 \times 4.65)$$
$$= 0.528 \text{ ft } [= 0.161 \text{ m}]$$

(c) Velocity of aerated mass (U_f),

$$U_f = 0.0267 q/h_f \phi D_f = [100 q/h_f \phi D_f] \tag{6-41, 6-42}$$

From Figure 6–9, $\phi = 0.20$, hence

$$U_f = 0.0267 \times 92.16/8.2 \times 0.2 \times 4.65 = 0.323 \text{ ft/s}$$
$$[= 100 \times 5.83 \times 10^{-3}/20.8 \times 0.2 \times 1.42 = 0.0985 \text{ m/s}]$$

(d) Reynolds modulus (R_{eh}),

$$R_{eh} = R_h U_f \rho_L/\mu_L \tag{6-43}$$
$$= 0.528 \times 0.323 \times 43.3/2.15 \times 10^{-4} = 34\,400$$
$$[= 0.161 \times 0.0985 \times 694/3.2 \times 10^{-4} = 34\,400]$$

(e) Friction factor (f). From Figure 6–10, $f = 0.075$.

(f) Calculate Δ.

$$L_f = 0.77D_T = 4.04 \text{ ft } [= 1.23 \text{ m}]$$
$$= 12 fU_f^2 L_f / R_h g \tag{6-44}$$
$$= 12 \times 0.075 \times (0.323)^2 \times 4.04/(0.528 \times 32.2)$$
$$= 0.0223 \text{ in.}$$
$$= [10^4 fU_f^2 L_f / R_h g] \tag{6-45}$$
$$[= 10^4 \times 0.075 \times (0.0985)^2 \times 1.23/(0.161 \times 981)$$
$$= 0.0565 \text{ cm}]$$

This liquid gradient is sufficiently small to be normally considered significant.

iv) Height of Aerated Mass in Downcomer

ssuming a clearance of 1.5 in. [38 mm] between the downcomer apron nd the tray,

$$_{da} = (1.5/12) \times 4.04 = 0.504 \text{ ft}^2 [= 0.0468 \text{ m}^2]$$
$$_{da} = 0.03(q/100A_{da})^2 = 0.03 (92.16/100 \times 0.504)^2 = 0.10 \text{ in.} \tag{6-19}$$
$$[= 16.5(q/A_{da})^2 = 16.5 \times (5.83 \times 10^{-3}/0.0468)^2 = 0.25 \text{ cm}] \tag{6-20}$$
$$\tag{6-18}, (6-46)$$
$$_{dc} = \Delta P_T + h_w + h_{ow} + \Delta + h_{da} = [\Delta P_T + 0.1h_w + h_{ow} + \Delta + h_{da}]$$
$$= 1.21 + 2.0 + 0.74 + 0.02 + 0.10 = 4.07 \text{ in.}$$
$$[= 3.07 + 0.1 \times 50 + 1.86 + 0.06 + 0.25 = 10.3 \text{ cm}]$$

Height of aerated liquid $= 4.07/0.60 = 6.78$ in. $[= 17.2$ cm$]$

his height is less than half the tray spacing and shows that the chosen pacing is satisfactory.

v) *Summary*

Type of tray	single-pass crossflow sieve tray	
Tower diameter	5.25 ft	[1.60 m]
Tray spacing	1.5 ft	[0.458 m]
Active area	16.0 ft²	[1.49 m²]
Hole area	2.16 ft²	[0.20 m²]
Downflow area	2.52 ft²	[0.236 m²]
Hole/tower area	0.10	
Hole/active area	0.135	
Hole size	$\tfrac{3}{16}$ in.	[4.75 mm]
Weir length	4.04 ft	[1.23 m]
Weir height	2.0 in.	[50 mm]
Downcomer clearance	1.5 in.	[38 mm]
Tray thickness	0.074 in.	[1.88 mm]

6.3 Valve Trays

The most common type of valve units have been described in Chapter 4
The design factors and methods for trays comprised of these units wil
now be considered in detail.

6.3.1 Flooding and Entrainment

It is recommended[18] that new columns should be sized so that the desigi
rates are no more than 82 per cent of the flooding rates. For vacuum tower
and for columns 3 ft [1 m] diameter the corresponding maximum figure
are 77 per cent and 75 per cent respectively. For towers less than 3 ft [1 m
diameter, the normal range of operation is between 65–75 per cent. Provide
these limits are adhered to, entrainment should not exceed 10 per cent i
the design procedure to be described is adopted, though Figure 6–4 fo
sieve trays may be used for valve trays in the absence of other publishe
data.

6.3.2 Tray Spacing

As a preliminary tray spacing estimate, the recommendations outlined fo
sieve trays may be adopted, though for very large diameter columns
spacing of 4 ft [1.22 m] is to be preferred.
    The tray spacing has an important effect on the cost of the column sinc
the overall height is directly dependent upon the chosen spacing. Whilst th
preliminary spacing may be selected as indicated, calculations of hydrauli
parameters may show that the initial figure can be reduced. This should b
done where possible provided that entrainment at the smaller tray spacin
remains acceptable.

6.3.3 Foaming

The design of a valve tray column incorporates a factor to allow for th
foaming characteristics of the system. These 'system factors' are enumerate
in Table 6–1.

TABLE 6–1. FOAMING SYSTEM FACTORS[1]

Service	Example	System factor
Non-foaming	Hydrocarbons, regular systems	1.00
Fluorine	Freon, BF_3	0.90
Moderate foaming	Oil absorbers, amine regenerators	0.85
Heavy foaming	Amine absorbers	0.73
Severe foaming	MEK units	0.60
Foam-stable systems	Caustic regenerators	0.30

6.3.4 Tray Type

Figure 6–11 shows sketches of trays ranging from single-pass trays to fiv
pass trays, and Table 6–2 may be used to select the appropriate type c
tray for the estimated column diameter. The use of multipass trays leads t

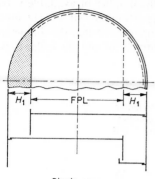

Single pass

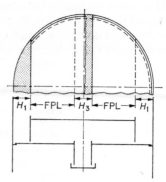

Two pass

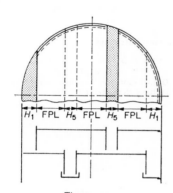

Three pass

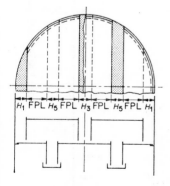

Four pass

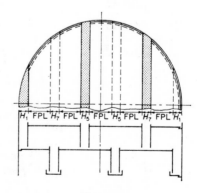

Five pass

FPL = Flow path length

▨ = Downflow area

H_1, H_3, H_5, H_7 = Downcomer widths –Table 6.3

FIGURE 6–11. Tray types[18]

smaller tower diameters but also decreases the number of available valve units and the flow path length, so that the separation may suffer. The pressure drop and downcomer backup may also increase so that the diameters given in Table 6–2 should be regarded as minimum for any particular duty.

TABLE 6–2. VALVE TRAY TYPES

Number of passes	Minimum diameter (ft)	Minimum diameter [m]	Preferred diameter (ft)	Preferred diameter [m]
2	5	1.5	6	1.8
3	8	2.5	9	2.8
4	10	3.0	12	3.7
5	13	4.0	15	4.5

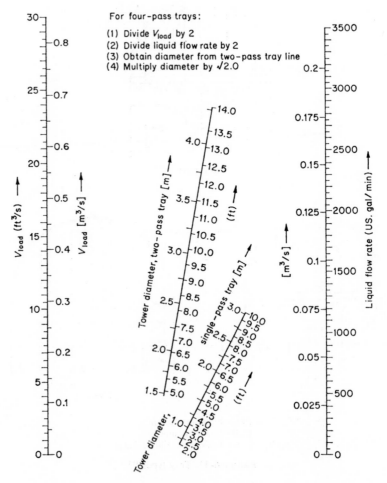

For four-pass trays:

(1) Divide V_{load} by 2
(2) Divide liquid flow rate by 2
(3) Obtain diameter from two-pass tray line
(4) Multiply diameter by $\sqrt{2.0}$

FIGURE 6–12. Nomograph for the estimation of valve tray diameter[18]

6.3.5 Tray Diameter and Layout

(i) Tray Diameter

A first estimate of the tray diameter may be obtained from Figure 6–12. The parameter 'V_{load}' is defined as

$$V_{load} = Q[\rho_V/(\rho_L - \rho_V)]^{0.5} \qquad (6\text{–}47)$$

where Q = volumetric vapour velocity (ft³/s) [m³/s],

ρ_L, ρ_V = liquid and vapour densities (lb/ft³) [kg/m³].

The resultant diameter obtained from the nomograph (Figure 6–12) can then be used to select the appropriate tray type from Figure 6–11 in conjunction with Table 6–2. The allocation of the tower area into the active and downcomer areas may now be considered.

(ii) Active Area

The active area is the area available for valve units between the inlet and outlet edges of the tray. It is a function of vapour and liquid loads, system

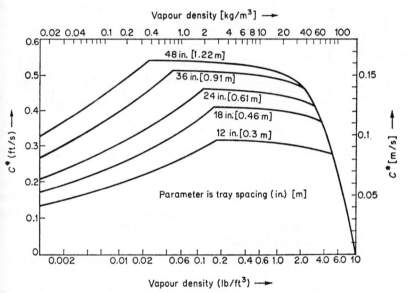

FIGURE 6–13. Variation of capacity factor with vapour density and tray spacing[18]

properties, flood factor and capacity factor, and its minimum value may be estimated from equations (6–48) or (6–49).

$$A_{a\,min} = \{Q[(\rho_V/(\rho_L - \rho_V)]^{0.5} + qL/13\,000\}/CF \qquad (6\text{–}48)$$

$$[A_{a\,min} = \{Q[(\rho_V/(\rho_L - \rho_V)]^{0.5} + 1.36qL\}/CF] \qquad (6\text{–}49)$$

$$C = C^*S \qquad (6\text{–}50)$$

$$L = 9.0D_t/N \qquad (6\text{–}51)$$

$$[= 0.75D_t/N] \qquad (6\text{–}52)$$

where C = vapour capacity factor (ft/s) [m/s],
S = system factor
F = flood factor

C^* is obtained from Figure 6–13.

Q = vapour flow rate (ft³/s) [m³/s],
q = liquid flow rate (US gal/min) [m³/s],
L = liquid flow path (in.) [m],
D_t = tower diameter (ft) [m].

(iii) *Downcomer Area* (A_d)

The downcomer area is a function of liquid rate, downcomer design velocity, and flood factor. The ideal downcomer design velocity (U^*) may be found as the smallest value from the graph of Figure 6–14 and equations 6–53a, b, c.

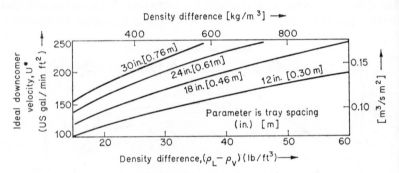

FIGURE 6–14. Ideal downcomer velocity as a function of tray spacing and density difference[18]

$U^* = 250 = [0.170]$ (6–53a)
$U^* = 41\ (\rho_L - \rho_V)^{0.5} = [0.007(\rho_L - \rho_V)^{0.5}]$ (6–53b)
$U^* = 7.5\ (T_s(\rho_L - \rho_V))^{0.5} = [0.008\ (T_s\ (\rho_L - \rho_V))^{0.5}]$ (6–53c)
U^* = ideal downcomer velocity (US gal/min ft²) [m³/sm²]
T_s = tray spacing (in.) [m]

The value of U^* is the *smallest* value obtained from graph and equations 6–53a, b, c.

The velocity from Figure 6–14 is corrected for the foaming tendency of the system by multiplying by the appropriate system factor S (Table 6–1) to obtain the downcomer design velocity, U.

$$U = U^* \times S$$ (6–54)

Then $$A_{d\ min} = q/U \cdot F$$ (6–55)

If the downcomer area calculated by equation (6–54) or (6–55) is less than 11 per cent of the active area, (A_a), the smaller of the following should be used:

$$A_{d\ min} = 0.11 A_a$$ (6–56)

or $$A_{d\ min} = 2\ (\text{calculated value of } A_{d\ min})$$ (6–57)

Having obtained the minimum downcomer and active area and selected the type of tray to be employed, the minimum tower area should be checked from equations (6–59) or (6–60) in (iv) which follows and compared with

$$A_T^* = A_{a\,min} + 2A_{d\,min} \tag{6-58}$$

(iv) *Tower Area*

The minimum tower area may be calculated from

$$A_{T\,min} = V_{load}/0.78FC \tag{6-59}$$

$$[= 1.28\, V_{load}/FC] \tag{6-60}$$

If $A_{T\,min}$ is larger than the area obtained from

$$A_T = A_{a\,min} + 2A_{d\,min} \tag{6-61}$$

then the larger value should be used and the active and downcomer areas increased proportionally. An example of this procedure is shown later in this chapter.

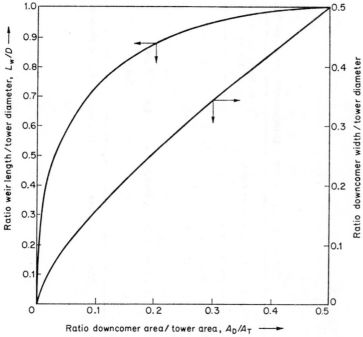

FIGURE 6–15. Downcomer sizing chart

Use may now be made of Table 6–3 or Figure 6–15 to determine the areas and widths of each individual downcomer. The width of side downcomers may be obtained from Figure 6–15 and an accurate estimate of all other downcomers obtained from the relevant width factors in Table 6–3, since

$$H = (WF)A_d/A_t \tag{6-62}$$

where
H = width of downcomer (in.) [m],
WF = width factor (—) [—].

TABLE 6-3. ALLOCATION OF DOWNCOMER AREAS AND WIDTH FACTORS

	Fraction of total Downcomer area				Width factors (A.A. Units)			Width factors [SI]		
Passes	A_{D1}	A_{D3}	A_{D5}	A_{D7}	H_3	H_5	H_7	H_3	H_5	H_7
2	0.50 each	1.00	—	—	12.0	—	—	1.0	—	—
3	0.31	—	0.69	—	—	8.63	—	—	0.72	—
4	0.21 each	0.58	0.50 each	—	6.9	6.78 each	—	0.58	0.57 each	—
5	0.16	—	0.46	0.38	—	5.66	5.5	—	0.47	0.46

NB. nomenclature for A_{D1}, A_{D3}, H_3, H_5 etc. are given in Figure 6-11.

The flow path length of a tower which is to be fitted with internal mainways should not be less than 16 in. [0.4 m]. In cases where a very high liquid load relative to the vapour load exists and internal mainways are required, it may be necessary to increase the tower diameter to accommodate this minimum flow path length.

With the downcomer widths established, the segmented downcomer areas may be obtained from Figure 6-15. For the remaining downcomers, the areas may be calculated from

$$A_{d1} = (SF) \times D_t \times H_1/12 \qquad (6\text{-}63)$$

$$[= (SF) \times D_t \times H_1] \qquad (6\text{-}64)$$

where A_{d1} = area of the individual downcomer (ft^2) [m^2],
H_1 = width of individual downcomer (in.) [m],
SF = span factor obtained from Table 6-4 (—) [—].

TABLE 6-4. DOWNCOMER SPAN FACTORS

Passes	Fraction of tower diameter		
	H_3	H_5	H_7
2	1.0	—	—
3	—	0.95	—
4	1.0	0.885	—
5	—	0.98	0.88

The downcomer areas of even-numbered trays may be somewhat different from that of odd-numbered trays for two or four passes. An average value may be used as the total downcomer area for further calculations.

(v) *Number of Valve Units*

As a first approximation, the number of valve units on a tray may be assumed to be within the range of 11–16 units/ft^2 of active area [120–175/m^2]. These figures are obtained with a unit spacing of 3 in. [76 mm], and should fewer be required the spacing may be increased to 4.5 in. or 6 in. [115 or 150 mm].

To obtain an accurate estimate of the number of units it is necessary to draw a detailed layout of the tray. One such typical drawing is shown in Figure 6-16.

(vi) *Hole Area*

The area used to calculate the hole velocity in subsequent pressure drop calculations is

$$A_h = \text{number of units}/78.5 \qquad (6\text{-}65)$$

$$[= \text{number of units}/845] \qquad (6\text{-}66)$$

where A_h = hole area (ft^2) [m^2].

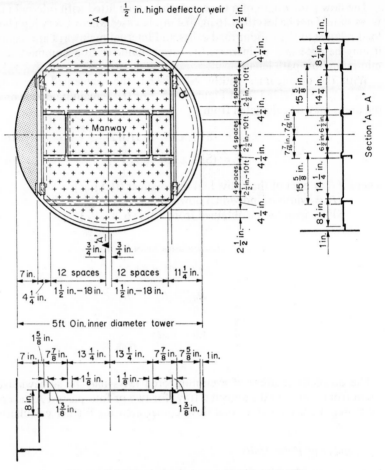

FIGURE 6–16. Typical tray layout (British Units)[18]

(vii) *Weirs*

Weir lengths may be obtained from Figure 6–15 and Table 6–3. An average length for odd and even numbered trays of two and four passes is used for calculating pressure drop.

A weir height of 2 in. [50 mm] is most commonly used, though as the weir height has a direct effect on the tray pressure drop this value may be reduced to $\frac{3}{4}$ in. [19 mm] for vacuum applications. If a long residence time on the tray is desired, the weir height may be increased to 6 in. [150 mm].

6.3.6 Hydraulic Parameters

(i) *Approach to Flooding*

The approach to flooding may be checked when the downcomer and active areas have been determined. Use is made of the term 'V_{load}' which has been defined earlier as

$$V_{load} = Q[\rho_V/(\rho_L - \rho_V)]^{0.5} \qquad (6\text{–}47)$$

The percentage of flood may then be calculated from equations (6–67) to (6–69) using the larger of the values obtained.

$$\% \text{ flood}/100 = (V_{\text{load}} + qL/13\,000)/A_a C \tag{6-67}$$

$$[= (V_{\text{load}} + 1.36qL)/A_a C] \tag{6-68}$$

or $$\% \text{ flood}/100 = V_{\text{load}}/0.78 A_T C \tag{6-69}$$

Where an existing column is to be rated for a new service and the down-comer area is small relative to the required downcomer area, the following equations may be used to estimate the approach to flooding:

$$\% \text{ flood}/100 = \left\{ \frac{V_{\text{load}}}{[(A_a C) - (A_d UL/13\,000)]D_{\text{LF}}} \right\}^{0.625} \tag{6-70}$$

$$\left[= \left\{ \frac{V_{\text{load}}}{[(A_A C) - (3.02 \times 10^{-3} A_d UL)]D_{\text{LF}}} \right\}^{0.625} \right] \tag{6-71}$$

where $$D_{\text{LF}} = \text{downcomer load factor} = (A_d U/q)^{0.6} \tag{6-72}$$

(ii) Pressure Drop

The pressure drop across a valve tray is a function of vapour and liquid rates, the number and the type of valve used, the metal density, the thickness of the valve, the weir height and the weir length. When the valves are not fully open, the dry tray pressure drop is proportional to the valve weight and virtually independent of the vapour rate. At higher vapour rates, which are sufficient to open the valves fully, the dry tray pressure drop is proportional to the square of the velocity of vapour through the holes.

(a) *Dry tray pressure drop.* The dry tray pressure drop may be calculated from the following equations with the correct figure being taken as the larger value from equations (6–73) or (6–74) and (6–75).

$$\Delta P_{\text{dry}} = 1.35 t_m \rho_m/\rho_L + K_1 V_h^2 \rho_v/\rho_L \tag{6-73}$$

$$[= 0.135 t_m \rho_m/\rho_L + K_1 V_h^2 \rho_v/\rho_L] \tag{6-74}$$

or $$\Delta P_{\text{dry}} = K_2 V_h^2 \rho_v/\rho_L \tag{6-75}$$

where $\Delta P_{\text{dry}} = $ dry tray pressure drop (in.) [cm],
$\rho_m = $ metal density (lb/ft^3) [kg/m^3],
$V_h = $ vapour velocity through valves (ft/s) [m/s],
$K_1, K_2 = $ coefficients obtained from Table 6–5 (—) [—].

Data for t_m and ρ_m for the most commonly used materials may be obtained from Table 6–6.

Referring to equations (6–73), (6–74) and (6–75) and Table 6–5, it will be seen that the use of the venturi-shaped holes of the type V-4 results in a lowering of the dry tray pressure drop.

(b) *Total pressure drop.* The total tray pressure drop may be calculated from equations (6–76) or (6–77).

TABLE 6–5. PRESSURE DROP COEFFICIENTS

Type of Unit	K_1	K_2 for deck thickness (A.A. Units) of			K_1	K_2 [SI Units]			
		14G (0.74 in.)	12G (0.104 in.)	10G (0.134 in.)	0.25		14 s.w.g. [1.88 mm]	12 s.w.g. [2.65 mm]	10 s.w.g. [3.3 mm]	¼ in. [6.35 mm]
V–1	0.2	1.05	0.92	0.82	0.58	5.46	28.7	25.1	22.4	15.8
V–4	0.1	0.5	0.39	0.38	—	2.73	13.6	10.6	10.4	—

$$\Delta P_{\mathrm{T}} = \Delta P_{\mathrm{dry}} + 0.4(q/L_{\mathrm{w}})^{0.67} + 0.4h_{\mathrm{w}} \qquad (6\text{-}76)$$

$$[= \Delta P_{\mathrm{dry}} + 55.4(q/L_{\mathrm{w}})^{0.67} + 0.04h_{\mathrm{w}}] \qquad (6\text{-}77)$$

where ΔP_{T} = total pressure drop (in.) [cm],
ΔP_{dry} = dry tray pressure drop (in.) [cm],
L_{w} = weir length (in.) [m],
q = liquid flow rate (US gal/min) [m³/s],
h_{w} = weir height (in.) [mm].

TABLE 6–6. DATA FOR MATERIALS OF CONSTRUCTION

Thickness			Density of Valve Materials		
Gauge	(in.)	[cm]	Metal	ρ_{m} (lb/ft)	ρ_{m} [kg/m³]
20	0.037	0.094	Mild steel	480	7 700
18	0.050	0.127	Stainless steel	510	8 180
16	0.060	0.152	Nickel	553	8 850
14	0.074	0.188	Monel	550	8 820
12	0.104	0.264	Titanium	283	4 540
10	0.134	0.340	Hastelloy	560	8 980
¼ in.	0.250	0.635	Aluminium	168	2 695
			Copper	560	8 980
			Lead	708	11 350

(iii) *Downcomer Back-up*

If the back-up in the downcomer exceeds 60 per cent of the tray spacing, flooding of the column may occur. The back-up may be calculated from equation (6–78) or (6–79).

$$h_{\mathrm{dc}} = 0.4(q/L_{\mathrm{w}})^{0.67} + h_{\mathrm{w}} + (\Delta P_{\mathrm{T}} + 0.65)(\rho_{\mathrm{L}}/(\rho_{\mathrm{L}} - \rho_{\mathrm{V}})) \qquad (6\text{-}78)$$

$$[= 55.4(q/L_{\mathrm{w}})^{0.67} + 0.1h_{\mathrm{w}} + (\Delta P_{\mathrm{T}} + 1.66)(\rho_{\mathrm{L}}/\rho_{\mathrm{L}} - \rho_{\mathrm{V}})] \qquad (6\text{-}79)$$

The value of h_{dc} should therefore be less than 0.6 times the tray spacing.

The lowest vapour rates at which V-1 and V-4 ballast trays can operate are given by equations (6–80) and (6–81) respectively.

$$V_{\mathrm{load}}/A_{\mathrm{h}} = 0.3 \ [= 0.10] \qquad (6\text{-}80)$$

$$V_{\mathrm{load}}/A_{\mathrm{h}} = 0.6 \ [= 0.20] \qquad (6\text{-}81)$$

To extend the lower operating limit of the tray, the following methods may be applied:

(a) Reduce the number of valves either by increasing the spacing or blanking the units at the inlet or outlet edge of the tray.

(b) Use selected rows of units with zero tab height which may be considered as inactive.

(c) Use a type of valve which has more inherent flexibility. These are normally only used in pilot plant investigations.

These design factors will now be used in an example to illustrate the method of approach in designing a valve tray.

Example 6–2

Using the same system as described in Example 5–5, design a valve tray for the given duty.

The approach to the solution will be similar to that adopted for sieve trays in Example 6–1.

(i) *Vapour*

$$\text{Mass flow rate} = 40\,000 \text{ lb/hr } [= 5.04 \text{ kg/s}]$$
$$\text{Vapour density, } \rho_V = 0.168 \text{ lb/ft}^3 \, [= 2.69 \text{ kg/m}^3]$$
$$\text{Volumetric flow, } Q = 66.08 \text{ ft}^3/\text{s } [= 1.87 \text{ m}^3/\text{s}]$$

(ii) *Liquid*

$$\text{Mass flow rate} = 32\,000 \text{ lb/hr} = [4.04 \text{ kg/s}]$$
$$\text{Liquid density, } \rho_L = 43.3 \text{ lb/ft}^3 = [694 \text{ kg/m}^3]$$
$$\text{Volumetric flow, } q = 739 \text{ ft}^3/\text{hr} = 12.32 \text{ ft}^3/\text{min}$$
$$= 92.16 \text{ US gal/min}$$
$$= [5.83 \times 10^{-3} \text{ m}^3/\text{s}]$$

iii) *Vapour Load*

$$\{(\rho_L - \rho_V)/\rho_V\}^{0.5} = \{(43.3 - 0.168)/0.168\}^{0.5}$$
$$= 16.0$$
$$V_{\text{load}} = Q/16 = 66.08/16.0 \qquad (6\text{–}47)$$
$$= 4.14 \text{ ft}^3/\text{s } [= 0.117 \text{ m}^3/\text{s}]$$

(iv) *Estimated Diameter*

From Figure 16–12,
$$D_T \simeq 3.75 \text{ ft } [= 1.14 \text{ m}]$$

From Table 6–2 a single pass crossflow tray is indicated.

(v) *System Factor*

From Table 6–1, a hydrocarbon system is non-foaming and the system factor is 1.

(vi) *Flood Factor*

As in the example for sieve trays, the column will be designed for 80 per cent of flooding, i.e. $F = 0.80$.

(vii) *Active Area*

The minimum active area is obtained from equation (6–48) or (6–49) using a flow path length L, where

$$L = 9.0D_T/N \tag{6-51}$$

$$[= 0.75D_T/N] \tag{6-52}$$

$$= 9.0 \times (3.75 \times 12)/1 = 33.8 \text{ in.}$$

$$[= 0.75 \times 1.14/1 = 0.855 \text{ m}]$$

Also $\qquad C = C^* \times S$

$$= 0.41 \times 1.0$$

from Figure 6–13, using a tray spacing $= 18$ in. [0.457 m]. Hence

$$C = 0.41 \text{ ft/s} = [0.125 \text{ m/s}]$$

$$A_{a \, min} = [(V_{load} + qL/13\,000)]/CF \tag{6-48}$$

$$= [4.14 + (92.16 \times 33.8)/13\,000]/0.41 \times 0.8$$

$$= 13.4 \text{ ft}^2$$

$$[= (V_{load} + 1.36qL)/CF \tag{6-49}$$

$$= (0.117 + 1.36 \times 5.83 \times 10^{-3} \times 0.855)/0.125 \times 0.8$$

$$= 1.24 \text{ m}^2]$$

(viii) *Downcomer Area* (A_d)

From Figure 6–14, the ideal downcomer design velocity is obtained as:

$$U^* = 209 \text{ US gal/min ft}^2 \text{ [0.142 m}^3/\text{m}^2\text{s] from the graph}$$

$$= 250 \text{ US gal/min ft}^2 \text{ [0.170 m}^3/\text{m}^2\text{s]} \tag{6-53a}$$

$$= 269 \text{ US gal/min ft}^2 \text{ [0.183 m}^3/\text{m}^2\text{s]} \tag{6-53b}$$

$$= 209 \text{ US gal/min ft}^2 \text{ [0.142 m}^3/\text{m}^2\text{s]} \tag{6-53c}$$

The smallest value of U^* is selected as 209 US gal/min ft² [0.142 m³/m²s]

$$U = 209 \times 1.0 = 209 \text{ US gal/min ft}^2 \tag{6-54}$$

$$\text{Minimum downcomer area} = q/UF \tag{6-55}$$

$$= 92.16/209 \times 0.8$$

$$= 0.552 \text{ ft}^2$$

$$[= \frac{5.83 \times 10^{-3}}{0.142 \times 0.8}$$

$$= 0.0512 \text{ m}^2]$$

As $A_{d \, min}$ is less than than 11 per cent $A_{a \, min}$, use the smaller of equations (6–56) or (6–57).

$$A_{d \, min} = 0.11A_a = 1.44 \text{ ft}^2 \text{ [}= 0.134 \text{ m}^2\text{]} \tag{6-56}$$

$$A_{d \, min} = 2(\text{calculated } A_{d \, min}) \tag{6-57}$$

$$= 1.104 \text{ ft}^2 \text{ [}= 0.1024 \text{ m}^2\text{]}$$

Hence $\qquad A_{d \, min} = 1.104 \text{ ft}^2 \text{ [}= 0.1024 \text{ m}^2\text{]}$

(ix) *Tower Area* (A_T)

$$A_T^* = A_{a\,min} + 2A_{d\,min} \qquad (6\text{–}58)$$

Check minimum tower area $= V_{load}/0.78FC \qquad (6\text{–}59)$

$$[= 1.28\, V_{load}/FC] \qquad (6\text{–}60)$$

$$A_{T\,min} = \frac{4.14}{0.78 \times 0.80 \times 0.41} = 16.0\ \text{ft}^2$$

$$\left[= \frac{1.28 \times 0.117}{0.8 \times 0.125} = 1.49\ \text{m}^2\right]$$

As the minimum tower area exceeds the calculated tower area from steps (viii) and (ix), use the former area, i.e.

$$A_T = 16.0\ \text{ft}^2\ [= 1.49\ \text{m}^2]$$

Hence   diameter $= (A_T/0.786)^{0.5}$

$$= 4.5\ \text{ft}\ [= 1.38\ \text{m}]$$

(x) *Tower Areas Check*

True active area $A_a = A_{a\,min} \times A_T/A_T^* \qquad (6\text{–}82)$

$$= 13.4 \times 16.0/15.6 = 14.0\ \text{ft}^2$$

$$[= 1.24 \times 1.49/1.45] = [1.28\ \text{m}^2]$$

True downcomer area $A_d = A_{d\,min} \times A_T/A_T^* \qquad (6\text{–}83)$

$$= 1.104 \times 16.0/15.6 = 1.13\ \text{ft}^2$$

$$[= 0.1024 \times 1.49/1.45 = 0.105\ \text{m}^2]$$

(xi) *Weir Sizing*

For a single-pass tray, the width H and the length L_w of the weir may be calculated from Figure 6–15.

$$\frac{A_d}{A_T N} = \frac{1.13}{16.0 \times 1}\left[= \frac{0.105}{1.48 \times 1}\right] = 0.0707$$

From Figure 6–15, $H/D = 0.123$, $L_w/D = 0.657$. Therefore

width of downcomer, $H = 0.123 \times 4.5 \times 12$

$$= 6.67\ \text{in.}\ [= 0.17\ \text{m}]$$

weir length $L_w = 0.657 \times 4.5 \times 12$

$$= 35.5\ \text{in.}\ [= 0.904\ \text{m}]$$

Liquid flow per pass $Z = q/N$

$$= 92.16\ \text{US gal/min}$$

$$[= 5.83 \times 10^{-3}\ \text{m}^3/\text{s}]$$

$Z/L_w = 92.16/35.5 = 2.59 < 10$

$[Z/L_w = 5.83 \times 10^{-3}/0.904 = 6.45 \times 10^{-3} < 0.025]$

Therefore the sizing is satisfactory.

(xii) *Number of Valve Units*

Assume for first estimate a layout of 11–16 units/ft^2 [118–173 units/m^2] of active area.

$$A_a = 14.0 \text{ ft}^2 \; [= 1.28 \text{ m}^2]$$

Number of units = 151–224

Suppose there are 180 units/tray.

$$\text{Hole area } A_h = 180/78.5 = 2.30 \text{ ft}^2$$
$$[= 180/845 = 0.213 \text{ m}^2]$$

(xiii) *Pressure Drop*

(a) Dry tray pressure drop. For the duty in this example carbon steel is adequate and equations (6–73) to (6–75) will be applied.

$$\Delta P_{\text{dry}} = 1.35 t_m(\rho_m/\rho_L) + K_1 V_h^2(\rho_V/\rho_L) \qquad (6\text{–}73)$$

$$[= 0.135 t_m(\rho_m/\rho_L) + K_1 V_h^2(\rho_V/\rho_L)] \qquad (6\text{–}74)$$

$$\Delta P_{\text{dry}} = K_2(V_h)^2(\rho_V/\rho_L) \qquad (6\text{–}75)$$

For 14g (14 gauge) trays (0.074 in. [or 1.88 mm]), V-1 units (14g)

$$V_h = 66.08/2.30 = 28.7 \text{ ft/s} \; [= 8.8 \text{ m/s}]$$

$$\Delta P_{\text{dry}} = 1.35 \times 0.074 \times 480/43.3 + 0.2 \times (28.7)^2 \times 0.168/43.3 \qquad (6\text{–}73)$$

$$= 1.74 \text{ in. } [= 4.32 \text{ cm}] \text{ of liquid}$$

or $\Delta P_{\text{dry}} = 1.05 \times (28.7)^2 \times 0.168/43.3 = 3.34 \text{ in.} \qquad (6\text{–}75)$

$$[= 28.7 \times (1.87/0.213)^2 \times (2.69/694)$$

$$= 8.50 \text{ cm of liquid}]$$

Taking the largest value of ΔP_{dry} gives a value equal to 3.34 in.; [8.50 cm] of liquid.

(b) Total pressure drop. ΔP_T. The total pressure drop is calculated from equation (6–76) or (6–77),

$$\Delta P_T = \Delta P_{\text{dry}} + 0.4(q/L)^{0.67} + 0.4 h_w \qquad (6\text{–}76)$$

$$[= \Delta P_{\text{dry}} + 55.4(q/L)^{0.67} + 0.04 h_w] \qquad (6\text{–}77)$$

Assuming a 2 in. [50 mm] weir height is used,

$$\Delta P_T = 3.34 + 0.4(92.16/35.5)^{0.67} + 0.4 \times 2.0 \qquad (6\text{–}76)$$

$$= 4.90 \text{ in. of liquid}$$

$$[= 8.50 + 55.4(5.83 \times 10^{-3}/0.904)^{0.67} + 0.04 \times 50 \qquad (6\text{–}77)$$

$$= 12.45 \text{ cm of liquid}]$$

(xiv) *Downcomer Back-up*

The downcomer back-up is calculated from equations (6–78) or (6–79).

$$h_{dc} = 0.4(q/L)^{0.67} + h_w + (\Delta P_T + 0.65)(\rho_L/(\rho_L - \rho_V)) \qquad (6\text{–}78)$$
$$[= 55.4\,(q/L)^{0.67} + 0.1h_w + (\Delta P_T + 1.66)(\rho_L/(\rho_L - \rho_V))] \qquad (6\text{–}79)$$
$$= 0.4(92.16/35.5)^{0.67} + 2.0(4.90 + 0.65)\{43.3/(43.3 - 0.168)\}$$
$$= 0.756 + 2.0 + 5.58$$
$$= 7.34 \text{ in. } [= 18.6 \text{ cm}]$$

The value of h_{dc} must be less than $0.6T_s$, i.e.

$$0.6T_s = 10.8 \text{ in. } [= 27.4 \text{ cm}]$$

$h_{dc} < 10.8$ in. and therefore the tray design is satisfactory.

(xv) *Discussion*

The pressure drop across the tray as calculated above is higher than that for the sieve tray example earlier in the chapter. The calculation below shows how the dry tray pressure drop may be reduced by using V-4 type units at a loading of 16 units/ft² [173 units/m²] of active area.

$$\text{Number of units} = 16 \times 14.0 = 224 \text{ units/tray}$$
$$\text{Hole area} = 224/78.5 \,[= 224/845]$$
$$= 2.85 \text{ ft}^2 \,[= 0.265 \text{ m}^2]$$
$$\text{Hole velocity} = 66.08/2.30 \,[= 1.87/0.265]$$
$$V_h = 23.2 \text{ ft/s } [= 7.07 \text{ m/s}]$$
$$V_h^2 = 538 \text{ ft}^2/\text{s}^2 \,[= 50.0 \text{ m}^2/\text{s}^2]$$

From Table 6–5, for V-4 units,

$$K_1 = 0.10 \,[= 2.73]$$
$$K_2 = 0.50 \,[= 13.6]$$

From equation (6–73)

$$\Delta P_{dry} = 1.35 \times 0.074 \times 480/43.3 + 0.10 \times 538 \times 0.168/43.3$$
$$= 1.32 \text{ in. } [= 3.32 \text{ cm}]$$

or from equation (6–75)

$$\Delta P_{dry} = 0.50 \times 538 \times 0.168/43.3$$
$$= 1.02 \text{ in. } [= 2.60 \text{ cm}]$$

Taking the larger value gives a dry tray pressure drop of 1.32 in. [3.32 cm].

Thus ΔP_{dry} has been reduced from 3.34 to 1.32 in. and, assuming that all other factors remain constant, the total pressure drop is calculated as before to give

$$\Delta P_T = 1.32 + 0.4(q/L)^{0.67} + 0.4h_w \qquad (6\text{–}76)$$
$$[= 3.32 + 55.4(q/L)^{0.67} + 0.04h_w] \qquad (6\text{–}77)$$
$$= 1.32 + 1.56 = 2.88 \text{ in.}$$
$$= [3.32 + 3.96 = 7.28 \text{ cm}]$$

This compares with the earlier value of 4.90 in. [12.45 cm], which is a considerable reduction in pressure drop. The final value is now the same as that obtained from the sieve tray design, though in this example pressure drop is unlikely to be a factor of controlling importance. Given the same pressure drop, a choice between the two tray types would probably depend upon the turndown ratio required. If any great degree of flexibility in operation were required, the ballast tray would be selected.

6.4 Nomenclature—Sieve and Valve Trays

A_a	active area	(ft²)	[m²]
A_d	downcomer area	(ft²)	[m²]
A_{da}	area under downcomer apron	(ft²)	[m²]
A_h	hole area	(ft²)	[m²]
A_n	net area	(ft²)	[m²]
A_p	perforated area	(ft²)	[m²]
A_T	tower area	(ft²)	[m²]
A_T^*	tower area, equation (6–58)	(ft²)	[m²]
C, C^*	vapour capacity factor	(ft/s)	[m/s]
C_{sb}	vapour capacity parameter	(ft/s)	[m/s]
C_{vo}	dry orifice coefficient	(—)	[—]
D_t	tower diameter	(ft)	[m]
d_h	hole diameter	(in.)	[mm]
f	friction factor	(—)	[—]
F	flood factor	(—)	[—]
F_{lv}	liquid flow parameter	(—)	[—]
g	acceleration due to gravity	(ft/s²)	[m/s²]
h_a	head loss due to aerated liquid	(in.)	[cm]
h_{da}	head loss under downcomer apron	(in.)	[cm]
h_{dc}	height of clear liquid in downcomer	(in.)	[cm]
h_e	equivalent height of clear liquid	(in.)	[cm]
h_{ow}	height of liquid crest over weir	(in.)	[cm]
h_w	weir height	(in.)	[mm]
h_σ	head loss due to bubble formation	(in.)	[cm]
H	width of downcomer	(in.)	[m]
$K_1 K_2$	coefficients in pressure drop equation from Table 6–5	(—)	[—]
l_w	weir length	(in.)	[m]
L	liquid mass flow rate	(lb/hr)	[kg/s]
L	flow path length	(in.)	[m]
L_t	distance between weirs	(ft)	[m]
L_w	weir length	(ft)	[m]
N	number of flow passes	(—)	[—]
ΔP_{dry}	dry tray pressure drop	(in.)	[cm]
ΔP_T	total tray pressure drop	(in.)	[cm]
q	volumetric liquid flow rate	(U.S. gal/ min)	[m³/s]
Q	volumetric vapour rate	(ft³/s)	[m³/s]
Q_p	aeration factor	(—)	[—]
R_h	hydraulic radius	(ft)	[m]
R_{eh}	modified Reynolds number	(—)	[—]
S	system factor, Table 6–1	(—)	[—]
SF	span factor, Table 6–3	(—)	[—]
t_m	valve metal thickness	(in.)	[mm]
U	downcomer design velocity	(ft/s)	[m/s]
U_f	froth velocity across plate	(ft/s)	[m/s]
U_h	vapour velocity through holes	(ft/s)	[m/s]
U_n	vapour velocity based on net area	(ft/s)	[m/s]
U_{nf}	flooding velocity based on net area	(ft/s)	[m/s]

U^*	ideal downcomer design velocity	(U.S. gal/ min ft^2)	[m^3/m^2 s]
V	mass flow of vapour	(lb/hr)	[kg/s]
V_d	downcomer liquid velocity	(ft/s)	[m/s]
V_h	hole velocity	(ft/s)	[m/s]
V_{load}	vapour load parameter, equation (6-47)	(ft^3/s)	[m^3/s]
WF	width factor, Table 6-3	(—)	[—]
X	entrainment correction factor, equation (6-11)	(—)	[—]
ρ	density	(lb/ft^3)	[kg/m^3]
ψ	fractional	(—)	[—]
Δ	liquid gradient	(in.)	[cm]
ϕ	relative froth density	(—)	[—]

Subscripts

L   liquid
V   vapour
min   minimum

REFERENCES

1. Smith, B. D. *Design of Equilibrium Stage Processes.* New York: McGraw-Hill, 1963.
2. Eduljee, H. E. *Brit. Chem. Eng.* 1958, **3**, 14.
3. Eduljee, H. E. *Brit. Chem. Eng.* 1959, **4**, 320.
4. Zenz, F. A., Stone, L., Crane, M. *H.C. Proc.* 1967, **46**, 12, 138.
5. McAllister, R. A., McGuinnis, P. H. Jnr., Plank, C. *Chem. Eng. Sci.* 1954, **9**, 25.
6. Fair, J. R. *Pet. Chem. Eng.* 1961, **33**, 10, 45.
7. Souders, M. and Brown, G. G. *Ind. Eng. Chem.* 1934, **26**, 98.
8. Norman, W. S. *Absorption, Distillation and Cooling Towers.* London: Longmans, 1961.
9. Mayfield, F. D., *et al. Ind. Eng. Chem.* 1952, **44**, 2238.
10. Zenz, F. A. *Pet. Ref.* 1954, **33·** 2, 99.
11. Khamdi, A. M., Skoblo, A. I., Molokanov, Y. K. *Khim. Tekhnol. Topliv Masel,* 1963, **8**, 2, 31.
12. Bernard, J. D. T., Sargent, R. W. H. *Tv. Inst. Chem. Eng.* 1966, **44**, T314-27.
13. Chase, J. D. *Chem. Eng.* 31 July 1967, **74**, 105.
14. Chan, B. K. C. Ph.D. Thesis, Univ. of Sydney, Aust. Dec. 1965.
15. Prince, R. G. H. *Inst. Sym. Distn. Ins. Chem. Eng.* 1960, 177-84.
15a. Foss, A. S. and Gerster, J. A. *Chem. Eng. Prog.* 1956, **52**, 28.
16. Chase, J. D. *Chem. Eng.* Aug. 28 1967, **74**, 139.
17. Hughmark, G. A. and O'Connell, H. E. *Chem. Eng. Prog.* 1957, **53**, 3, 127-32.
18. *Ballast Tray Manual,* Bulletin No. 4900 (revised), Dallas, Texas: Fritz Glitsch and Sons Inc., 1970.

Chapter 7: Mechanical Design

7.1 Introduction

The chemical engineer is normally required to specify the main dimensions of the pieces of equipment for which he is responsible and throughout this book emphasis has been placed on the specification and sizing of both major and ancillary items of chemical plant. This information by itself, however, is normally insufficient to permit drawings to be made to allow fabrication and erection to proceed. Whilst the detail required for complete specifications to be completed normally lies in the hands of a specialist mechanical engineer, the process chemical engineer should be sufficiently aware of the problems and design methods involved to ensure that his own calculations do not produce unduly difficult and expensive problems in later stages.

In this chapter, an attempt will be made to outline some of the problems involved and to suggest suitable design methods for preliminary equipment specifications. Emphasis will be placed upon the mechanical design of heat exchangers and columns, since these are most frequently encountered by the chemical engineer. At the end of this chapter, a short section is included on the estimation of various parameters by 'rules of thumb' based on plant design and operating experience, which may prove useful for the quick estimations often needed in the field of plant design.

7.2 The Mechanical Design of Heat Exchangers

7.2.1 General

The mechanical design of exchangers, particularly shell and tube units, involves the determination of the thickness of the component parts of the exchanger and enables an engineering drawing of the unit to be prepared from which construction can be carried out. In simple terms, the chemical engineer is mainly concerned with the sizing of the unit, the number and layout of tubes and optimization procedures, while the actual design and preparation of working drawings lies more within the realm of mechanical engineering. It is important, however, that the chemical engineer should be aware of the limitations of mechanical design and should not, for example, specify a tube sheet thickness which is impracticable or indeed impossible to fabricate. The complete design of a shell and tube unit is essential to any second-order cost estimation and, although an approximate cost can be extrapolated from the heat transfer area, accurate sizing of the component parts is most important where the cost of the unit is a significant feature in the optimization of a section of plant.

Mention has already been made of the advantages of standardization in heat exchanger construction, by way of standard tube length, tube layout, sizing and so on, and exchangers are normally designed to a standard code of practice evolved over many years. Such codes include those of the American Society of Mechanical Engineers,[1] the Tubular Exchanger Manufacturers Association,[2] and, in the U.K., the British Standard Specification on Pressure Vessels (B.S. 1500).[3] The term 'pressure vessel' accurately describes the shell of an exchanger. Such design codes should be consulted and adhered to, and the following notes serve to provide only a summary of the more important relationships.

Before considering the relationships for calculation of the thickness of the various components, several terms common to many of the equations must be mentioned. Firstly, the vessel is usually designed with a 10 per cent safety factor with respect to pressure. That is, design pressure = 1.1 × working pressure. Obviously, where the pressure is likely to fluctuate in a process the working pressure is taken as the maximum pressure encountered. Materials of construction can only withstand certain limiting stresses without deformation or indeed complete destruction, and the term permissible stress (f) features in many of the relations which follow. Values of this stress for some common materials are given in B.S. 1500 (1958),[3] Table 4 and *ASME Boiler and Pressure Vessel Code*, Section viii, 'Unfired Pressure Vessels' (1959)[1] gives tables of recommended permissible stresses for various materials at different temperatures. Typical values for carbon steel are given in Table 7-1.

TABLE 7-1. PERMISSIBLE STRESSES FOR CARBON STEEL UP TO 650°F [616 K][3]

| Grade | Stress (f) | |
	(lb/in.^{2})	[MN/m^2]
A	13 400	92.33
B	14 500	99.9
C	15 700	108.2

A further term to be considered is the welded joint factor, J, which is an indication of the strength of a weld in relation to the strength of the parent material. Owing to possible defects in the weld J is usually less than unity, and the actual value is dependent on the stress relief and the extent of weld examination carried out. Values of J are given in Table 1 of BS 1500.[3]

The question of corrosion allowance is dealt with later in this chapter, and it need only be noted at this stage that a certain amount is added to the calculated thickness of a component to allow for the anticipated reduction in thickness during the life of the exchanger. It may be added that such corrosion may not be uniform across the surface of a plate, for example, and the extent of corrosion depends largely on the fluids involved and the conditions under which the unit is operated. If no data are available from past experience, an allowance of at least $\frac{1}{16}$ in. [1.59 mm] should be made.

7.2.2 Thickness of Various Components

(i) *Shell and Channel*

The thickness of the shell and channel is given by B.S. 1500 (1958)

$$t = P(D + 2c)/(2fJ - P) + c \qquad (7-1)$$

where P = design pressure (lb/in.^{2}) [MN/m^2],
  D = shell internal diameter (in.) [mm],
  f = permissible stress (lb/in.^{2}) [MN/m^2],
  J = welded joint factor (—) [—],
  c = corrosion allowance (in.) [mm],
  t = thickness (in.) [mm].

The value calculated in this way should be compared with the minimum thickness given in Table 7–2 and the *greater* of these two values should be specified.

TABLE 7–2. MINIMUM PLATE THICKNESS (CARBON STEELS)

Shell diameter		Minimum plate thickness	
(in.)	[m]	(in.)	[mm]
8–12	0.203–0.305	see B.S. 1600	see B.S. 1600
13–24	0.330–0.610	$\frac{3}{8}$	9.5
25–29	0.635–0.737	$\frac{3}{8}$	9.5
30–39	0.762–0.990	$\frac{7}{16}$	11.1
40–60	1.016–1.524	$\frac{1}{2}$	12.7

Where the shell diameter is less than 24 in. [0.61 m], it is usually more economic to use standard pipe rather than plate and the thickness in this case is specified in B.S. 1600 'Wrought Steel Pipe for the Petroleum Industry'.[4] It must be appreciated that, in order to accommodate the flange of the floating head cover in this type of exchanger, the diameter of the cylinder of the shell cover is always greater than the diameter of the main shell.

(ii) *Shell Cover*

This component is usually in the shape of an ellipse with the major axes in the ratio 2:1. The thickness is specified in ASME,[1] section viii as

$$t = P(D + 2c)/(2fJ - 0.2P) + c \qquad (7\text{–}2)$$

where the symbols are as defined previously.

(iii) *Channel Cover*

The thickness of the channel cover may be calculated from the following equation quoted in TEMA:[2]

$$t = (5.7P(G/100)^4 + 2h_\text{G}A_\text{B}(G/100)/\sqrt{d_\text{B}})^{0.33} + g \qquad (7\text{–}3)$$

$$[= 0.0254(0.02PG^4 + 7.65 \times 10^3 h_\text{G}A_\text{B}G/\sqrt{d_\text{B}})^{0.33} + g] \qquad (7\text{–}4)$$

where G = diameter at location of gasket load reaction (see Figure 7–1) (in.) [m],

$h_G = 0.5 (PCD - G)$ (in.) [m],

A_B = cross sectional area of bolts ($n \times a_B$) (in.^{2}) [m^2],

d_B = nominal bolt diameter (in.) [m],

n = number of bolts (—) [—],

a_B = cross sectional area of one bolt ($\pi d_B^2/4$) (in.^{2}) [m^2],

g = depth of groove (in.) [m],

t = thickness (in.) [m].

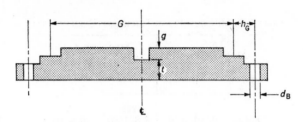

FIGURE 7–1. Dimensions of channel cover

If the cover is grooved—as is obviously the case with multipass exchangers—there is no need to add a corrosion allowance to the thickness providing $g \geqslant 0.188$ in. [4.77 mm].

(iv) *Tubes*

As outlined in Chapter 3, tubes of the following standard lengths should be used wherever possible: 8, 12, 16 or 20 ft [2.44, 3.66, 4.88 or 6.10 m]. The thicknesses of typical standard tube sizes are given in Table 7–3.

TABLE 7–3. THICKNESSES OF STANDARD TUBES

Outside diameter		Wall thickness					
		Non-ferrous			ferrous		
(in.)	[mm]	s.w.g.	(in.)	[mm]	s.w.g.	(in.)	[mm]
$\frac{3}{4}$	19.1	14	0.080	2.03	16	0.060	1.52
1	25.4	12	0.104	2.64	14	0.080	2.03
$1\frac{1}{4}$	31.7	10	0.128	3.25	12	0.104	2.64

Further data on tube dimensions, flow area/tube and surface area per unit length of tube are given in Kern[5] Table 10, whilst Table 9, 'Layout of tube sheets', can save considerable time at the drawing board. Approximate tube counts based on these data are included as Figure 7–2. As a general rule, the minimum distances between tube centres should be 1.25 × tube outer diameter. In condensers the number of tubes which can be fitted in a

given shell is reduced somewhat, as space must be allowed for the impingement plates to be fitted beneath the nozzle inlet, thereby preventing direct
injection of the vapours onto the top rows of tubes.

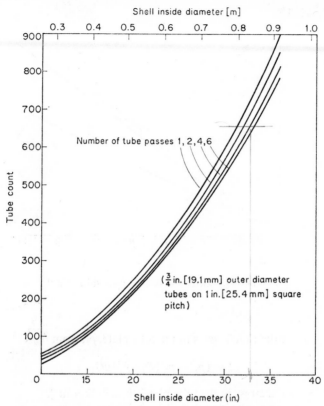

FIGURE 7–2. Approximate tube sheet layouts[5]

(v) *Tube Plates*

The calculation of tube plate thickness is a relatively tedious operation in
that trial and error procedures are involved; however, suitable charts are
available which greatly simplify the calculation. The method outlined here
is based on B.S. 1500,[3] pages 50–6 and involves the following steps:

(a) Calculate k from

$$k = \sqrt{\{1.32 \ \sqrt{(na/\eta Lh')}D/h'\}} \qquad (7\text{--}5)$$

where D = internal diameter of shell (in.) [cm],

h' = an assumed plate thickness (in.) [cm] (1 in. [2.54 cm] is a good
guess),

a = cross-sectional area of the metal in one tube (in.²) [cm²],

L = effective tube length (in.) [cm],

η = deflection efficiency $(= (A - C)/A)$ (—) $[\text{—}]$,

n = number of tubes (—) $[\text{—}]$,

A = cross-sectional area of bore of shell (in.²) [cm²],

C = cross-sectional area of tube holes in tube plate (in.²) [cm²].

(b) With this value of k, obtain G_1, a non-dimensional function of K from Figure 7-3 or B.S. 1500.[3]

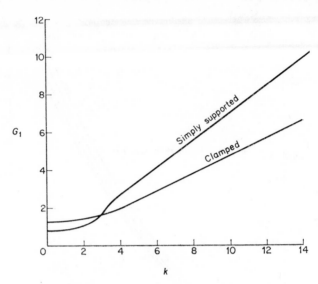

FIGURE 7-3. Calculation of tube plate thickness[3]

(c) Calculate the radial stress in the tube plate, p_r, from

$$p_r = (p_1 - p_2)(D/h')^2/4\mu G_1 \qquad (7\text{-}6)$$

where p_r = radial stress in the tube plate (lb/in.²) [kN/m²],
　　p_1 = pressure outside tubes (lb/in.²) [kN/m²],
　　p_2 = pressure inside tubes (lb/in.²) [kN/m²],
　　μ = factor depending on the number of passes (—) [—].

(d) Assess the value of p_r.

If $p_r < f$, the maximum permissible stress, then the thickness of the tube plate, h, is given by

$$h = h' + c + g \text{ (in.)} \qquad (7\text{-}7)$$

$$[= 0.01h' + c + g] \text{ [m]} \qquad (7\text{-}8)$$

bearing in mind that the corrosion allowance is only necessary for one side of the plate providing $g \geqslant 0.188$ in. [4.77 mm], thus allowing a sufficient margin on the other side.

If $p_r \geqslant f$, then an increment is added to h' and the calculation is repeated starting again at step (a). As noted previously the minimum value of h is usually 1 in. [25.4 mm] as this thickness is necessary to permit the tubes to be expanded into the plate.

(vi) Floating Head Cover

The following equation is given in ASME,[1] section viii:

$$t = 5PL'/6fJ + 2c \qquad (7\text{-}9)$$

where L' = the internal radius of the cover (in.) [m],
t = the thickness of the cover (in.) [m].

A chart is available in ASME for checking this value, which gives the shell thickness of cylindrical and spherical vessels under external pressure. If the pressure on the outside of the floating cover (the convex side) is greater than that on the inside (concave side) then the design pressure P is taken as 1.67 × external design pressure.

(vii) Tie Rods and Spacers

The minimum number and diameter of tie rods for various shell sizes are shown in Table 7–4, which is taken from TEMA.[2]

TABLE 7–4. NUMBER AND SIZE OF TIE RODS

Nominal shell size		Number of tie rods	Diameter of tie rods	
(in.)	[m]		(in.)	[mm]
8–15	0.203–0.381	4	$\frac{3}{8}$	9.53
16–27	0.406–0.686	6	$\frac{3}{8}$	9.53
28–33	0.711–0.838	6	$\frac{1}{2}$	12.70
34–48	0.864–1.219	8	$\frac{1}{2}$	12.70

(viii) Partition Plates, Baffles and Tube Support Plates

Suggested thicknesses of pass partition plates in carbon steel shells are:

nominal shell diameter up to 24 in. [0.610 m]: $\frac{3}{8}$ in. [9.5 mm]

nominal shell diameter above 24 in. [0.610 m]: $\frac{1}{2}$ in. [12.7 mm]

TEMA[2] gives values for the minimum thicknesses of baffles and support plates which are included as Table 7–5.

TABLE 7–5. BAFFLE AND SUPPORT PLATE THICKNESSES

Distance between adjacent segmental plates		Nominal shell inside diameter							
		8–14 in. [0.203–0.356 m]		15–28 in. [0.381–0.711 m]		28–38 in. [0.737–0.965 m]		> 39 in. [> 0.99 m]	
(in.)	[mm]	(in.)	[mm]	(in.)	[mm]	(in.)	[mm]	(in.)	[mm]
< 12	< 0.305	$\frac{1}{8}$	3.2	$\frac{3}{16}$	4.8	$\frac{1}{4}$	6.4	$\frac{1}{4}$	6.4
12–18	0.305–0.457	$\frac{3}{16}$	4.8	$\frac{1}{4}$	6.4	$\frac{5}{16}$	7.9	$\frac{3}{8}$	9.5
18–24	0.457–0.610	$\frac{1}{4}$	6.4	$\frac{3}{8}$	9.5	$\frac{3}{8}$	9.5	$\frac{1}{2}$	12.7
24–30	0.610–0.762	$\frac{3}{8}$	9.5	$\frac{3}{8}$	9.5	$\frac{1}{2}$	12.7	$\frac{5}{8}$	15.9
> 30	> 0.762	$\frac{3}{8}$	9.5	$\frac{1}{2}$	12.7	$\frac{5}{8}$	15.9	$\frac{5}{8}$	15.9

(ix) Supports and Flanges

Heat exchangers and condensers are usually supported on saddles fabricated from girders. Their design is relatively specialized and certainly beyond the scope of the present text. For a useful treatment of the subject, Zick[6] should be consulted in the first instance. The various design codes include information on, for example, ancillary components such as vents and thermowells. One feature relating to the support of condensers is that saddles are provided which incline the unit at 1–5° to the horizontal, thus ensuring complete drainage on shutdown.

The connecting nozzles for fluid inlet and outlet are usually designed as in the case of standard piping, though the cost of these items is normally insignificant compared to the cost of the exchanger and the use of tables of economic pipe diameters such as given in Perry[7] is quite acceptable. Flanges for nozzles and the shell of exchangers also require specialized treatment, though B.S. 2041[8] gives flange sizes for most normal sizes of heat exchanger.

Example 7–1. Design of a heat exchanger

Design scantlings for a 29 in. [737 mm] internal diameter condenser having cooling water as the tube-side fluid in two passes. The following data is available:

Shell cover: 33 in. [838 mm] internal diameter
Crown radius of floating head cover: 19 in. [483 mm]
Depth of pass-partition grooves: $\frac{3}{16}$ in. [4.8 mm]
Tubes: 526, $\frac{3}{4}$ in. [19 mm] 14 BWG [2 mm wall] 16 ft 0 in. [4880 mm] long
      on 1 in. [25 mm] square pitch
Corrosion allowance: $\frac{1}{8}$ in. [3.2 mm]
Materials: tubes—steel to B.S. 1508–240
          bolts—steel to B.S. 1506–613 B6
          other components—steel to B.S. 1501–151 grade C.

Operating conditions:
    pressure 132 lb/in.^{2} [910 kN/m^2] shell, 50 lb/in.^{2} [345 kN/m^2] tube
    temperature: 146°F [336 K] shell, 105°F [314 K] tube

Control of quality of the exchanger material to comply with Class I requirements of BS 1500 (1958).

The components of tubular heat exchangers are normally designed for one of the following pressures: 75, 150, 300, 450 or 600 lb/in.^{2} g [517, 1034, 2068, 3100, 4134 kN/m^2]. In this case,

shell design pressure $= 1.1 \times 132 = 145$ lb/in.^{2} [$= 1000$ kN/m^2]

tube design pressure $= 1.1 \times 50 = 55$ lb/in.^{2} [$= 379$ kN/m^2]

Therefore, to follow the normal practice, the shell will be designed for 150 lb/in.^{2} [1034 kN/m^2] and the channel 75 lb/in.^{2} [527 kN/m^2] pressure.

The maximum permissible stress for the materials used is obtained from
B.S. 1500,[3] Table 2:

Material	Permissible Stress	
	(lb/in.^{2})	[MN/m^2]
B.S. 1508–240	14 000	96.5
B.S. 1501–151 gr C	15 700	108.2
Bolt material		
B.S. 1750 (1951)	20 000	137.8

The condenser will be tested to Class I requirements of B.S. 1500,[3] and
hence joint factor $J = 0.95$.

(i) *Shell*

$$t = 150(29 + 2 \times \tfrac{1}{8})/(2 \times 15\,700 \times 0.95 - 150) + \tfrac{1}{8} = 0.273 \text{ in.}$$

$$[t = 1.034\,(0.737 + 2 \times 0.0032)/(2 \times 108.2 \times 0.95 - 1.034)$$
$$+ 0.0032 = 0.006\,93 \text{ m}]$$

The nearest practical size is $\tfrac{5}{16}$ in. [7.9 mm]. However, from Table 7–2,
the minimum thickness required is $\tfrac{3}{8}$ in. plate [9.5 mm], which is greater
than the calculated and hence $\tfrac{3}{8}$ in. [9.5 mm] plate will be specified.

(ii) *Channel*

$$t = 75\,(29 + 2 \times \tfrac{1}{8})/(2 \times 15\,700 \times 0.95 - 75) + \tfrac{1}{8} = 0.199 \text{ in.}$$

$$[t = 0.517\,(0.737 + 2 \times 0.0032)/(2 \times 108.2 \times 0.95 - 0.577)$$
$$+ 0.0032 = 0.005\,05 \text{ m}]$$

The nearest practical size is $\tfrac{1}{4}$ in. [6.4 mm]. However, from Table 7–2, the
minimum thickness required is $\tfrac{3}{8}$ in. [9.5 mm] plate, i.e. greater than the
calculated value, and hence $\tfrac{3}{8}$ in. [9.5 mm] plate will be specified.

(iii) *Shell Cover (2:1 elliptical)*

$$t = 150(33 + 2 \times \tfrac{1}{8})/(2 \times 15\,700 \times 0.95 - 0.2 \times 150) + \tfrac{1}{8} = 0.292 \text{ in.}$$

$$[t = 1.034(0.838 + 2 \times 0.0032)/(2 \times 108.2 \times 0.95 - 0.2 \times 1.034)$$
$$+ 0.0032 = 0.007\,41 \text{ m}]$$

The nearest practical size is $\tfrac{5}{16}$ in. [7.9 mm] and this will be specified.

(iv) *Shell Cover Cylinder*

$$t = 150(33 + 2 \times \tfrac{1}{8})/(2 \times 15\,700 \times 0.95 - 150) + \tfrac{1}{8} = 0.166 \text{ in.}$$

$$[t = 1.034(0.838 + 2 \times 0.0032)/(2 \times 108.2 \times 0.95 - 1.034)$$
$$+ 0.0032 = 0.004\,22 \text{ m}]$$

The nearest practical size is $\tfrac{3}{16}$ in. [4.8 mm]; however, from Table 7–2, the
minimum thickness required is $\tfrac{7}{16}$ in. [11.1 mm] and this would be specified.
In many cases, it is cheaper to fabricate both the shell cover and the shell

cover cylinder from the same plate thickness, the greater of the two calculated values being used—in this case $\frac{7}{16}$ in. [11.1 mm].

(v) Channel Cover

In order to calculate the thickness of this component, it is necessary to know the values of G, h_G, A_B, all of which depend on the flange to which the cover is bolted. B.S. 2041 [8] gives flange details for various sizes of exchanger, and for the present unit the following data are appropriate:

$$\text{bolt circle diameter} = 34 \text{ in. } [0.864 \text{ m}]$$
$$\text{number of bolts} = 28$$
$$\text{bolt diameter} = \tfrac{7}{8} \text{ in. } [22.2 \text{ mm}]$$
$$\text{assumed value of } G = 30.5 \text{ in. } [0.775 \text{ m}]$$

$$\text{total core area of bolts, } A_B = 28 \times 0.422 = 11.82 \text{ in.}^2$$
$$[= 28 \times 0.000\,272 = 0.007\,62 \text{ m}^2]$$
$$h_G = \tfrac{1}{2}(34 - 30.5) = 1.75 \text{ in.}$$
$$[= \tfrac{1}{2}(0.864 - 0.775) = 0.045 \text{ m}]$$
$$t = (5.7 \times 75(30.5/100)^4 + 2 \times 1.75$$
$$\times 11.82 (30.5/100)/0.875)^{0.33} + \tfrac{3}{16}$$
$$= 2.763 \text{ in.}$$
$$[= 0.0254(0.02 \times 0.517 \times 0.775^4$$
$$+ 7.65 \times 10^3 \times 0.045 \times 0.007\,62$$
$$\times 0.775/0.0222)^{0.33} + 0.0048$$
$$= 0.0703 \text{ m}]$$

Hence $2\frac{3}{16}$ in. [71.7 mm] thick plate will be specified.

(vi) Tube Plates

As a first approximation assume the plate thickness, $h' = 1$ in. [2.54 cm].

$$\text{Internal diameter of shell} = 29 \text{ in. } [73.7 \text{ cm}]$$
$$\text{Cross-sectional area of metal in one tube, } a = 0.1683 \text{ in.}^2 [1.09 \text{ cm}^2]$$
$$\text{Effective tube length, } L = 188 \text{ in. } [478 \text{ cm}]$$
$$\text{Number of tubes, } n = 526$$
$$\text{Cross-sectional area of bore of shell, } A = \pi \times 29^2/4 = 660.0 \text{ in.}^2$$
$$[= \pi \times 73.7^2/4 = 4270 \text{ cm}^2]$$
$$\text{Cross-sectional area of tube holes, } C = 526 \times 0.75^2 \times \pi/4$$
$$= 232.2 \text{ in.}^2$$
$$[= 526 \times 1.91^2 \times \pi/4$$
$$= 1490 \text{ cm}^2]$$
$$\text{Deflection efficiency} = (660.0 - 232.2)/660.0$$
$$= 0.648$$
$$[= (4270 - 1490)/4270$$
$$= 0.648]$$

Hence,

$$k = \sqrt{\{1.32 \sqrt{(526 \times 0.1683/(0.648 \times 188 \times 1))} \times 29.0/1.0)} = 5.72$$

$$[= \sqrt{\{1.32 \sqrt{(526 \times 1.09/(0.648 \times 478 \times 2.54))} \times 73.7/2.54)} = 5.72]$$

Assuming the plate is simply supported, then $G_1 = 3.95$ from either Figure 7-3 or B.S. 1500 and hence the radial stress in the tube plate, p_r is given by

$$p_r = (150 - 75)(29.0/1.0)^2/(4 \times 0.5 \times 3.95) = 7981 \text{ lb/in.}^2$$

$$[= (1034 - 517)(73.7/2.54)^2/(4 \times 0.5 \times 3.95) = 54\,989 \text{ kN/m}^2$$

$$= 55.0 \text{ MN/m}^2]$$

This is less than f, the maximum permissible stress, and hence the assumed value of h', 1.0 in. [2.54 cm] is acceptable.

Thus the specified plate thickness becomes

$$h = 1.0 + 0.125 + 0.1875 = 1.3125 \text{ in.}$$

$$[= 0.01 \times 2.54 + 0.0032 + 0.0048 = 0.0334 \text{ m}]$$

The nearest standard size is $1\frac{5}{16}$ in. [0.0334 m] and this will be specified.

(vii) *Floating Head Cover*

The shell side pressure is greater than the concave or tube side pressure and hence P becomes 1.67 × external design pressure = 25.1 lb/in.^{2} [1.73 MN/m^2]. Thus,

$$t = 5 \times 257 \times 19.0/(6 \times 15\,700 \times 0.95) + 2 \times \frac{1}{8} = 0.576 \text{ in.}$$

$$[= 5 \times 1.73 \times 0.483/(6 \times 108.2 \times 0.95) + 2 \times 0.0032 = 0.0131 \text{ m}]$$

The nearest standard plate thickness is $\frac{9}{16}$ in. [0.0143 m] and this will be specified.

(viii) *Ancillary Components*

The minimum number and diameter of tie rods is shown in Table 7-4 and this will be specified, i.e. 6 tie rods, $\frac{1}{2}$ in. [12.70 mm] diameter. Similarly, the thicknesses of tube support plates (baffles are not required in a condenser) are taken from Table 7-5. $\frac{5}{16}$ in. [7.9 mm] thick support plates will be specified, 18 in. [0.457 m] apart.

(ix) *Layout of Specification*

The sizes of the various components as calculated in this example may be laid out in a specification as shown in Table 7-6—data which of course should be supplemented with an engineering drawing of the unit. (In this calculation, the equivalent SI dimension is taken as the exact conversion of the A.A. value. It is obviously more practical to work to the nearest SI standard size (25 mm nominal pipe, for example) though until materials and codes of practice are available in standard SI sizes, the standard A.A. size is deemed more appropriate.)

214   PROCESS PLANT DESIGN

TABLE 7–6. LAYOUT OF HEAT EXCHANGER SPECIFICATION

Design pressure: shell 150 lb/in.^{2} [1.034 MN/m^2], tube 75 lb/in.^{2} [0.517 MN/m^2]

Design temperature: shell 146°F [336 K], tube 105°F [314 K]

Tubes steel to B.S. 1508–240 number: 526 outer diameter: $\frac{3}{4}$ in. [19 mm] BWG: 14 [2 mm wall] length: 16 ft [4880 mm] pitch: 1 in. [25.4 mm] square

Shell steel to B.S. 1501–151 gr C inner diameter: [737 mm] thickness: $\frac{3}{8}$ in. [9.5 mm]

Shell cover M.S. Tema Standard inner diameter: 33 in. [838 mm] thickness: $\frac{7}{16}$ in. [11.1 mm]

Floating head cover M.S. crown radius: 19 in. [483 mm] thickness: $\frac{9}{16}$ in. [14.3 mm]

Channel steel to B.S. 1501–151 gr C. 24 in. × 29 in. inner diameter [610 mm × 737 mm inner diameter] thickness: $\frac{3}{8}$ in. [9.5 mm]

Channel cover M.S. diameter 35$\frac{1}{8}$ in. [892 mm] thickness: 2$\frac{13}{16}$ in. [71.7 mm]

Tube sheets stationary: M.S. $\frac{5}{16}$ in. [33.4 mm] thick floating: M.S. 1$\frac{5}{16}$ in. [33.4 mm]

Tube supports M.S. 7 off $\frac{5}{16}$ in. [7.9 mm] thick

Tie rods M.S. 6 off $\frac{1}{2}$ in. [12.7 mm] outer diameter sleeve: M.S. $\frac{3}{4}$ in. [19.1 mm] outer diameter

Gaskets metal-faced asbestos Tema Standard R.65

Connections shell   in: M.S. 8 in. [205 mm] standard ASME
                out: M.S. 5 in. [130 mm] standard ASME
        channel   in: M.S. 8 in. [205 mm] standard ASME
                 out: M.S. 8 in. [205 mm] standard ASME

Corrosion allowance shell: $\frac{1}{8}$ in. [3.2 mm] tube: $\frac{1}{8}$ in. [3.2 mm]

Weight   shell: 2200 lb [1000 kg] bundle: 4500 lb [2040 kg]
        total: Dry 10 500 lb [4770 kg]

Remarks   *Code requirements:* B.S. 1500 class 1
          2 M.S. impingement plates $\frac{1}{2}$ in. [12.7 mm] thick 10 in. [254 mm] dia.
          Vents on shell and floating head. Drain on shell and channel.
          2 Lifting lugs 4 in. [100 mm] dia.
          Temperature and pressure connections on all inlet nozzles
          Unit to be mounted on saddle at 2° to horizontal

7.3 The Mechanical Design of Columns

The mechanical design of columns may be classified under six broad headings:[9]

1. vessel design,
2. support design,
3. foundation design,
4. flange and connection design,
5. vessel internal design,
6. materials of construction.

These headings will be considered in turn with the greatest emphasis being placed on the internals of columns which are closer to the needs of the chemical engineer than the remainder which are usually the province of the specialized mechanical or civil engineer.

7.3.1 Vessel Design

B.S. 1500[3] classifies pressure vessels into four categories:

(i) Severe-duty vessels which are to contain toxic or lethal substances at a temperature above their self-ignition temperature or fired vessels for pressures exceeding 100 lb/in.^{2} [689 kN/m^2]. These are equivalent to what are known as 'Class I vessels'.

(ii) Light duty vessels are for relatively light duties with plate thicknesses not exceeding $\frac{3}{8}$ in. [15.9 mm] built for working pressures not exceeding 50 lb/in.^{2} [344 kN/m^2] vapour pressure or 250 lb/in.^{2} [1.72 MN/m^2] hydrostatic pressure, at temperatures not exceeding 300°F [421 K] and unfired. They are generally rated as 'Class III vessels'.

(iii) Medium duty vessels are those which are not covered by (i) and (ii) above and are rated as 'Class II vessels'.

(iv) Vessels for service at sub-atmospheric pressure and which are liable to embrittlement at temperatures below 32°F [273 K] should be subjected to impact testing at their lowest operating temperatures.

The design codes for each of these classes are set out in great detail in B.S. 1500[3] and while some relevant equations have been abstracted in the section on heat exchangers, further detail at this point cannot attempt to replace use of the standard to which further reference should be made. With regard to loadings on the vessel, note should be made of the following factors and the final loading should be a combination of these factors, acting together, which will produce the worst stress/temperature condition.

(a) the internal or external operating pressure plus a factor to produce the design pressure to avoid unnecessary blowing of safety devices,

(b) shock loads, including rapid pressure fluctuations and surge loads due to rapid movement of the contents of the vessel,

(c) the weight of the vessel and its contents at operating temperature and pressure,

(d) bending moments which may be caused by the eccentricity of the centre of the working pressure relative to the central axes of the vessel,

(e) superimposed loads,

(f) wind loads which may be estimated from a procedure of Molineux[9] based on methods from several sources,[10]–[14]

(g) local stresses due to supporting lugs, internal structures or connecting piping,

(h) temperature differences,

(i) additional pressure due to static head of contained fluids.

The minimum shell thickness for a light duty vessel is given by B.S. 1500[3] as

$$t = PD/(2fJ - P) + c \qquad (7\text{–}10)$$

where t = shell thickness (in.) [mm],

P = maximum working pressure (lb/m^2) [MN/m^2],

D = wide diameter of shell (in.) [mm],

f = maximum allowable working stress (lb/m^2) [MN/m^2],

J = weld efficiency factor (maximum = 0.8) (—) [—],

c = corrosion allowance (in.) [mm].

Example 7–2

Estimate the minimum thickness of the shell of the distillation column of Example 6–1 which is 5.25 ft [1.60 m] diameter. The tower is to operate at atmospheric pressure and will be made of 18/8 stainless steel.

From Table 1a, B.S. 1500,

$$f = 19\ 600\ \text{lb/in.}^2\ [= 135\ \text{MN/m}^2]$$
$$D = 63\ \text{in.}\ [= 1600\ \text{mm}]$$

The design pressure will be taken as 50 lb/in.^{2} [0.345 MN/m^2] = P. $J = 0.8$ (assumed), $c = \frac{1}{16}$ in. [1.59 mm] (assumed). Hence

$$t = 50 \times 63/(2 \times 19\ 600 \times 0.8 - 50) + 0.0625 = 0.164\ \text{in.}$$
$$[= 0.345 \times 1600/(2 \times 135 \times 0.8 - 0.345) + 1.59 = 4.03\ \text{mm}]\quad(7\text{–}10)$$

This is the minimum shell thickness and would almost certainly be increased for a tower of this diameter and height. Nelson[15] has suggested that the thickness of tower shells allowing for rigidity and corrosion should range from 0.25 in. [6.35 mm] for towers of approximately 3 ft [1 m] diameter to 0.5 in. [12.7 mm] for 5 ft [1.5 m] diameter vessels.

In extremely tall towers, the thickness based on internal pressure alone will be insufficient to withstand the combined stresses due to wind and weight. As these stresses increase from top to bottom of the tower, the thickness is often increased for the tower sections of the column. Bergmann[14] has suggested that as a general guide the thickness of the lower sections of the tower be increased by $\frac{1}{16}$ in. [1.59 mm] for each 20 ft [6.1 m] height over 50 ft [15.2 m].

To enable an estimate of the cost of a vessel to be made, Table 7–7 is included and its use is illustrated by means of two examples.

Example 7–3

Estimate the cost of the mild steel shell of a distillation column which weighs 50 000 lb [22 700 kg].

From Table 7–7, the cost is given by

$$10^3\ \$ = 2.01\ (X)^{0.68}$$

where X = weight in lb $\times$ 10^3 lb. As the weight is 50 000 lb, $X = 50$. Hence

$$\text{cost} = 2.01(50)^{0.68} \times 10^3 = \$28\ 500$$
$$[= 26.4(X)^{0.68}$$
$$= 26.4(22\ 700)^{0.68} = \$28\ 500]$$

Example 7–4

Estimate the cost of a 10 ft [3.048 m] diameter column in mild steel including heads, nozzles and mainways if the shell is 1 in. [25.4 mm] thick.

From Table 7–7, the cost is given by

$$£/\text{ft} = 1.50(10 + 10)^{1.76} = 292\ £/\text{ft}$$

or
$$[£/\text{m} = 40.1(3.048 + 3.048)^{1.76} = 960\ £/\text{m}]$$

TABLE 7-7. COST DATA FOR TOWER SHELLS (1970 COSTS). COST $= C[f(X)]^n$

Type	Thickness (in.) [mm]	X (A.A.) [SI]	Limits of X (A.A.)	Limits of X [SI]	Cost units	f(X) (A.A.)	f(X) [SI]	C (A.A.)	C [SI]	n	Reference
Mild steel shell only		diameter ft m	2–9	0.61–2.7	$/ft or [$/m]	X	X	11.8	216	1.44	16
Mild steel shell only		weight 10³ lb kg	1–100	450–45 000	10³$ or [$]	X	X	2.01	26.4	0.68	17
Mild steel shell including heads, nozzles and manholes	2 — 50.8	diameter ft [m]	3–16	0.92–4.9	£/ft or [£/m]	(X + 10)	(X + 3.048)	1.19	45.8	2.06	18
	1 3/16 — 30.2		2–16	0.61–4.9		(X + 10)	(X + 3.048)	1.19	37.1	1.89	18
	1 — 25.4		2–16	0.61–4.9		(X + 10)	(X + 3.048)	1.50	40.1	1.76	18
	3/4 — 19.1		2–16	0.61–4.9		(X + 10)	(X + 3.048)	1.52	37.7	1.70	18
	5/8 — 15.9		2–16	0.61–4.9		(X + 10)	(X + 3.048)	1.24	31.6	1.72	18
	1/2 — 12.7		2–13	0.61–4.0		(X + 10)	(X + 3.048)	1.30	33.6	1.73	18
	5/16 — 7.96		1.5–10	0.46–3.0				0.59	16.8	1.81	18

7.3.2 Vessel Supports

The design of column supports is the province of a specialist mechanical engineer as the variables involved are complex. Account must be taken of the size and weight of the vessel, the operating temperature and pressure, the surrounding structure (if any) and the stresses imposed on the vessel by the supports themselves. Figure 7–4 shows three typical support geometries

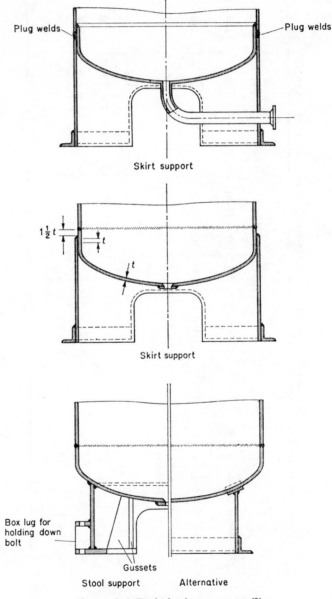

FIGURE 7–4. Typical column supports[3]

which may be used as an alternative to a simple bracket support between the vessel and the framework.

The skirt support plate should not be less than 0.25 in. [6.35 mm] thick and the compressive stress should not exceed 10 000 lb/in.² [68.9 MN/m²], with openings provided to permit inspection of the vessel bottom if this is not readily visible through the supporting framework. Steel supporting structures which do not form part of the vessel should comply with B.S. 449—'Use of Structural Steel in Building'.[19]

7.3.3 Foundations

A design procedure for foundations has been produced by Marshall,[20] whereby a foundation is assumed and the soil loading and stability checked in a series of trial and error calculations until a design results in which the soil below the foundation is always under compression at all points, and the loading does not exceed the maximum permissible for the particular subsoil. Typical values have been included in Section 11.5.2.

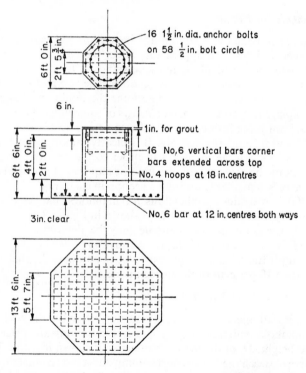

FIGURE 7–5. Typical pedestal slab for a fractionating column (British Units)[19],[16]

A typical foundation design for a fractionating column is shown in Figure 7–5, taken from Molineux[9] and originally based on Marshall.[16] The design was based on a tower 4 ft [1.22 m] diameter, 54 ft [16.5 m] high with a total weight of 172 000 lb [78 000 kg]. The pressure as a result of the wind velocity, z, is calculated from equations (7–11) or (7–12) to enable the bending moment and the maximum soil loading due to overturning to be estimated.

$$z = 0.0025(V_w)^2 \text{ (lb/ft}^2) \tag{7–11}$$

$$[= 0.061(V_w)]^2 \text{ [kg/m}^2] \tag{7–12}$$

where V_w = wind velocity (mph) [m/s].

The sum of the loadings due to the weight of the tower and the overturning load due to the wind must be less than the allowable maximum soil loading. The dimensions of Figure 7–5 are based on a wind load of 100 m.p.h. [44.7 m/s].

The octagonal shape is recommended for this type of foundation as it combines features of stability and ease of construction with minimum material requirements. The foundation base should be located below the frost line for the particular climate in which the column is located. Further information on foundation design has been presented by Brown[21,22] who considers octagonal, rectangular and circular reinforced structures.

7.3.4 Manholes and Flanges

Manholes are provided in a column for inspection and maintenance, and their number and size can appreciably affect the cost of the tower. The inside diameter of the manhole must be sufficient to allow access to the column and also passage of the tray sections. The size and the number of sections comprising each tray will determine the minimum diameter, and if, for example, a tray contains 5, 6 or 7 rows of valve units at row centres of 2.5 in. [63.5 mm], the inside manhole diameter will be 16, 18.5 and 21 in. [0.41, 0.47 and 0.53 m] respectively. It is clearly important in the case of large towers to use large manways in order to reduce the complexity of construction of the trays. Manholes are not required between each tray if means of access is available through a section in each tray.

B.S. 1500[3] provides specifications for manholes, flanges and branch pipes, and a typical circular manhole is shown in Figure 7–6.

The wall thickness of the manhole may be determined from either equation (7–10) or, for thick shells, from equation (7–13) below, which may also be used for the calculation of vessel thickness where the shell thickness is greater than 10 per cent of the inside diameter.

$$t = (D/2)\{[(fJ + P)/(fJ - P)]^{0.5} - 1\} + c \tag{7–13}$$

Minimum wall thicknesses are specified and frequently the wall thickness of both manhole and shell will be equal. Limitations are placed on the maximum height, H, of the manhole opening as shown in Table 7–8.

The force between the flanges on a column and a pipe connection must be in excess of the load due to the pressure in the pipe in order to avoid leakage. The flanges must also carry any bending moment on the pipe due to expansion or contraction of lengths joined by bends to the point of anchorage.[9] The flanges should be sufficiently wide and should be neither too thick nor too thin to enable the bolts when tightened to make a good seal with the gasket material. Molineux[9] has detailed a design procedure which should be followed for flange design.

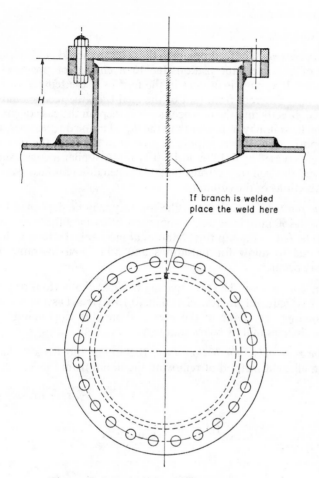

If branch is welded
place the weld here

FIGURE 7–6. Typical circular manhole and cover

TABLE 7–8. MAXIMUM HEIGHTS OF MANHOLE OPENINGS

Diameter of manhole		Height H	
(in.)	[mm]	(in.)	[mm]
14	0.37	10	0.25
15	0.38	11	0.28
16	0.41	12	0.31
18	0.46	12	0.61
24	0.61	42	1.07

7.3.5 Vessel Internals

(i) Plate Columns

(a) *The feed tray.* The feed stream to the column should be checked for an estimate of possible vaporization due to a change of pressure and/or temperature. If vapour is present in the feed or if the temperature of the feed is greatly different from that of the feed tray, the feed should not be introduced directly into the downcomer[23] except in the case of circulating reflux which, as it may be cooler than the liquid in the downcomer, will not cause vaporization. Where vapour is present in the feed, the tray spacing should be increased by 6–12 in. [0.15–0.31 m]. A similar increase should be made where the feed tray is transitional between different flow paths in the adjacent sections of the column.

(b) *The top tray.* Reflux fed to the top tray may be deposited behind a 4–6 in. [100–150 mm] inlet weir without any other modifications, but if the feed is to be fed to the top tray, additional mechanical strength should be incorporated to allow for pulsations and additional hydraulic stresses which may occur.

(c) *The bottom tray.* The tray spacing between the bottom tray and the one above is frequently increased by 6 in. [0.15 m] to act as a safety factor to avoid flooding the tower in the event of pressure fluctuations, vapour surges or changes in flow below that tray.

(d) *The zone below the bottom tray.* Kitterman and Ross[23] have suggested an efficient method of removing the liquid from the bottom of the

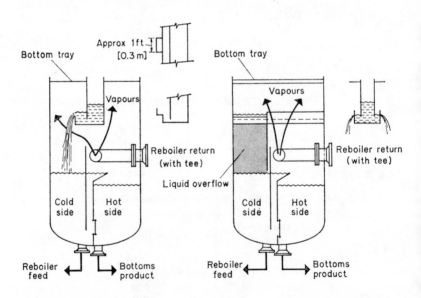

(a) Conventional configuration   (b) The configuration suggested by Kitterman and Ross (23)

FIGURE 7–7. The zone below the bottom tray of a distillation column[23]

tower and introducing the return from the reboiler using a system of preferential baffles. This idea is illustrated in Figure 7–7, where the more commonly specified arrangement is shown for comparison. In the conventional configuration (a) the liquid from the bottom tray's seal pan overflows to only one side of the preferential baffle (the cold side), while the liquid portion of the reboiler return flows to, and is drawn from, the other side (the hot side). The baffle runs parallel to the long dimension of the seal pan so that half the reboiler vapour has to pass through a curtain of liquid before reaching the first tray. In the preferred arrangement (b) the baffle is now perpendicular to the long dimension of the seal pan and the overflow weirs of the pan are designed to allow liquid to flow only to the permitted area. Thus the vapour does not have to pass through the liquid curtain to reach the first tray and entrainment and the risks of premature flooding are reduced.

(e) *Downcomer types.* Some of the many variations of downcomer types are illustrated in Figure 7–8[24] which is self-explanatory.

(f) *Tray flatness.* The floor of the tray should remain essentially flat under all design conditions. Deflection will be caused by the weight of the tray itself, its liquid loading, thermal expansion and vapour thrust. The support for the tray is normally designed to take the tray weight plus 20–25 lb/ft² [98–122 kg/m²] of uniform load with a maximum deflection of $\frac{1}{8}$ in. [3.2 mm] or $\frac{3}{16}$ in. [4.8 mm] for towers less than and greater than 12.5 ft [3.66 m] in diameter respectively.

Small towers usually have a tray support ring to which the tray is fixed, no additional support being required. For trays of diameter greater than approximately 12 ft [4 m], one or more major support beams are usually required. These will be designed to meet the criterion of flatness just described and are usually installed parallel to the liquid flow.

(g) *Inlet and draw-off sumps.* The tray spacing should be increased to allow for a sump depth of two-thirds the normal tray spacing as indicated in Figure 7–8. In this way, liquid entering the tray has a vertical component of velocity which results in better aeration at the inlet edge of the tray and increases both tray efficiency and capacity and decreases downcomer back-up.[24]

(ii) *Packed Towers*

Consideration will now be given to the internals of packed towers, the choice of which can greatly affect the performance of the selected packing. With the increased accuracy of design data, the 100 per cent safety factors commonplace ten years ago are now no longer necessary, although a greater degree of importance is now attached to the design of individual items of equipment within the tower. For example, a support plate is now accepted as a more important component than a mere support for the packing above it—its design can vitally affect the performance of the tower. Similar consideration must be given to distributors and redistributors if full benefit is to be obtained from modern high efficiency packings, and these items will now be considered in turn.

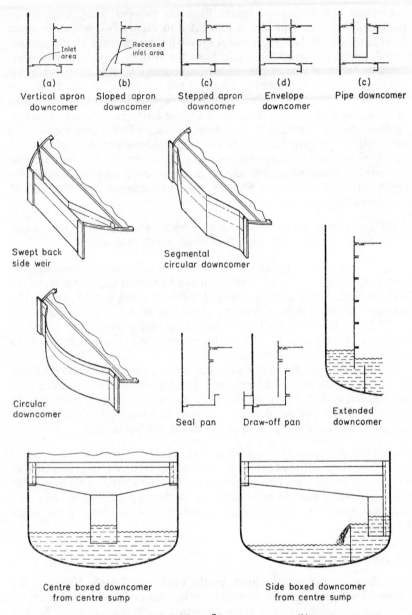

FIGURE 7–8. Different downcomer types[24]

(a) *Packing support plates.* As mentioned above, the support plate is more than a packing support; it must possess a high percentage free area in order to permit relatively unrestricted flow of downcoming liquid and to allow free upward flow of gas; it must be able to redistribute the liquid uniformly if required; it must possess high mechanical and thermal strength and be available in a range of corrosion resistant materials and finally, for ease of construction and handling, the plate should be available in sections.

The use of thick, flat ceramic plates which had a free area of between 15 and 25 per cent has now been superseded by the gas injection type of plate[25] which is illustrated diagrammatically in Figure 7–9. The free

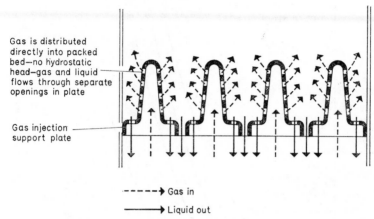

Gas is distributed directly into packed bed—no hydrostatic head—gas and liquid flows through separate openings in plate

Gas injection support plate

‑‑‑‑‑▶ Gas in

──────▶ Liquid out

FIGURE 7–9. The gas injection type of support plate[25]

space of the gas injection plate may exceed 100 per cent of the column cross-sectional area and it also provides separation between outgoing liquid at the base of the support and the gas which is injected directly into the packing. These factors mean that such a device will handle high liquid and gas loadings whilst maintaining a low pressure drop across the plate. Figure 7–10 shows the relationship between liquid and vapour flows and pressure drop for the Type 804–R2 plate of the Norton Chemical Process Products Division of the Norton Co. of Ohio.[25] This type of plate is suitable for use in towers ranging from 4–10 ft [1.2–3 m] diameter and variations on the design are available for tower diameters in the range 4–66 in. [0.1–1.67 m].[25]

(b) *Hold-down plates.* The use of a hold-down plate on top of a packed section serves an important purpose by resting directly on top of the packing and restraining the bed under conditions of high gas rates or fluctuating gas flows when, without restraint, ceramic or carbon packings may break and fall back through the bed, thus reducing its capacity. The plate normally weighs about 20 lb/ft² [approximately 100 kg/m²] which is insufficient to cope with a major disturbance in the tower, but is able to function efficiently in normal circumstances. The free area is a compromise between as great a value as possible and the need to prevent the packing passing through the plate. This normally results in a figure of 70 per cent of the free area of the packing being used.

For plastic packings, the function of the hold-down plate is no longer that of the prevention of breakage, but is the prevention of loss of pieces of packing by carry-over with the gas. For this reason, the plate is referred to as a bed-limiter. These weigh approximately half of the weight of the hold-down plate and are fixed to the tower wall. Their free space may be

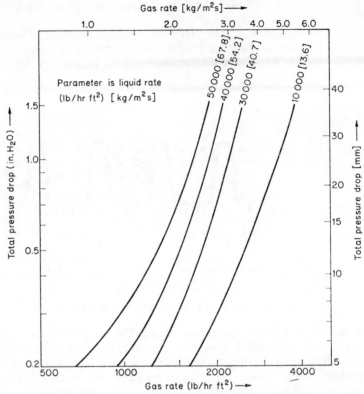

FIGURE 7–10. Pressure drop characteristics of the packing support plate, type 804–R2[25]

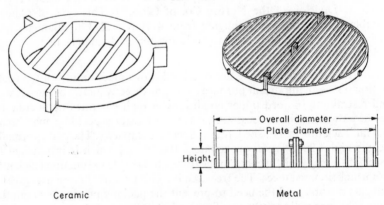

Ceramic   Metal

FIGURE 7–11. Hold-down plates[25]

higher, as wire or plastic meshes fixed to a lighter weight frame form a less bulky construction. Typical hold-down plates and bed-limiters are shown in Figure 7–11.

(c) *Liquid distributors.* The high performance obtainable from modern tower packings depends upon equally high performance from the tower

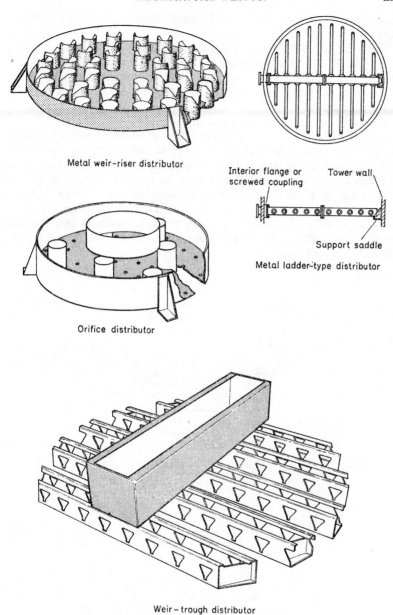

Metal weir-riser distributor

Interior flange or   Tower wall
screwed coupling

Support saddle

Metal ladder-type distributor

Orifice distributor

Weir-trough distributor

FIGURE 7–12. Types of liquid distributors for packed towers[25]

internals. A one-point liquid distributor located above a packed bed has been shown to produce a distribution through the bed in which 67.5 per cent of the liquid is confined to the centre of the tower and only 6.5 per cent reaches the area near the walls.[25] This performance is so poor that the use of an efficient packing is clearly useless. If, instead, the one-point distributor is replaced by a four-point one, the liquid flow to all areas in the tower is nearly equal, so that with proprietary distributor designs, multi-point liquid distribution is always provided.

There are two basic types of distributors—the weir type and the orifice type, both of which are low pressure drop or gravity devices. High pressure devices are rarely used as their small orifices are easily blocked by dirty liquids and the fine particle spray which they produce is easily entrained in the gas steam.

The main types of distributor are illustrated in Figure 7–12.

For small diameter towers of less than approximately 3 ft [1 m] diameter, the pan-type orifice distributor is to be preferred, where the liquid is fed to the centre of the plate and develops a uniform head over all the orifices. The ladder-type orifice distributor may be used for towers of between 18 and 120 in. [0.46–3.1 m] diameter, where the uniform distribution is achieved with a high free area for gas flow. Limitations of the orifice distributors include a low turndown ratio and a tendency to block in dirty service, though they are available in any material which is capable of being fabricated in pipe form.

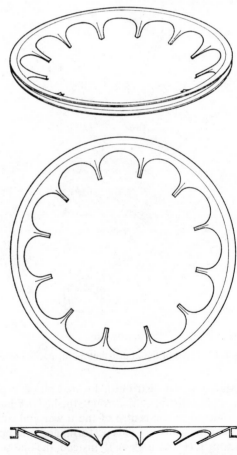

FIGURE 7–13. The Rosette type of wall wiper liquid redistributor[25]

The most popular type of distributor is the weir type, as it overcomes the disadvantages of the orifice type while possessing a high free area. Weir-type distributors are available for towers greater than 20 in. [0.51 m] diameter and are of sectional construction in a wide range of materials. They provide uniform controlled distribution of flow rates between 2 and 50 gal/min ft² [0.0014–0.034 m³/s m²].

As the liquid distributor has little weight to support, fixing may be made to a continuous support ring around the inside of the column, but care must be taken to ensure that the final assembly is completely level.

(d) *Liquid redistributors.* The need for liquid redistribution has been noted in Chapter 5, where recommendations were made for maximum bed heights. The liquid redistributor collects the downcoming liquid from the bed above and distributes it uniformly to the bed below to correct maldistribution. Since the redistributor has to be used in conjunction with a gas injection support plate, the free area is reduced at that point in the column. If this combination is likely to cause excessive pressure drop, use may be made of the simple wall wiper shown in Figure 7–13, which has a free area of 72 per cent and removes liquid from the wall of column and transports it to the interior of the bed. Wall wipers are most frequently used for small diameter columns and may be either welded to the tower wall or flange mounted.

A redistributor used in conjunction with a support plate, shown in Figure 7–14, may be used in a variety of ways: to correct maldistribution, to act as a draw-off point for the liquid or to act as a feed pipe in a packed distribution column. The redistributor should be located 6–18 in. [0.15–0.46 m] above the next packed section.

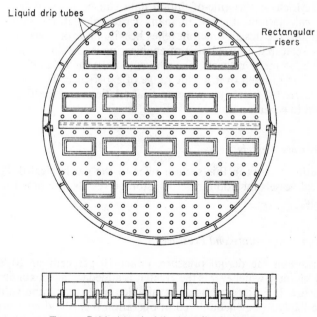

FIGURE 7–14. A typical liquid redistributor[25]

7.3.6 Materials of Construction

The choice of the materials of construction for any piece of equipment is an economic balance between capital cost and corrosion resistance. Materials of construction for many pieces of equipment have been mentioned in this text and it is important to consider the following characteristics in making the final selection:

    (i) Sufficient physical strength to carry the required load.
    (ii) Sufficient thermal shock resistance to withstand sudden and possibly extreme temperature variations.
    (iii) Resistance to the chemicals with which the material will make contact.
    (iv) Cost.

7.4 Practical Rules of Thumb

Although a great many sophisticated equations have been developed in engineering design there are several practical limitations, and considerable data has emerged largely as a result of experience. For example, in optimizing the velocity of liquid through the tubes in a shell and tube unit, no matter how complicated the operating conditions and equipment geometry the design value for most situations is between 2 and 5 ft/s [0.61–1.53 m/s]. Similarly the economic velocity of gases in heat exchange equipment is of the order of 50 ft/s [15.3 m/s] and it is surprising how often this figure emerges even as a result of quite complicated optimization techniques. A great deal of this information has been accumulated by design organizations, and in many cases it forms an important part of their design manuals and procedures. Much of this rule-of-thumb data can be justified by economic balance calculations, though in many cases calculations would be too involved and the design function relies heavily on experience. In this way it will be seen that a broad area of engineering is still an art, and the best designer is, in many respects, one who can make the best informed guess, utilizing data gleaned over a wide range of experience. This section serves to summarize the better known empirical relations and limiting principles at present in common use.

7.4.1 Pressure Vessels

The limiting factors in this section refer to the mechanical design of all cylindrical vessels including reactors, storage vessels, fractionating towers and so on.

(i) *Design Temperature and Pressure*

In determining the design pressure, either 10 per cent or 10–25 lb/in.^{2} [70–170 kN/m^2], whichever is the greater, is added to the maximum operating pressure. If no information is available on the maximum value of the pressure likely to be encountered in operation then the design pressure is taken as the normal operating pressure plus 25 lb/in.^{2} [170 kN/m^2]. For vessels operating at pressures of 0–10 lb/in.^{2} g [0–70 kN/m^2] and 600–1000°F [590–800 K], the design pressure is specified as 40 lb/in.^{2} [275 kN/m^2].

For vessels operating between −20°F [245 K] and 650°F [615 K], the design temperature may be taken as the operating temperature plus 50°F [28 K]. Below −20°F [245 K] special steels are required, and above 650°F [615 K] the allowable design stress falls very sharply.

(ii) *Wall Thickness*

The wall thickness of a pressure vessel may be approximated from the relation

$$t = PR/(SE - 0.6P) + c \qquad (7\text{-}14)$$

where P = design pressure (lb/in.^{2}) [kN/m^2],
  R = internal diameter (in.) [m],
  S = allowable working stress (lb/in.^{2}) [kN/m^2],
  E = welded joint efficiency (usually 0.8–0.95),
  c = corrosion allowance (in.) [m],
  t = wall thickness (in.) [m].

Where materials of construction are in contact with corrosive fluids, the corrosion allowance is usually $\frac{1}{4}$ in. [6.4 mm] on exposed surfaces. For non-corrosive fluids, this value may be reduced to $\frac{1}{8}$–$\frac{3}{16}$ in. [3.2–4.8 mm] and where the duty is known precisely, for example in steam drums and air receivers, values as low as $\frac{1}{16}$ in. [1.6 mm] may be specified.

There is obviously a minimum value for the wall thickness below which the vessel would not be rigid structurally. This value is given by

$$t = (R + 100)/1000 \qquad (7\text{-}15)$$
$$[= (2.54 + R)/1000] \qquad (7\text{-}16)$$

The appropriate corrosion allowance must be added to this minimum thickness.

7.4.2 Reactor Design Temperature

It is usual to assume an external surface temperature of 200°F [365 K] and thence calculate the convection and radiation losses from the surface. The design temperature, i.e. the inner wall temperature, is then estimated by means of a heat balance. This approximation applies to vessels operating above 900°F [750 K] and which are lined internally to provide insulation and erosion protection.

A very rudimentary way of obtaining heat losses from a vessel is to estimate the outside surface temperature by touch, subtract 60°F [288 K] from the result and multiply this difference by a factor of 2 [11.4] multiplied by the approximate surface area in ft^2 [m^2]. This gives the total heat loss in BTU/hr [W].

7.4.3 Drums

Cylindrical drums are installed in process plant for mainly two purposes; firstly to effect the separation of entrained phases, and secondly to provide surge capacity between various sections of the plant. Drums dealing with liquids are normally installed horizontally and gas drums vertically.

(i) Dimensions

The most economic length to diameter ratio L/D is in the range 2.5–5, a value of 3 being fairly common. For settlers, L/D is normally 4–5. In the case of gas drums it should be noted that for higher operating pressures, L/D is increased.

(ii) Separation Space

Where a liquid contains solids or, more commonly, droplets of an immiscible liquid and a separation into the two phases is to be effected, space must be allowed in the vessel for this to be achieved and the design is normally based on the relative velocity between the two phases. The general relation for a spherical particle moving under gravity may be re-arranged to give the velocity of the particle (or droplet) relative to the surrounding medium as

$$v = 6.55(D_p/C)^{0.5}\{(\rho_s - \rho)/\rho\}^{0.5} \text{ ft/s} \tag{7-17}$$

$$[= 3.60(D_p/C)^{0.5}\{(\rho_s - \rho)/\rho\}^{0.5} \text{ m/s}] \tag{7-18}$$

where D_p = particle diameter (ft) [m],
           C = constant (for Re < 2, $C = 24/\text{Re}$
                     for Re > 500, $C = 0.44$)
          ρ_s = density of particle or droplet (lb/ft³) [kg/m³],
           ρ = density of continuous medium (lb/ft³) [kg/m³],
           v = relative velocity of the particle (ft/s) [m/s].

In the case of liquid droplets settling in a continuous phase, D_p may be assumed to have a minimum value of 0.004 in. [0.10 mm]. The settling velocity used in the design of the drum should at no time exceed 10 in./min [4.5 mm/s]. For vapour–liquid separation, the equation

$$K' = 6.55(D_p/C)^{0.5} \tag{7-19}$$

$$[= 3.60(D_p/C)^{0.5}] \tag{7-20}$$

may be used in conjunction with the general relationship, where K varies from 0.1 for low entrainment to 1.0 for severe entrainment. For vertical drums, K is usually 0.2 providing the vapour space exceeds 3 ft [1 m].

(iii) Surge Capacity

Where a drum is included on a plant to provide a buffer or a hold-up between sections of the plant, the size of the vessel depends on its location and the extent to which a failure in flow is undesirable. The design is usually based on the hold time when the vessel is half full. For reflux drums a value of 5 min [300 s] is acceptable, and for a drum in the line feeding material to a column the hold time is usually in the range 5–10 min [300–600 s] when half full. Where it is important that interruptions in the flow should be avoided if at all possible, a drum with a hold time of 30 min [1.8 ks] is specified.

It is often necessary to separate liquid droplets from a gas stream as, for example, in the inlet line to a compressor, and the vessel in which this is achieved is termed a 'knock-out drum'. The volume of such a vessel is usually about ten times the volumetric flow/min [volume $[m^3] = 600$ (m^3/s)].

7.4.4 Fractionating Towers

Methods for the precise design of fractionating towers have already been presented in other sections of this text and the following data provides only a rough guide, useful in the preliminary sizing of equipment. As such they are ideally suited to first approximation costing procedures and preliminary screening of process alternatives.

(i) Flow Conditions

The proportion of the feed which is vaporized may be taken as being directly proportional to the ratio of total overhead product to feed to the column. The optimum reflux ratio in many columns, R, is given by

$$R = 1.10 - 1.50R_{min} \qquad (7-21)$$

where R_{min} is the minimum allowable reflux ratio. Where reflux is expensive, for example in refrigerated towers, R is taken as $1.2R_{min}$ or less.

(ii) Tower Diameter

Except in cases where a very large liquid load is involved, the maximum allowable superficial velocity is given by

$$v = K(TSh/MP)^{0.5} \qquad (7-22)$$

$$[= 6.75K(TSh/MP)^{0.5}] \qquad (7-23)$$

where $T =$ vapour temperature ($^\circ$R) [K],
$s =$ specific gravity of liquid at plate temperature,
$h =$ plate spacing (in.) [m],
$M =$ molecular weight of vapour (—) [—],
$v =$ maximum superficial velocity (ft/s) [m/s],
$P =$ operating pressure (lb/in.^{2}a) [kN/m^2],
$K =$ tower constant from Table 7–9.

TABLE 7–9. VALUES OF K FOR USE ON EQUATIONS (7–22) AND (7–23)

Operating pressure		K
(lb/in.^{2} g)	[kN/m^2]	
15–50	104–350	1.25
50–200	350–1400	1.15
200	1400	0.85–1.05
vacuum towers		0.90

and the tower cross section is given by

$$A_t = 10.7 W/K(T/MPSh)^{0.5} \qquad (7\text{–}24)$$
$$[= 1.22 \ W/K \ (T/MPSh)^{0.5}] \qquad (7\text{–}25)$$

where W = vapour flow rate (lb/hr) [kg/s],
A_t = tower cross section (ft²) [m²].

This approximation also applies to absorbers, except that in this case the tower constant is given by

$$K = 0.15 + 0.75 \ (\mu_1 + 0.30) \qquad (7\text{–}26)$$
$$[= 0.15 + 0.75/(1000\mu_1 + 0.30)] \qquad (7\text{–}27)$$

where μ_1 = viscosity of liquid (cP) [Ns/m²]; the equations apply where $\mu_1 > 0.5$ cP [5 × 10⁻⁴ Ns/m²].

7.4.5 Heat Exchangers

The design of heat exchange equipment has been dealt with in Chapter 3 and worked examples on a variety of designs have been presented in Holland's text *Heat Transfer*.[26] Where a suggested process is being costed to a first approximation it is not usually necessary to design an exchanger precisely, and certainly, optimization procedures are not required. In such cases, an informed guess as to a likely overall coefficient of heat transfer, based largely on past experience, permits an estimation of exchanger area and hence cost, providing reasonable temperature ranges can be selected and limitations as to maximum size of exchanger are observed. The following notes serve as a guide to the selection of these parameters.

(i) *Operating Temperatures*

Where a fluid is being cooled, the optimum temperature difference between the outlet fluid and that of the incoming cooling water is, at least, 10–20°F [5–10 K], though this range may be reduced in the case of clean fluids. Similarly 30°F [15 K] is the optimum difference in temperature between that of condensing steam and the outlet fluid being heated in a boiler. It should be noted that in this type of equipment the maximum flux is 10 000 BTU/hr ft² [31.5 kW/m²]. For a conventional heat exchanger, 30°F [15 K] should be the difference between the temperatures of the incoming hot and the outgoing cold streams.

A further point in connection with coolers is the limitation of the cooling water outlet temperature in relation to the problem of scale formation. With fresh water the outlet temperature is usually limited to 120–125°F [320–325 K] and the maximum which can be tolerated with salt water is 110°F [315 K]. In any exchanger, the design is unacceptable economically if the correction to the logarithmic mean temperature difference to allow for imperfect cross-flow is less than 0.75.

(ii) *Heat Transfer Area*

For a given shell diameter the smaller the tube diameter, the larger the surface area, but the cleaning of the exchanger becomes more difficult. In general the use of ¾ in. outer diameter [19 mm] tubes is limited to clean fluids, and, where scale and frequent cleaning may be anticipated, in. outer diameter [25–37 mm] tubes are specified. Similarly, tub positioned on a triangular pitch only for clean fluids, and where a u. fluid is passed through the shell square pitch is employed. The use of finned tubes has only received scant attention in the present text, but these should be considered where, fairly obviously, the outside film coefficient is much lower than the inside and also where the estimated surface area is less than 200 ft² [20 m²].

The maximum area which can be accommodated in a single shell with 16 ft [4.88 m] tubes is shown in Table 7–10.

TABLE 7–10. MAXIMUM HEAT TRANSFER AREA PER UNIT

Tube outer diameter		Maximum surface area	
(in.)	[mm]	(ft²)	[m²]
$\frac{3}{4}$	19	3500	322
1	25	3200	294
$1\frac{1}{2}$	37	2000	186

7.4.6 Pipelines and Pumps

The power required to drive a centrifugal pump for a given duty is approximated by

$$P = \Delta H \, Gs/3960 \qquad (7\text{--}28)$$
$$[= \Delta H \, Gs/123] \qquad (7\text{--}29)$$

where ΔH = required head (ft of liquid) [m],
$\quad s$ = specific gravity of liquid (—) [—],
$\quad G$ = volumetric flowrate (gal/min) [cm³/s],
$\quad P$ = power requirement (hp) [W].

In sizing a pump, it is usual to assume an efficiency of 60 per cent. In the above relation, the required head is specified as

$$\Delta H = (\text{discharge pressure} - \text{suction pressure})$$

The optimization of fluid velocity in a heat exchanger has been dealt with in section 3.5; however, Table 7–11 gives a rough guide to the economic velocity of fluids in pipelines and also the maximum allowance friction loss per unit length of pipe.

TABLE 7–11. ECONOMIC FLUID VELOCITY

Economic fluid velocity (ft/s) [m/s]		Pressure drop allowable	
Pipe diameter d (in.)	d [mm]	(lb/in².)/100 ft	[(kN/m²)/100 m]
Pump discharge tube			
$(d/3 + 5)$ft/s	$(d/250 + 1.5)$m/s	2.0	45.2
Pump suction line			
$d/6 + 1.3$	$d/500 + 0.4$	0.4	9.0
Steam or gas line $20d$	$0.24d$	0.5	11.3

For liquids, a velocity in excess of 20 ft/s [6 m/s] should never be specified in order to avoid erosion of the pipe. On suction lines to pumps, the pressure drop should always be less than 50 per cent of the total head developed by the pump.

The determination of the optimum pipe diameter has been considered in Chapter 2, though Figure 7–15 is included for a rapid estimation of the approximate optimum size.

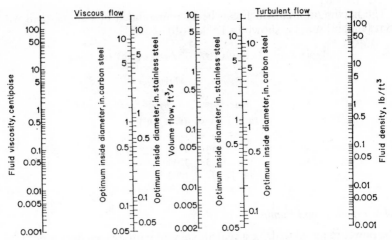

FIGURE 7–15. Nomograph for the rapid determination of the approximate economic pipe diameter (British Units)[27]

This nomograph has been produced by British Chemical Engineering[27] using the methods of Peters and Timmerhaus[28] and taking account of the method or pricing employed by the British Steel Corporation in the U.K.

7.5 Nomenclature

a	cross-sectional area	(cm²)	[m²]
a_B	cross-sectional area of a bolt	(in.²)	[m²]
A	cross-sectional area equation (7–5)	(in.²)	[m²]
A_B	total cross-sectional area of bolts	(in.²)	[m²]
A_t	tower area	(ft²)	[m²]
c	corrosion allowance	(in.)	[mm, m]
C	cross-sectional area of tube holes equation (7–5)	(in.)	[cm²]
C	factor in cost equations	(—)	[—]
d_B	bolt diameter	(in.)	[m]
D	internal diameter	(in.)	[mm, m]
D_p	particle diameter	(ft)	[m]
f	maximum allowable stress	(lb/in.²)	[kN/m² or MN/m²]
g	depth of groove, Figure 7–1	(in.)	[m]
G	diameter equations (7–3) and (7–4)	(in.)	[m]
G	volumetric flow	(gal/min)	[cm³/s]
h	plate spacing	(in.)	[m]
h_G	defined in equations (7–3) and (7–4)	(in.)	[m]
ΔH	head of liquid	(ft)	[m]
J	welded joint factor	(—)	[—]
k	thickness, equation (7–5)	(in.)	[cm]
K'	constant in equations (7–19) and (7–20)	(—)	[—]

K	constant, equations (7–22), (7–23)	(—)	[—]
L	tube length	(in.)	[cm]
L'	internal radius of cover	(in.)	[m]
M	molecular weight	(—)	[—]
n	number of bolts or tubes	(—)	[—]
n	exponent in cost equations	(—)	[—]
p	pressure	(lb/in.^{2})	[kN/m^2, MN/m^2]
P	power	(h.p.)	[W]
P_r	radial stress	(lb/in.^{2})	[kN/m^2]
P_1	pressure outside tubes	(lb/in.^{2})	[kN/m^2]
P_2	pressure inside tubes	(lb/in.^{2})	[kN/m^2]
R	internal diameter	(in.)	[m]
R_e	Reynolds number	(—)	[—]
R_{min}	minimum reflux ratio	(—)	[—]
s	specific gravity	(—)	[—]
S	allowable working stress	(lb/in.^{2})	[kN/m^2]
t	thickness	(in.)	[mm, m]
T	temperature	(°R)	[K]
V	velocity	(ft/s)	[m/s]
V_w	wind velocity	(m.p.h.)	[m/s]
W	vapour flow rate	(lb/hr)	[kg/s]
X	factor in cost equations	(—)	[—]
z	pressure due to wind velocity	(lb/ft^2)	[kg/m^2]
η	deflection efficiency	(—)	[—]
μ	factor in equation (7–6)	(—)	[—]
μ_L	viscosity	(cP)	[N s/m^2]

REFERENCES

1. *Unfired Pressure Vessels*, ASME Boiler and Pressure Vessel Code, 1959.
2. Standards of the Tubular Exchanger Manufacturers Association, New York, 1959.
3. British Standards Institution, 'Fusion-welded Pressure Vessels', B.S. 1500, 1958.
4. British Standards Institution, 'Dimensions of Steel Pipe for the Petroleum Industry', B.S. 1600, 1970.
5. Kern, D. Q. *Process Heat Transfer*. New York: McGraw-Hill Book Co., 1950.
6. Zick, L. P. *Stresses in Large Horizontal Cylindrical Pressure Vessels in Two Saddle Supports*, Welding Research Supplement. Sept. 1957.
7. Perry, J. H. Ed. *Chemical Engineers' Handbook*, 4th ed. New York: McGraw-Hill Book Co., 1963.
8. British Standards Institution, 'Tubular Heat Exchangers for use in the Petroleum Industry'. B.S. 2041, 1953.
9. Molineux, F. *Chemical Plant Design* 1. London: Butterworths, 1963.
10. Donnell, L. H. *Trans. A.S.M.E.* 1934, **56**, 795.
11. Jorgensen, S. M., *Pet. Ref.* 1945, **24**, 381.
12. Brownell L. and Young, E. H. *Process Equipment Design*. New York: Wiley, 1959.
13. Rase, H. P. and Barrow, M. H. *Project Engineering of Process Plant*. New York: Wiley, 1957.
14. Bergman, E. O. *Trans. A.S.M.E.* 1955, **77**, 863.
15. Nelson, W. L. *Petroleum Refinery Engineering*. New York: McGraw-Hill Book Co., 1956.
16. Zimmerman, O. T. *Cost Eng.* 1968, **13**, 4, 9.
17. Billet, R. and Raichle, L. *Chem. Eng.* 1967, **74**, 4, 148.
18. Stogens, M. *Brit. Chem. Eng.* 1961, **6**, 3, 182.
19. British Standards Institution, 'Use of Structural Steel in Building', B.S. 449, 1959.
20. Marshall, V. O. *Pet. Ref.* 1943, **22**, 8, 101.
21. Brown, A. A. *Hydrocarbon Proc.* 1966, **45**, 7, 109.
22. Brown, A. A. *Hydrocarbon Proc.* 1966, **45**, 12, 109.
23. Kitterman, L. and Ross, M. *Hydrocarbon Proc.* 1967, **46**, 5.

24. 'Ballast Tray Design Manual', Bulletin No. 4900 (revised). Dallas: F. W. Glitsch and Sons Inc., 1970.
25. 'Packed Tower Internals' Manual TA–70. Akron, Ohio: Norton Chemical Process Products Division.
26. Holland, F. A., Moores, R. M., Watson, F. A. and Wilkinson, J. K. *Heat Transfer*. London: Heinemann Educational Books Ltd, 1970.
27. Anon. *Brit. Chem. Eng.* 1971, **16**, 4/5, 313.
28. Peters and Timmerhaus. *Plant Design and Economics for Chemical Engineers*. New York: McGraw-Hill Book Co., 1968.

Part 3: Process Plant Selection

Chapter 8: Physical Separation Processes

8.1 Introduction

The field of solid–liquid separation processes has attracted considerable attention in the published literature; it is less common to read of gas–solid separations and rarer still to read of solid–solid separation equipment or problems associated with selection procedures. In this chapter, these three classes will be considered in reverse order. The intention is to provide a guide to selection rather than a descriptive catalogue of available equipment.

8.2 Solid–Solid Separations

Before attempting to separate mixtures of solids, it is usually necessary to provide a feed of material which lies within a limited particle size range. A screening or classification step is therefore likely to be the first step in a separation process.

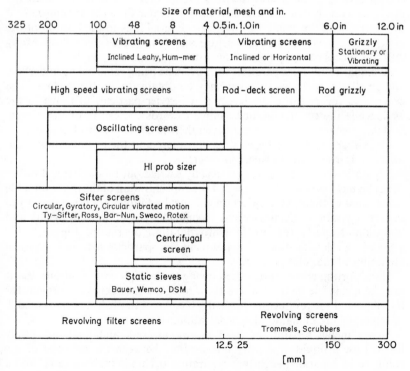

FIGURE 8–1. Screen selection guide[1]

Screens are devices with surfaces which are able to effect separation by size, the screen acting as the separating medium with apertures for the passage of undersize materials. The normal range of screen size lies between 0.005 in. and 10 in. [0.12–250 mm] though it is possible to separate boulders as large as 6 ft [2 m] diameter.[1] Screen materials most commonly used are woven wire mesh and perforated plates, the latter usually being restricted to severe surface conditions and the handling of larger materials.

The most common problem arising with screens is 'blinding'. This may occur in one of two ways: either by plugging of the apertures when screening a large volume of near-size material, or by coating when undersize particles adhere firstly to the wires and then to each other, finally closing the aperture. The blinding condition is most often cured by increasing the frequency and stroke of the screen, thus providing a more vigorous motion. Blinding due to coating with small damp particles may be prevented by the use of electrically heated decks which alter surface tension effects between the damp particles and the heated wire.

Figure 8–1[1] may be used to select the most suitable type of screen for a given range of particle size. Description of the items of equipment is available elsewhere[2] and is not included in this text.

Whilst screening can effect particle concentrations over a wide range of sizes, other methods are more applicable to particles up to 0.005 in. [0.12 mm] diameter. In the range 65–200 mesh, classification is often carried out by means of a hydroseparator which is an overfed thickener. One typical arrangement is illustrated in Figure 8–2.[3] The feed rate to the thickener is such that the overflow rate is greater than the settling rate of the slurry, with the result that oversize material falls to the bottom of the tank, where it is raked to the centre and removed as underflow. Many differing pieces of equipment all operate on the same principle.

A similar resulting separation with oversize being removed as underflow and smaller particles as overflow may be obtained with the cyclone. The centrifugal forces cause the larger and heavier particles to move to the wall, while the smaller particles move towards the centre of the vortex. Suitable mechanical arrangements can then separate the two fractions.

Centrifuges may also be used as classifiers and their applicability and selection is discussed elsewhere in this text.

Separation of solid mixtures is usually carried out by mechanical rather than chemical means and is aimed at the concentration of the more valuable constituent of the mixture. Certain chemical additives are often employed in wet separation techniques and include the use of wetting agents and 'flotation' reagents for use in froth flotation operation. Figure 8–3[3] attempts a classification of equipment by particle size and physical attributes of the solid product.

Hand sorting is very rarely used owing to the high cost of labour, but it may be a suitable method where the size of the product is convenient for handling and the product itself is valuable. An example has been quoted[3] of hand picking in diamond concentration.

Heavy media separations (also known as 'sink and float' or 'heavy liquid') use a liquid of density intermediate between the densities of the solids to be separated such that separation occurs between a 'sink' and a 'float' fraction when the solids are added to the liquid. The difficulty lies

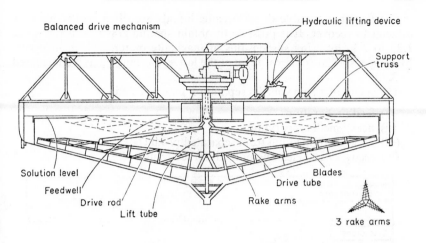

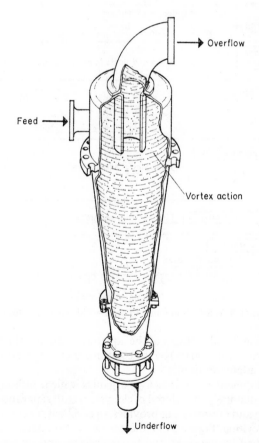

FIGURE 8–2. A typical hydroseparator and a liquid cyclone[3]

in the fact that suitable dense organic liquids are often costly, toxic or difficult to recover. It is possible to obtain specific gravities in the range 1.25 to 3.40 by using suspensions of magnetite or ferrosilicon in water; by using a special spherical form of the latter, the range may be increased to 3.70.

The stages in a heavy media separation may be listed as:

(i) feed preparation,
(ii) heavy media separation,
(iii) removal of medium from separated products,
(iv) recovery and cleaning of medium for re-use.

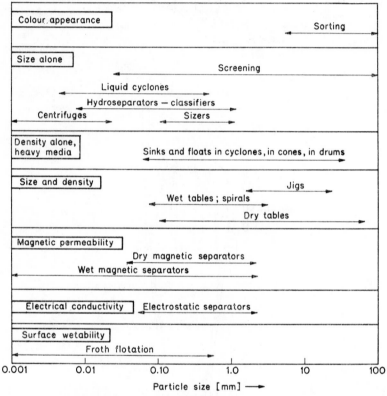

FIGURE 8–3. Selection guide to solid–solid separation techniques

One of the great advantages of the aqueous magnetite slurry is the ease of performing steps (iii) and (iv) by the use of magnetic separators which will be discussed in a later section.

The development of inert heavy organics which, although costly, can be re-used indefinitely in a closed cycle system with separation and recovery by scrubbing and evaporation, provides an excellent medium for separation in the hydrocylone. Figure 8–3 indicates the wide material size range which can be handled by the sinks and floats process, and by use of the appropriate physical equipment the range can be extended to material up to 12 in. [300 mm] diameter.

The use of jigs finds wide application in coal and mineral preparation. The jig consists essentially of a submerged screen which can either be vibrated (in the case of a movable screen jig) or can remain fixed with pulses of liquid forcing a way through the screen. The frequency of pulsation is usually 2–5 strokes/s with a stroke length of 0.2–2 in. [5–50 mm]. After each pulsation, the solids mixture settles and stratifies with light material on the top and heavy material at the bottom of the bed. As the layers build

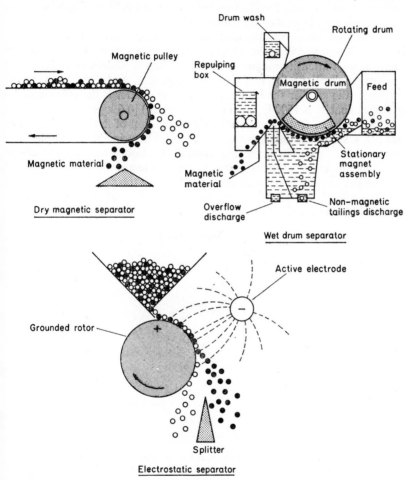

FIGURE 8–4. Diagrammatic representation of magnetic and electrostatic separators[3]

up each can be removed separately. Whilst jigs can handle a wide size range, it is usual for sinks and floats to handle large material and for shaking tables to deal with the range less than 10 mesh.

The riffled table is a flat table inclined at about 3° to the horizontal. Parallel to the top of the table runs a series of slats or riffles about ¼ in. [6 mm] high. Water flows down the table and the whole assembly reciprocates logitudinally. The material is fed to the table at a top corner, and under the influence of the table motion and the water velocity the larger and less dense particles are carried downwards and the small heavy particles move parallel to the riffles. Thus a number of fractions may be separated with optimum efficiency obtained using a classified feed between 8 and 100 mesh [2–0.15 mm]. The normal table size is between 4 and 6 ft [1.2–1.8 m] wide and 12 and 16 ft [3.6–4.9 m] in length, requiring a power consumption of 0.75–1.0 hp [0.56–0.76 kW] per table and using a water rate of between 40 and 350 gal/ton [180–1300 cm³/kg] of solids. Feed rates per table of between 8 and 10, and 2 and 3 ton/hr [2.2–2.8 and 0.56–0.85 kg/s] are used for roughing and cleaning operations respectively.

Separations based on magnetic and electrostatic properties are simple in principle and typical applications are illustrated in Figure 8–4.[3] Development of high intensity magnetic separators has considerably widened the field of applications for this type of equipment, and for both types of separator the smaller the feed size range, the greater is the efficiency. An average capacity is 1500 lb/hr (ft of rotor length) [0.6 kg/s m].

The subject of froth flotation for the separation of finely divided solids has attracted much detailed attention.[4],[5] The principle is simple—if a mixture is suspended in an aerated liquid the gas bubbles will tend to adhere preferentially to one of the constituents (the one less easily wetted) such that its effective density is reduced and it will rise to the surface. If a frothing agent is added, the particles may be held in the surface and eventually discharged over a weir. The technique is illustrated in Figure 8–5.[3]

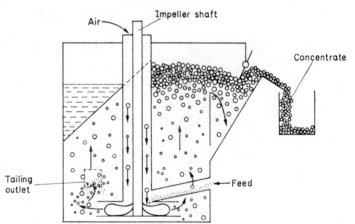

FIGURE 8–5. Froth flotation technique for the separation of fine particles[3]

Froth flotation is useful for small particles and for mixtures where density differences are too small for gravity separation. Chemical additives are often used to make the product more hydrophobic and the waste hydrophilic. These compounds are often expensive and rather exotic for particular applications, although they are only required in very small concentrations.

8.3 Gas–Solid Separations

The necessity for gas cleaning may arise from one or more of the following considerations:

(i) Health—to reduce the concentration of toxic or combustible dust either within a plant for the benefit of personnel or to reduce emission to the surrounding population.

(ii) Explosion—a number of wide-ranging materials when present in small concentrations form explosive mixtures in air.

(iii) Economic—the dust particles may themselves be valuable and worth recovering from a financial viewpoint.

(iv) Social—with public emphasis focused so much on pollution of the environment, it may be desirable to reduce the amount of material discharged to the atmosphere as mists or plumes even though they may be completely harmless.

The problem of gas cleaning may be difficult with a wide spectrum of solid materials in a size range which can vary from large molecules to visible pollutants such as large soot particles. The range of equipment for the treatment of solid–gas mixtures is therefore wide and an attempt will be made to classify the available plant by considering two basic categories— dry collectors and wet scrubbers. Inevitably, a third category arises with any attempted classification—miscellaneous.

The size range of the solid particles will normally be within the range of 0.1–100 microns and their concentration between 0.1 and 100 grains/ft^3 [3.5–3500 grains/m^3]. (The grain per unit volume is a measure of dust concentration where 7000 grains = 1 lb [1 grain = 65 mg].)

To make preliminary selection of suitable gas cleaning plant, it is necessary to know the dust concentration, particle size, gas flow, and the permissible emission rate. More detailed requirements for complete specification are available elsewhere.[6],[7]

8.3.1 Dry Collectors

Various types of dry collection equipment are shown in Figure 8–6. They are useful where the product is valuable as it requires no further processing, though the cleaned gas will not be cooled or completely free of fines. Although generally low in efficiency, they are relatively cheap and operating costs are low.

The gravity settling chamber (Figure 8–6(a)) is the simplest and least efficient form of collector. By passing the gas through an enlarged section where the velocity falls, the particles can settle out by gravity. This type of unit requires a lot of space and is little used. Variations exist in an attempt to minimize space requirements and improve efficiency, such as the baffle collector (Figure 8–6(b)).

The louvre separator (Figure 8–6(c)) and the skimming chamber effect a sudden change in the direction of the gas flow while the solid particles are carried by inertia to another section of the equipment where they may be removed.

The most commonly used of the dry collectors is one of the varieties of the cyclone. Gas enters the cyclone tangentially at the top of the cylindrical section and spirals downwards to the bottom section, which is usually conical in shape. The centrifugal force holds the solid particles against the vessel wall and they may be separated from the gas at the bottom of the

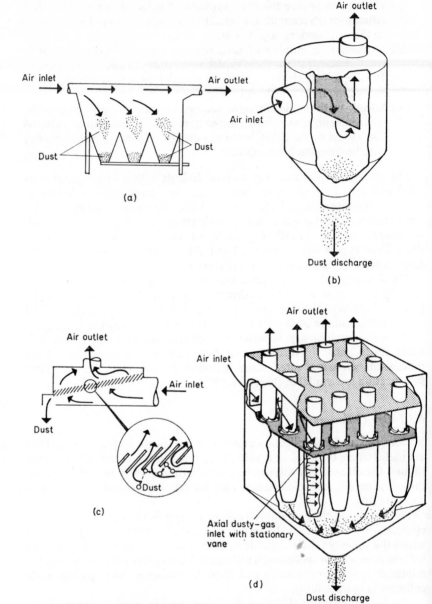

FIGURE 8–6. Types of dry collectors, (a) gravity settling chamber, (b) baffle collector, (c) louvre collector, (d) multiple cyclone collector[6]

section. The cost of these units is low, they have no moving parts and can withstand high temperatures. As small diameter cyclones are more efficient than large ones, they are frequently arranged in parallel with gas feed from a surrounding chamber (Figure 8–6(d)).

Two further types of collectors, impingement and dynamic, provide a high

TABLE 8–1. CHARACTERISTICS OF DRY GAS COLLECTORS[6]

Type of equipment	Minimum particle size (μ)	Minimum loading (grains/ft³)	[mg/m³]	Approximate maximum efficiency (%)	Gas pressure loss (in. water gauge)	[mm water]	Gas velocity (ft/min)	[m/s]	Maximum capacity (ft³/min × 10³)	[m³/s]	Relative space required
Settling chamber	50	5	11 500	50	0.2	5	300–600	[1.5–3]	None	None	Large
Baffle chamber	50	5	11 500	50	0.1–0.5	2.5–12.5	1000–2000	[5–10]	None	—	Medium
Skimming chamber	20	1	2 300	70	1	25	2000–4000	[10–20]	50	23	Small
Louvre	20	1	2 300	80	0.5–2	12.5–50	2000–4000	[10–20]	30	14	Medium
Cyclone	10	1	2 300	85	0.5–3	12.5–75	2000–4000	[10–20]	50	23	Medium
Multiple cyclone	5	1	2 300	95	2–6	50–150	2000–4000	[10–20]	200	93	Small
Impingement	10	1	2 300	90	1–2	25–50	3000–6000	[15–30]	None	None	Small
*Dynamic	10	1	2 300	90	Provides head	—	—	—	50	30	—

* Usually requires 1–2 hp/1000 ft³/min [1.6–3.2 kW/(m³/s)]

velocity by a venturi or a centrifugal fan respectively which accelerates the particles on to a surface of an obstacle while the gas is diverted around the obstacle and led way in a cleaned condition.

Table 8–1 may be used to summarize the main characteristics of dry dust collectors and should aid equipment selection.

8.3.2 Wet Scrubbers

The following points should be noted about the properties of wet scrubbers:[6]

(i) The gas is cooled and washed.

(ii) The solids are recovered as either slurry or a solution which may then need further processing.

(iii) Corrosive gases may be neutralized by the correct choice of scrubbing medium.

(iv) Efficiency of collection is generally higher than dry collectors, but mists and plumes may be formed.

(v) Explosion hazards are reduced.

(vi) Inlet gas conditions are virtually unlimited.

(vii) Capital costs are usually reasonable but operating costs may be high.

(viii) Freezing conditions may have to be considered.

Figure 8–7 illustrates some of the types of available wet scrubbers.

The gravity spray scrubber (Figure 8–7(a)) is the simplest and least efficient of the wet scrubbers. It is, however, cheap to operate, as liquid is pumped to the nozzles from which the collecting droplets fall by gravity. This type of collection is useful for coarse solids present in high concentrations, and is often installed directly in a stack to avoid using an additional fan and separate scrubber. The gas velocity must be low to minimize entrainment, and this may give rise to large diameter equipment. Insertion of a packed section into a spray tower forms the basis of the packed bed scrubber (Figure 8–7(b)). In order to minimize blockage in the interstices of the packing, the packed tower should only be used for very dilute mixtures and the dust collection aspect of the operation may be secondary in relation to either cooling or gas absorption duties.

A gas stream carrying both dust particles and water droplets may be passed at high velocity through the holes of a perforated plate to impinge upon a baffle on the other side. Flow is countercurrent and the solids are removed in the liquid overflow from the impingement plate. The centrifugal scrubber (Figure 8–7(c)) is available in many different forms, each one having the common feature that liquid is sprayed into a rising vortex of gas. Here the gas and liquid droplets combine and are accelerated to the wall where they are collected.

In the dynamic wet scrubber, liquid is sheared mechanically to break it into droplets with which the solids will combine. The same principle is employed in the submerged nozzle scrubber where the gas atomizes the liquid, and in the jet type where water flow is used in a jet ejector (Figure

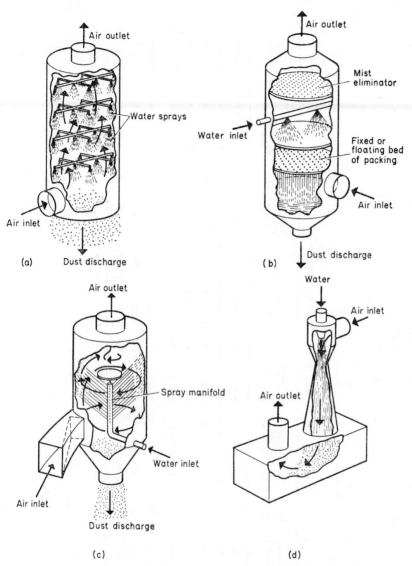

FIGURE 8–7. Wet scrubbers, (a) gravity scrubber, (b) packed bed scrubber, (c) centrifugal scrubber, (d) jet scrubber[6]

8–7(d)). In the venturi scrubber, the liquid is again atomized by the gas and very high collection efficiencies can be achieved even for the collection of submicron dusts.

A summary of these types of scrubbers is presented in Table 8–2.

8.3.3 Miscellaneous Equipment

Fabric filters could be included in the classification 'dry collectors', but as they can deal with very small particles at very high efficiencies they are best

TABLE 8–2. CHARACTERISTICS OF WET SCRUBBERS[6]

Type of equipment	Minimum particle size (μ)	Minimum loading (grains/ ft³)	[mg/m³]	Approx. max. efficiency (%)	Gas pressure (in. H₂O)	[mm H₂O]	Liquid pressure loss (lb/ in.²)	[kN/m²]	Gas velocity (ft/min)	[m/s]	Maximum capacity (ft³/min ×10³)	[m³/s]	Liquid rate (g.p.m./ 1000 ft³ min)	[kg/s]/ (m³/s)	Space required
Gravity spray	10	1	2300	70	1	25	20–100	140–700	100–200	0.5–1	100	47	0.5–2	0.06–0.27	Medium
Centrifugal	5	1	2300	90	2–6	50–150	20–100	140–700	2 000–4 000	10–20	100	47	1–10	0.13–1.3	Medium
Impingement	5	1	2300	95	2–8	50–200	20–100	140–700	3 000–6 000	15–30	100	47	1–5	0.13–0.67	Medium
Packed bed	5	0.1	230	90	1–10	25–250	5–30	35–210	100–300	0.5–1.5	50	23	5–15	0.67–2.0	Medium
*Dynamic	1	1	2300	95	Provides lead		5–30	35–210	3 000–4 000	15–20	50	23	1–5	0.13–0.67	Small
Submerged nozzle	2	0.1	230	90	2–6	50–150	None		3 000	15	50	23	No pumping		Medium
Jet	0.5–5 only	0.1	230	90	Provides lead		5–100	35–700	2 000–20 000	10–100	100	47	50–100	6.7–13.5	Small
Venturi	0.5	0.1	230	99	10–30	250–750	5–30	35–700	12 000–42 000	60–210	100	47	3–10	0.40–1.35	Small

* Dynamic scrubber usually requires 3–20 hp/1000 ft³/min [4.8– 32 kW/(m³/s)]

considered separately. Their use is limited by the temperature of the gas stream, though materials are available[6] which permit operation up to 550°F [560 K]. The dirty gas flows through a porous medium of woven or felted fabric and deposits the solids in the voids. The pressure drop increases with time and eventually reaches a point where the solids have to be removed.

Typical gas rates for fabric filters are 1.5–15 ft³/min ft² of filter area [0.0075–0.075 m³/s m²]. Details of physical properties of various media can be found in the literature.[2]

Electrostatic precipitators have the same properties as dry fabric filters in that they will collect very small particles at very high efficiencies, but normally, unless the gas is hot and corrosive, the fabric filter would be the first choice. Under difficult conditions, however, the high cost of the electrostatic precipitator becomes justifiable as the only high efficiency device capable of performing the duty. The range of operation is normally between 0 and 700°F [255–644 K], though this may be extended further beyond each limit if necessary.

The principle of the precipitator is simple; the gas between a high voltage electrode and an earthed electrode is ionized, the dust particles are charged by the gas ions and then migrate to the earthed electrode, where they collect. Removal of the solids may be effected by vibration, scraping or washing. The limitation of the technique is that the electrical resistivity of the dust should lie between 10^4 and 10^{10} Ω/cm. Dusts with a low resistivity will not adhere to the collector, and if the resistivity is above the upper limit the grounded electrode is likely to become covered with an insulating film of dust and lose its effectiveness. A high resistivity may be reduced by the addition of conditioning agents such as moisture or acid mist in order to improve the collection process.

Table 8–3 compares the range of operation of the electrostatic precipitator and the fabric filters with devices discussed earlier.

TABLE 8–3. CHARACTERISTICS OF ELECTROSTATIC PRECIPITATORS AND FABRIC FILTERS

		Fabric filter	Electrostatic precipitator
Particle size		0.2 μ	2 μ
Minimum loading	(grains/ft³)	0.1	0.1
	[mg/m³]	230	230
Approximate efficiency		99%	99%
Gas pressure loss	(in. water)	2–6	0.2–1.0
	[mm water]	50–150	5–25
Power requirement	(kW/(1000 ft³/min))	—	0.1–0.6
	[kW/(m³/s)]	—	0.21–1.3
Gas velocity	(ft/min)	1–20	100–600
	[m/s]	0.005–0.10	0.5–3
Maximum capacity	(10³ ft³/min)	200	10–2000
	[m³/s]	94	4.7–940
Relative size		Large	Large

Cost data based on the volumetric gas flow rate has been compiled for several of the types of gas–solid separation discussed in this section, and these data are presented in Table 8–4.

TABLE 8-4. COST DATA FOR DUST COLLECTORS. COST $(1970) = CX^n$

Type	X (A.A.)	[SI]	C (A.A)	[SI]	n	Reference
Mild Steel cyclone	ft³/min	m³/s	3.36	1 550	0.80	41
Cloth filter	ft³/min	m³/s	28	5 170	0.68	41
Electrostatic precipitator	ft³/min	m³/s	403	130 000	0.75	41
Gas scrubbers						
(a) Packed tower	10³ft³/min	m³/s	615	1 060	0.73	42
(b) Centrifugal collector	10³ft³/min	m³/s	868	1 500	0.73	42
(c) Venturi	10³ft³/min	m³/s	503	860	0.73	42

Example 8-1

Compare the cost of treating 1000 ft³/min [0.472 m³/s] of dust laden gas in a mild steel cyclone and in a packed tower scrubber.

From Table 8–4, for a mild steel cyclone,

$$X = 1000 \; [= 0.472]$$

Hence
$$\text{cost (\$)} = 3.36(1000)^{0.80}$$
$$[= 1550(0.472)^{0.80}]$$
$$= \$845$$

For a wet packed tower scrubber,

$$X = 1 \; [= 0.472]$$

Hence
$$\text{cost (\$)} = 615(1)^{0.73}$$
$$[= 1060(0.472)^{0.73}]$$
$$= \$615$$

8.4 Liquid–Solid Separation

The most important consideration in the selection of plant for solid–liquid separation may be listed as follows:[8]

(i) The cake dryness which is required. The value selected will influence both the costs of the separation equipment and the subsequent drying process. Material efficiencies and purities may also be affected.

(ii) Cake washing efficiency, which affects the quantity of wash liquor, material efficiencies and purity.

(iii) Filtrate clarity, which could be a problem in purity and/or product appearance.

(iv) Crystal breakage, which is often important from considerations of product appearance and particle size distribution.

Additional factors to be considered include toxicity and flammability of the solvent, vapour losses from the solvent, and materials of construction.

Figure 8–8[9] presents the relationship between particle size and solids concentration for several kinds of equipment, while the factors listed above

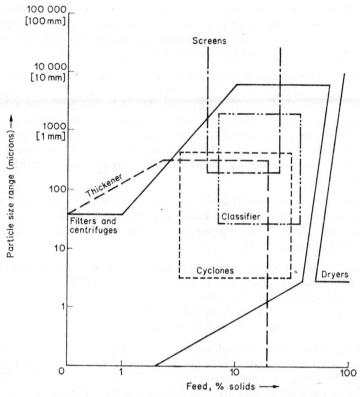

FIGURE 8–8. Equipment selection on the basis of particle size and solids concentration[9]

are considered as a function of the different items of separation and equipment in Table 8–5.

TABLE 8–5. QUALITATIVE GUIDE TO SEPARATION EQUIPMENT[19]

Type of equipment	Overflow or filtrate clarity	Solids dryness	Washing efficiency	Resistance to crystal breakage
Screens	2–4	2	3	5
Gravity settlers	5–7	5	—	—
Gravity filters	5–7	3	6	7
Rotation equipment	2–4	2	—	—
Vacuum filters	6	3–5	6	7
Pressure filters	6–7	3–5	5–7	7
Hydrocyclones	3–5	2	—	3
Centrifugal settlers	3–6	2–5	3	—
Centrifugal filters	3–7	7	4–6	3–6

Key

1 = Very poor	3 = Fair	6 = Very good
2 = Poor	4 = Quite good	7 = Excellent
	5 = Good	

It will be seen that the choice of methods for the separation of a liquid–
solid mixture is wide, though this may be limited depending on whether
a clear liquid or a dry solid is of prime importance. Each of the main
methods of separation will be considered in turn.

Gravity sedimentation is employed in two situations—where the separation is easy or where the volumetric flow rates are so large that the cost of
any other method is prohibitive. The larger the solid particles and the greater
the density difference between solid and liquid, the easier will be the separation, though the percentage of solids in the mixture should be low to avoid
difficulties in solid removal. Although clear liquor can be obtained, the
solids remaining are wet and further dewatering by filtering or centrifuging
may be necessary.

Gravity filtration is usually associated with high volumetric flow rates and
the removal of traces of minute solids, as in the case of water purification.
The filtration media is often a porous material such as crushed coke or a
granular bed such as sand.

Filtration in one of many forms (e.g. vacuum and pressure filtration) will
be discussed in a later section. It has the advantage of producing a clear
liquor and a reasonably dewatered solid at an economic rate providing the
feed contains greater than about 0.5 weight per cent solids and the filtration rate exceeds 5 gal/hr ft^2 [6.8 × 10^{-5} m^3/s m^2].

Hydrocyclones have been discussed by Bradley[40] and are worthy of
consideration when a slurry is to be concentrated, two immiscible liquids
have to be separated, or when a clarified liquid overflow is required. Their
advantages and limitations are fully discussed in the reference quoted.

Centrifugal sedimentation is used for the recovery of very fine to medium
size solids from slurries containing less than 0.5 per cent solids, where high
temperature and pressure operation is necessary, and where toxic materials
are handled. As mentioned earlier, centrifugal sedimentation is also used
for liquid–liquid separations.

Centrifugal filtration may be considered as an alternative to vacuum
filtration where a solid product with a low moisture content is required.

These last two classes of equipment will be considered in greater detail in
the next section. Table 8–5 is included as a guide to the preliminary selection
of separation equipment.

8.4.1 Sedimentation

Figure 8–8 shows that thickening and classification are appropriate methods
of separation over a wide range of particle size and concentration. These
two operations are most frequently thought of as sedimentation processes
but it is important to realize that the primary purpose of thickening is to
concentrate a relatively large quantity of suspended solids in a feed stream,
while classification attempts the removal of small quantities of fine particles
in order to produce a clear effluent. The size range of this type of equipment
is very wide, ranging from 3 to 400 ft [1–120 m] diameter.

Satisfactory operation of the thickener depends upon the presence of a zone of negligible solid content near the top. This zone must be sufficiently deep to enable any particles carried upwards from the feed position to settle against the upward flow of liquid. The volumetric flow rate of liquid up through the classification zone is equal to the difference between the feed rate and the rate of removal in the underflow.

The mass rate of discharge of liquid in the overflow is given by

$$(F' - U')W \tag{8-1}$$

where F' = ratio of liquid to solid in the feed (lb/lb) [kg/kg],
       U' = ratio of liquid to solid in the underflow (lb/lb) [kg/kg],
      W = mass feed rate of solid (lb/hr) [kg/s].

The upward velocity in the classification zone is given by

$$(F' - U')W/A\rho \tag{8-2}$$

where A = cross sectional area of the tank (ft^2) [m^2],
      ρ = density of the liquid (lb/ft^3) [kg/m^3].

As the upward velocity must not exceed the rate of sedimentation of the slurry, u_c, the area for successful operation is given by

$$A = (F' - U')W/\rho u_c \tag{8-3}$$

The rate of sedimentation has been the subject of considerable investigation and the following relationships may be used to calculate u_c:[12],[13]

$$u_c = e^2 d^2(\rho_s - \rho)g10^{-1.82(1-e)}/18\mu \tag{8-4}$$
$$[= 100e^2 d^2(\rho_s - \rho)g10^{-1.82(1-e)}/18\mu]$$

or

$$u_c = ed^2(\rho_s - \rho)g/18\mu_c \tag{8-5}$$
$$[= 100ed^2(\rho_s - \rho_c)g/18\mu_c]$$

where u_c = settling velocity of suspension (cm/s) [m/s],
      d = diameter of particle (cm) [mm],
ρ_c, ρ_s, ρ = density of suspension, solid and fluid respectively (g/cm^3) [kg/m^3],
    u_c, u = viscosity of suspension and fluid respectively (poise) [N/s m^2]
      g = acceleration due to gravity (cm/s^2) [m/s^2],
      e = voidage of suspension (dimensionless).

It can be shown[14] that the depth of the thickening region is given by

$$h' = \frac{Wt_r}{A\rho_s}(1 + X\rho_s/\rho) \tag{8-6}$$

where h' = depth (ft) [m],
      t_r = required retention time of solids as determined experimentally (hr) [s],
      x' = mean mass ratio of liquid to solids in thicknening section (—) [—].

The area should be calculated for the whole range of concentrations present in the thickener and the design based on the maximum value obtained. The depth calculated from equation (8–6) should be regarded as approximate and a safety factor of 3 ft [1 m] should normally be added.[14]

Typical designs of thickeners of various sizes are presented in Figure 8–9, together with the different rake designs for particular applications.

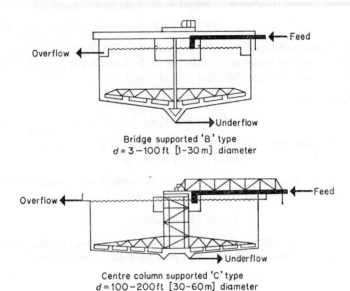

Bridge supported 'B' type
$d = 3 - 100$ ft [1-30 m] diameter

Centre column supported 'C' type
$d = 100 - 200$ ft [30-60 m] diameter

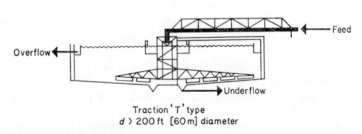

Traction 'T' type
$d \rangle 200$ ft [60 m] diameter

FIGURE 8–9. Types of thickener[9]

Classifiers share the same characteristics as thickeners in that large throughputs may be handled at low cost per unit volume. However, since the classifier is designed to handle dilute suspensions to produce a clear effluent, the power requirement is lower owing to reduced torque. Type B is usually employed for diameters between 8 and 45 ft [3–15 m] and Type C for all other sizes up to 400 ft [60 m].

Classifiers are often rectangular in shape with a length three times the width. Solids are removed by scrapers on endless chains which discharge into a sludge hopper at one end. Circular classifiers are more efficient, however, and are used where optimum performance is required. Where mixing and flocculation are necessary in the classification process, a reactor-classifier may be preferred. This unit has a combination drive to turn the rate mechanism at a slow speed and, by means of a concentric shaft, rotate a turbine in the reaction well. This application is particularly useful in water treatment for turbidity, colour removal and lime softening.

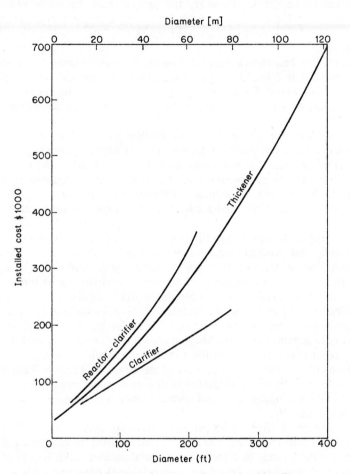

FIGURE 8–10. Cost data for sedimentation plant[9]

Cost data for sedimentation equipment is presented in Figure 8–10.[9] The costs include site preparation and erection of the tanks (which are assumed mild steel), with concrete bottoms in diameters exceeding 100 ft [30 m]. The costs of pumps, piping, electrical work and instrumentation are not included.

Operating costs may be estimated by considering that a 200 ft [60 m] diameter thickener operates with a 15 hp [11 kW] motor to give a peripheral speed of 25 ft/min [0.12 m/s]. This power figure is reduced to 1 hp [0.75 kW] for a similar size of classifier with a rotation speed of 0.02 rpm [0.3 mHz]. Labour requirements are very modest.

8.4.2 Filters

Filtration is by nature a discontinuous operation. The solids cannot be collected indefinitely without stopping to clean or renew the filter media. In addition, solvents may have to be recovered from the cake or the cake has to be dried. However, the process may be made essentially continuous by the exposure of fresh filter surface to the slurry. In this way the process has only to be stopped when the filter medium has to be replaced.

Experience has shown that continuous filtration should not normally be attempted if $\frac{1}{8}$ in. [3 mm] of cake cannot be formed under vacuum in less than about 5 min [300 s]. Use of the cake formation rate as a classification enables a guide to filter selection to be compiled, Table 8–6.[10]

The choice of equipment may be further narrowed by consideration of other factors. For example, temperature or vapour pressure limitations may eliminate vacuum operation, and if either of these factors is combined with a high production rate the choice may be reduced to a continuous pressure filter. Chemical resistance considerations may limit the range of suitable filter media which may in turn influence the type of equipment.

Referring to Table 8–6, fast filtering slurries are usually found in mineral processing and product streams from crystallization units. The choice of suitable filtering plant for this type of slurry is wide, and if the throughput is high the alternative methods of screening, centrifuging or settling may prove more economic. Medium filtering slurries are also capable of being handled in many forms of equipment, including the vacuum belt or, where the production rate is low, with one of the forms of the batch pressure filter. For the remaining slow filtering applications, emphasis tends to move away from continuous operation towards the batch pressure filters. Descriptions of the mechanical aspects of the filters in Table 8–6 are widely available[2],[14],[10],[15] and details of the more specialized topics such as ultrafiltration processing[16] and special filtration techniques[17] have been published recently.

The choice of filter media for any particular duty is very wide and is being constantly increased. The ratio of cost between the cheapest and most expensive media may be a factor of several thousand and it is important to realize that cost figures based on superficial area have very little meaning when compared with each other. Table 8–7, adapted from an excellent series of articles on filtration media,[18] attempts to classify the media according to a degree of rigidity. The table shows the range of costs involved together with the particle retention characteristics and the porosity. It immediately becomes apparent that a larger area of a low porosity medium will be needed to perform the same duty as one with a high porosity. Hence the price differential will be narrowed accordingly. Similarly, differences will occur in the filtration rate when considered per unit area under a defined pressure drop and hence the filtration area will depend upon the medium. Cost of fabrication distorts the cost figures and an example may be considered where made-up costs are compared with the cost of the cheapest medium alone.

TABLE 8–6. FILTER SELECTION[10]

Slurry characteristics	Fast filtering	Medium filtering	Slow filtering	Dilute	Very dilute
Cake formation rate	in./s [cm/s]	in./min [mm/s]	0.05–0.25 in./min [0.02–0.12 mm/s]	0.05 in./min [0.02 mm/s]	no cake
Normal concentration	20%	10–20%	1–10%	5%	0.10%
Settling rate	Very rapid	fast	slow	slow	—
Leaf test rate (lb/hr ft²) [kg/s m²]	500 0.7	50–500 0.07–0.7	5–50 0.007–0.07	5 0.007	— —
Filtrate rate (g.p.m./ft²) [m³/s m²]	5 3.4×10^{-3}	0.2–5 $(0.13\text{–}3.4) \times 10^{-3}$	0.01–0.02 $(0.007\text{–}0.013) \times 10^{-3}$	0.01–2 $(0.007\text{–}1.3) \times 10^{-3}$	0.01–2 $(0.007\text{–}1.3) \times 10^{-3}$

FILTER APPLICATION

Continuous vacuum filters

Multicomponent drum
Single compartment drum
Scroll discharge
Belt
Disc
Precoat
Continuous pressure precoat
Batch vacuum leaf
Batch nutsche

Batch pressure filters

Plate and frame
Vertical leaf
Tubular
Horizontal Plate
Cartridge edge

TABLE 8-7. FILTER MEDIA COST DATA[18]

Type	Example	Minimum particle size retained (μ)	Porosity (% free area)	Cost (£/ft²)	Cost [£/m²]
Rigid porous media	Ceramics	1	30–50		
	Sintered metal	3		4–40	42–420
Metal sheets	Perforated	100	20		
	Woven wire	5	15–35	2–10	20–100
Porous plastics	Pads, sheets	3	30–50	0.15	1.6
	Membranes	0.005		1.25	13.5
*Woven fabrics	Natural fibres			0.03	0.3
	Synthetic fibres	10	50–60	0.06–0.10	0.65–1.1
Cartridges	Yarn wound	2	50–60	0.25–1.90	2.7–1
	Spools, fibres				
Non-woven sheets	Felts	10	60–95	0.10	1.1
	Paper cellulose	5		0.003–0.01	0.03–0.11
	Glass	2		0.02–0.03	0.2–0.3
	Sheets and mats	0.5		0.02–0.05	0.2–0.5
Loose solids	Fibres–asbestos	Sub-micron	80	0.005	0.05
	Powders		80–90	0.01	–0.11

* Further details of filter fabric characteristics for 11 natural and synthetic fibres may be found in reference 11.

Example 8–2[18]

5 gal/min [3.8 × 10^{-4} m³/s] of a medium viscosity oil is to be filtered in a cartridge filter assembly. The choice of materials lies between impregnated paper, felt, woven wire and sintered bronze. The costs of the media on an area basis are shown in Table 8–8, together with the costs of fabrication. How would the choice of medium be made?

TABLE 8–8. COMPARISON OF COSTS OF ELEMENT TO FILTER OIL[18]

| | Type of medium in elements | | | |
	Impregnated paper	Felt	Woven wire	Sintered bronze
Cost of medium (£/ft²)	0.0125	0.10	3.875	4.20
[£/m²]	0.135	1.08	41.7	45.2
Ratio of cost to paper	1	8	309	336
Cost of made-up element (£)	0.875	1.50	32.0	12.5
Ratio to paper element	1	1.7	36.5	14.3

The wide difference in the original costs has narrowed considerably in the course of fabrication, though paper remains the cheapest medium. However, the life of the paper filter is relatively short and it will have to be replaced, while the woven wire and the sintered bronze can be reused many times. In order to make the correct selection, it is necessary to know the life of each element and to compare the time for cleaning the expensive materials with the downtime for replacing the paper element. In this way the optimum medium may be selected.

An example where these factors are included is given below.[18]

Example 8–3

A choice is to be made between a natural fibre and a fibreglass cartridge filter for the filtration of paint. The latter is more expensive initially but lasts three times as long as the natural fibre cartridge. The costs are included in Table 8–9, which shows how the fibreglass cartridge can show a very substantial saving over its cheaper rival in this particular example. This is a particular illustration of the economic considerations and the result is not a general trend.

Filter cakes which are normally too thin to be discharged continuously from conventional filters may be removed from a continuous precoat filter. This is simply a rotary drum filter which can be operated either as an open or vapour-tight vacuum or as a pressure filter. A 2–6 in. [50–150 mm] thick layer of precoat is put on to the filtering surface before the process starts. During the filtration cycle an advancing knife edge shaves off a thin layer of the precoat together with the deposited solids. It is useful, therefore, in applications where the cake would ordinarily be too thin to discharge or where the cake formed is impervious. In the latter case it is removed as it is formed.

TABLE 8–9. SELECTION OF FILTER CARTRIDGES[18]

	Wound natural fibre	Fibreglass cartridge
Volume of paint/annum (gal)	5×10^6	5×10^6
[m³]	18 900	18 900
Volume filtered/cartridge (gal)	111	333
[m³]	0.42	1.26
Number of cartridges/annum	45 000	15 000
Cost per cartridge	£0.25	£0.46
Annual cost of cartridges	£11 250	£6900
Annual saving with fibreglass	£4350 = 38.7%	
Labour cost per filter change	£0.56	£0.56
Number of filter changes	5000	1666
Annual cost of filter changes	£2800	£934
Annual saving in labour cost	£1866	
Total annual savings with fibreglass cartridges	£6216 = 44.2%	

The precoat, which is most commonly diatomaceous earth, has to be replaced at intervals, thus making the process not strictly continuous. The cycle may be typically several days and depends upon the depth of precoat cut and the rotation speed. Maintenance costs are generally low and the product will be contaminated with the precoat material.

Filter aids act as both a precoat and as a 'body mix'. Examples include diatomaceous earth and asbestos or cellulose fibres. They are added to the system as a powder which forms a filter aid in the filter. When added as a 'body mix' they are deposited with the solids and form a porous structure that holds the solids but permits liquid to flow through the cake. They are especially useful where the solids are slimy, but it must be remembered that there is an optimum amount to add to the system—too little will allow the contaminants to plug the filter and bed; too much will increase pressure drop and shorten the filtration cycle.

8.4.3 Centrifuges

Introduction

The principles of gravity separation are well known and if, for example, a mixture of oil, sand and water is allowed to settle in a vessel, the familiar situation arises with the sand layer eventually at the bottom of the tank, oil at the top, and the water layer between the two.

If the vessel were to be rotated rapidly about its axis, centrifugal force would be developed at right angles to the axis of rotation. In an analogous manner to the above example, sand would be forced to the wall of the tank, water would form the middle layer and oil the inner layer. The density of each component determines its ultimate position, and the substitution of centrifugal force for gravity allows the very rapid separation of the mixture.

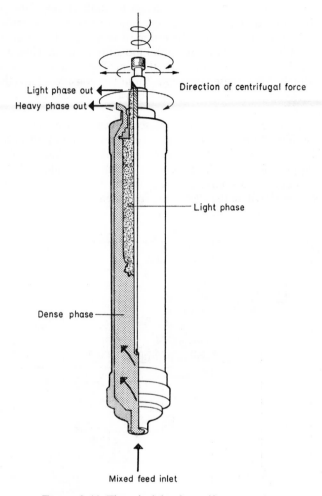

Light phase out

Heavy phase out

Direction of centrifugal force

Light phase

Dense phase

Mixed feed inlet

FIGURE 8–11. The principle of centrifuge operation

The feed of the mixture to the vessel may be either batch or continuous, and Figure 8–11 illustrates the principle of the latter mode of operation. To complete the separation, a suitable means of product discharge has to be devised, and this part of the design leads to the large number of centrifuge types which are available on the market. The other main factors influencing design are the feed and product specifications.

A centrifuge is basically a device for separating phases which may be either two immiscible liquids, with or without the presence of an insoluble solid, or a liquid and an insoluble solid. In addition, the centrifuge may be used for the breaking of emulsions and for the classification of solids by size or density. Equipment selection for centrifuging is frequently dependent upon the operations which precede and follow it. Holley[20] has quoted the example of the three consecutive steps of crystallization, centrifuging and drying. To economize on crystallizer costs, the product size may be reduced. This may alter the optimum type of centrifuge and will certainly alter its capacity. In order to minimize drying costs, it is common practice to specify a low initial moisture content. This, however, will increase the separation problem in the centrifuge and will increase costs correspondingly. Thus these operations should be considered together, and not each in isolation, as only in this way can the most economical process selection be effected.

When considering separation processes it was shown that centrifuges may be considered under two classes—sedimentation and filtration—and these will now be considered in turn, remembering that the former is generally applicable when the product is a clarified liquid and the latter type is used to produce a pure, dry solid. Figure 8–12 shows a classification by particle size[21] where the classes are indicated.

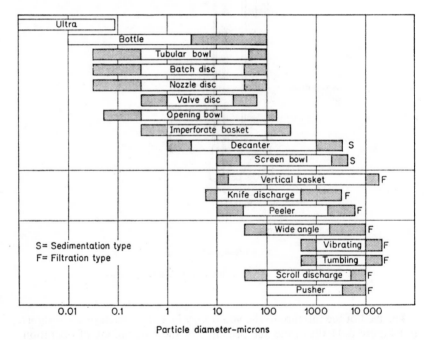

FIGURE 8–12. Centrifuge classification by particle size[21]

Sedimentation centrifuges

There are four main types of sedimentation centrifuge:

    (i) the tubular bowl,
    (ii) the disc,
    (iii) the solid bowl scroll discharge,
    (iv) the solid bowl batch basket.

Concise descriptions of these types have been given by Morris[22] and will not be repeated here. Unless flow rates are extremely small, the tubular bowl centrifuge is rarely used for solids concentrations greater than 1 per cent. For concentrations between 1 and 15 per cent any of the three remaining types may be used, and for concentrations exceeding 15 per cent either the solid bowl scroll discharge or the batch basket type is employed, depending whether or not the operation is continuous or batch.

Where a choice is to be made between the different types, use is frequently made of the *sigma* (Σ) *theory*. This theory, with its uses and limitations has been discussed by Ambler[23] and Trowbridge,[24] the overflow rate (i.e. feed rate minus the solids discharge rate) and its ratio to sigma are known, the most suitable machine may be selected, sized and costed. The theory is commonly worked in CGS units and this practice will be followed in this text.

The diameter of the smallest particle it is required to separate must be known together with the density difference between the solid and liquid phase and the liquid viscosity.

$$\Sigma = Q/2V_g \qquad (8\text{-}7)$$

where
$V_g = \Delta\rho d^2 g/18\mu$
Σ = sigma value (cm^2)
Q = overflow rate (cm^3/s),
V_g = settling velocity (cm/s),
$\Delta\rho$ = density difference (g/cm^3),
d = particle diameter (cm),
g = acceleration due to gravity (cm/s^2),
μ = liquid viscosity (centipoise).

Knowing the properties of the system, Q and Q/Σ may be calculated and compared with the values in Table 8–10 to enable a suitable selection to be made. The value of sigma as calculated should be divided by the relevant value of the efficiency as quoted in Table 8–10 to obtain a true value of Σ.

TABLE 8–10. SELECTION OF SEDIMENTATION CENTRIFUGES[19]

Centrifuge type	Efficiency (%)	Normal operating range Q (cm^3/s), Q/Σ (cm/s)
Tubular bowl	90	$Q = 115$ at Q/Σ $= 5 \times 10^{-6}$ to $Q = 1050$ at Q/Σ $= 3.5 \times 10^{-5}$
Disc	45	$Q = 26.5$ at Q/Σ $= 7 \times 10^{-6}$ to $Q = 3$ at Q/Σ $= 4.5 \times 10^{-5}$
Solid bowl scroll discharge	60	$Q = 230$ at Q/Σ $= 1.5 \times 10^{-4}$ to $Q = 38\,000$ at Q/Σ $= 1.5 \times 10^{-3}$
Solid bowl basket	75	$Q = 105$ at Q/Σ $= 5 \times 10^{-4}$ to $Q = 30\,000$ at Q/Σ $= 1.5 \times 10^{-2}$

Its typical significance is that its value represents the area (cm^2) of a gravity settling tank which is capable of the same separating ability for continuous flow operation as the centrifuge. It is also entirely a function of the physical characteristics of the centrifuge[22] so that the sigma value may be quoted to a manufacturer who can then quote for the appropriate size equipment.

It is clearly important that the value of Σ is as accurate as possible and it may be necessary to run a pilot-plant study and then follow the scale-up technique which has been presented by Trowbridge.[24] The sigma value may also be used to obtain an average cost on a 1970 basis for solid bowl scroll discharge centrifuges.[25] If X is the separation capacity ($\Sigma \times 10^{-5}$ cm^2 units) and $50 \leqslant X \leqslant 600$, then the installed cost with stainless steel contact parts is

$$\text{cost (£)} = 675X^{0.57} \tag{8-8}$$

or, capable of pressures to 5 p.s.i.g. [136 kN/m^2],

$$\text{cost (£)} = 775X^{0.57} \tag{8-9}$$

or, capable of pressures up to 15 p.s.i.g. [207 kN/m^2],

$$\text{cost (£)} = 875X^{0.57} \tag{8-10}$$

Use of these formulae and of the sigma value will now be illustrated by means of an example.

Example 8-4

The output from a crystallizer is a 10 per cent by volume slurry of crystals which have a density of 2.2 g/cm^3. The acceptable crystal loss from the centrifuge is such that it is necessary to retain a crystal size of 40 μ. Obtain an estimate of the cost of the most suitable type of centrifuge. Would it be worth improving the crystallizing process to narrow the product size distribution so that the acceptable centrifuge loss then corresponds to a crystal size of 80 μ?

$$\text{Volume of overflow} = 100 - (0.1 \times 100)$$
$$= 90 \text{ g.p.m.}$$
$$[= 6820 \text{ cm}^3/\text{s}]$$
$$\text{Density difference } \Delta\rho = (2.2{-}1.2)$$
$$= 1.0 \text{ g/cm}^3$$
$$\text{Viscosity, } \mu = 5.0 \text{ cP}$$
$$\text{Settling velocity } V_g = \rho d^2 g / 18 \mu$$
$$= 1.0(0.004)^2 \times 981/(18 \times 5.0)$$
$$= 1.75 \times 10^{-4} \text{ cm/s}$$

From equation (8–7),

$$Q/\Sigma = 2V_g$$
$$= 3.50 \times 10^{-4} \text{ cm/s}$$

From Table 8–10, using the values of Q and Q/Σ, the solid bowl scroll discharge centrifuge is suitable. Equation (8–8) may now be used to find the capital cost of the equipment.

$$\Sigma = Q/(\eta \times 2V_g) \qquad (8\text{–}11)$$

where η = efficiency of centrifuge = 0.60. Hence

$$\Sigma = 6820/(0.60 \times 2 \times 1.75 \times 10^{-4})$$

$$= 325 \times 10^5 \, cm^2$$

$$\text{Installed cost} = 675 X^{0.57} \qquad (8\text{–}8)$$

$$X = 325$$

Hence the total installed cost is £18 200.

If now the size of retained particle is changed to 80 μ, the following analysis results.

$$V_g = 7.0 \times 10^{-4} \, cm/s$$

$$Q/\Sigma = 1.4 \times 10^{-3} \, cm/s$$

Hence the type of centrifuge remains unaltered.

$$\Sigma = 81 \times 10^5$$

$$X = 81$$

$$\text{Installed cost} = 675 X^{0.57} \qquad (8\text{–}8)$$

$$= £8240$$

The installed cost of the centrifuge has been reduced by 54 per cent and this saving must be compared with the increased cost of improving the crystallization process.

Centrifugal Filtration

In centrifugal filters, as opposed to centrifugal sedimentors which have just been discussed, use is made of centrifugal force to increase the passage of liquid through the pores in the solids as they are held against the rotating bowl wall. Selection is more of an art when compared with the sigma method for sedimentation centrifuges, and pilot plant or other experience is usually necessary to make certain of correct selection.

In general the centrifugal filter is applicable to a coarser solids size range than the sedimentation type as has been indicated in Figure 8–12. The concentration of the feed slurry is of prime importance since the mechanism of separation is by liquid flow through a bed of solids and this liquid flow is the rate controlling step. It can be shown[26] that if the slurry is preconcentrated from 10 to 50 per cent by weight, the load time is reduced by a factor of ten. The capacity of the centrifuge, and therefore its cost, may then be reduced—not by the factor of ten, as the load time may represent only some 25 per cent of the cycle time—but by approximately 13 per cent. Figure 8–13[25] may be used to estimate the cost of manually unloaded and fully automatic cycling perforate basket centrifuges.

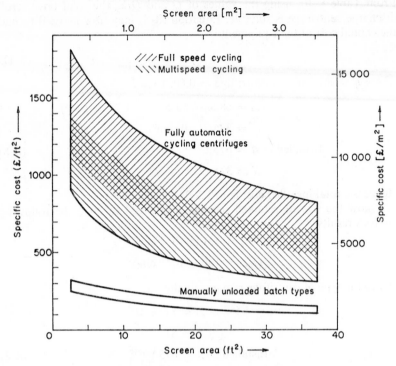

FIGURE 8–13. Centrifuge cost data[25]

Classification of centrifugal filters may be conveniently grouped into two main classes. In fixed-bed filtration, the centrifuge cake remains on the wall of the bowl until it is removed, either manually or automatically, using a knife mechanism. In moving-bed filtration, the solids forming on the bowl wall move along the bowl through washing and drying zones under the action of a scroll-, push-, or vibration-mechanism or by choice of bowl angle. The two classes may be combined in the form of a family tree of filters as shown in Figure 8–14.[26] To avoid description of the many variations, Figure 8–15 should be self explanatory.

The basket type of centrifuge ((a)–(c)) is the simplest and cheapest, and forms the basic design from which the many variants have arisen. The more automatic models are more desirable for continuous production but the batch types are invaluable in small-scale production and for pilot-plant work. The basic design consists of a perforated basket lined with a filtering medium. The bowl may rotate on a vertical, horizontal or inclined axis and the feed may come from either direct discharge from a pipe or by means of an accelerating cone which moves axially in the basket in an attempt to hold up a uniform filter cake. Solids discharge may be achieved by a variety of means—by bowl removal, manual shovelling, scrapers and ploughs.

Discharge by means of a knife mechanism is a feature of types shown in (d)–(h) in the figure, and of these types (d) and (e) are capable of multispeed, automatic operation at a proportionally higher cost. They are

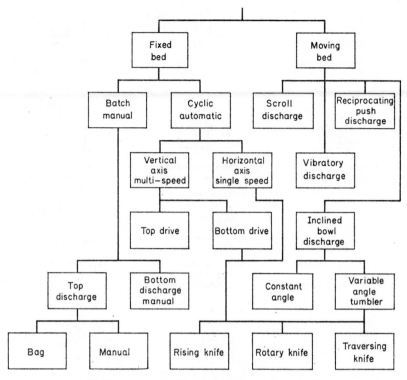

FIGURE 8–14. Classification of centrifugal filters[26]

particularly useful when dealing with fragile materials or needle-shaped crystals, when it is desirable to feed and remove at low speed to ensure efficient removal of mother liquid and to minimize crystal breakage.

The single, high speed centrifugal filter ((f), (g) and (h)) is useful where cakes are thin and short cycle times are required. Operation can easily be carried out under pressure and at extremes of temperature.

The remaining centrifuges shown in Figure 8–15 may be rated as continuous in that the solid is discharged continuously. In (i), (j) and (k), discharge of solids is achieved by virtue of the inclined wall, by vibrating the wall, and by a complex variation of the wall angle respectively. These three types give a variable degree of dryness from a concentrated slurry with a preferably viscous mother liquor. Under these conditions the centrifuges provide low cost separation up to capacities of 150 ton/hr [4.2 kg/s] with a specific energy consumption of 0.1–0.2 kWh/ton of product [0.0035–0.007 MJ/kg].

The scroll discharge centrifuge[20] provides low-cost, flexible operation over a wide range of applications. Although it is not suitable for handling fragile crystals owing to breakage, it is used for medium and coarse crystalline solids where some fines loss can be tolerated. Costs of this type of machine are shown in Table 8–11.

(a) Bottom drive batch
    basket with bag

(b) Top drive bottom
    discharge batch
    basket

(c) Bottom drive bottom
    discharge batch
    basket

(d) Bottom drive automatic
    basket, rising knife

(e) Bottom drive automatic
    basket, rotary knife

(f) Single–speed
    automatic rising
    knife

(g) Single-speed
    automatic
    rotary knife

(h) Single–speed
    automatic
    traversing knife

(i) Inclined wall
    self discharge

(j) Inclined vibrating
    wall self discharge

(k) Inclined 'tumbling'
    wall self discharge

(l) Inclined wall
    scroll discharge

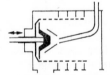

(m) Traditional single–
    stage pusher

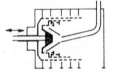

(n) Traditional
    multi–stage
    pusher

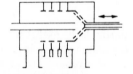

(o) Conical pusher with
    de–watering cone

FIGURE 8–15. Schematic sections of different types of centrifugal filters[26]

TABLE 8–11. COST DATA FOR CENTRIFUGAL FILTERS

Total Installed Cost of Perforate Basket Scroll Discharge Centrifuges

Total installed cost $= CX^n$

$X = 1000$ lb/hr dry solids

$[= $ kg/s dry solids$]$

Average particle size	Exponent n	Wt. % (dry basis) solids					
		15%		35%		80%	
		Values of Constant C					
		(A.A.)	[SI]	(A.A.)	[SI]	(A.A.)	[SI]
100 μ	0.26	7450	12 800	5580	9550	4840	8280
150 μ	0.31	5700	10 800	4460	8470	3960	7520
250 μ	0.32	5200	10 100	4340	8420	3350	6500

Example 8–5

Estimate the cost of a perforate basket scroll discharge centrifuge to handle 10 000 lb/hr [1.26 kg/s] of material on a dry solids basis. The slurry contains 35 per cent solids and the average particle size is 150 μ.
In this case,

$$X = 10 = [1.26]$$

Hence from Table (8–11), total cost (£) = $4460 \, (10)^{0.31}$

$$[= 8470 \, (1.26)^{0.31}]$$

$$= £9100$$

The remaining machines ((m), (n) and (o)) use the pusher principle of removing the solid cake. The field of application of these centrifuges is similar to the scroll discharge type, but they are capable of better separation of mother and wash liquors and the fines loss is reduced. They are more expensive than the scroll discharge type and cost data is presented in Table 8–12.

TABLE 8–12. TOTAL INSTALLATION COST OF PERFORATE BASKET PUSHER DISCHARGE CENTRIFUGES

Total installed cost = $CX^{0.41}$

$X = 1000$ lb/hr dry solids

[= kg/s dry solids]

Wt. % (dry basis) solids	Values of Constant, C			
	85% solids > 150 μ		70% solids > 150 μ	
	(A.A.)	[SI]	(A.A.)	[SI]
10	6750	15 800	11 300	26 500
20	5300	12 400	8850	20 700
30	4600	10 800	7700	18 000
40	4250	9950	7100	16 600
50	4130	9650	6900	16 100
60	4000	9350	6680	15 600
70	3890	9100	6500	15 200

Example 8–6

Estimate the cost of a perforated basket pusher discharge centrifuge to handle a feed containing 40 per cent solids with 85 per cent > 150 μ. The throughput on a dry solids basis is 10 000 lb/hr [1.26 kg/s]. Compare this cost with a 40 per cent solids feed with 70 per cent > 150 μ.
In the first case,

$$X = 10 = [1.26]$$

$$\text{cost (£)} = 4250 \, (10)^{0.41}$$

$$[= 9950 \, (1.26)^{0.41}]$$

$$= £10 \, 900$$

In the second case, the cost is given by

$$(\pounds) = 7100 \, (10)^{0.41}$$
$$[= 16\,600 \, (1.26)^{0.41}]$$
$$= \pounds 18\,200$$

Tables 8–11 and 8–12 both show that costs for these last types of machine are particularly dependent upon particle size and slurry concentration. As discussed earlier with regard to centrifugal sedimentors, consideration should be given to the possibility of increasing particle size by alteration of crystallizer conditions or to the installation of some means of preconcentration of the slurry. Since all costing data is subject to limitations and accuracy, the equations presented should give an idea of the order of costs involved but may be used with certainty for comparison purposes when considering alteration to process conditions.

8.4.4 Drying and Dryers

Drying of solid materials is frequently the last operation in a manufacturing process and is employed for a number of reasons. Handling may be facilitated and transport costs reduced, free flowing products may be required or corrosion problems minimized by removal of moisture.

A wide range of drying equipment is available and a choice must be made to provide the most economic solution for any particular application. The material to be dried will have a great influence on the final choice and a classification according to physical state is shown in Table 8–13.

TABLE 8–13. CLASSIFICATION OF DRIED MATERIALS

Type of Product	Examples
Loose bulk solids	Granules, chips, pellets, crystals, powders
Cohesive products	Centrifuges and filter cakes, thick pastes and slurries
Liquids	Chemical solutions, thin pastes and slurries
Gases	To remove moisture before reaction or to minimize corrosion, e.g. benzene and chlorine respectively

Unfortunately, classification of product in this way does not provide a direct access to the optimum type of dryer. It is necessary to consider the drying characteristics of the particular substance, remembering that moisture to be removed may include not only free surface moisture but internal moisture, water of crystallization or any combination of these. The theory of drying is well treated elsewhere[14] and will not be considered here, but consideration of the combination of theory and experience may enable the drying problem to be considered in terms of residence time requirements.[27]

The removal of surface moisture alone is the simplest drying problem and, providing the product and the drying gases are well distributed, only short to medium residence times are required. The diffusion mechanism of internal moisture removal means that residence times must be longer,

though particle size plays an important part. Removal of water of crystallization is a problem which can only be resolved by determination of the drying
characteristics of each application.

Figure 8–16 shows typical drying curves for two different types of
material.[14]

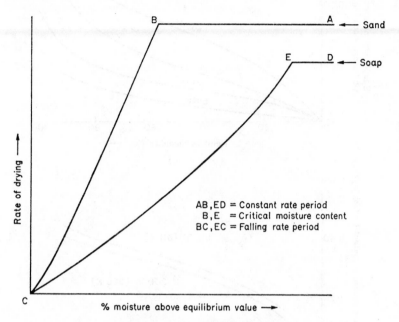

FIGURE 8–16. Typical drying curves[14]

The equilibrium moisture content of any substance will depend upon
the temperature and humidity of the surroundings and will vary according to
the material. For example, non-porous, insoluble solids such as sand will
have a very low equilibrium moisture content, while organic materials such
as wood, leather and textiles may have equilibrium values as high as 20
per cent at 60 per cent relative humidity at 20°C as shown in Figure 8–17,
where the typical dependence of the equilibrium moisture content with
temperature is also indicated. Above the equilibrium value the rate of drying
will assume the form shown above depending upon the material.

Reference to the drying curves shows that for the drying of materials
in slab or filter cake form, the rate controlling factor is the rate of diffusion
of moisture from the wet mass to the surface. To improve the drying process,
therefore, consideration should be given to preforming the feed to increase
the effective surface area for heat and mass transfer. This technique may be
extended in some cases to total dispersion drying in (say) a fluidized bed
dryer where each particle can be brought into contact with the drying
medium. Details of preforming equipment for normal applications are
available elsewhere[28] but the simplest means of achieving the increased
surface area is by the use of shredding machines. Equipment becomes more
complex with granulators and extruders for use with wet filter cakes and
clays.

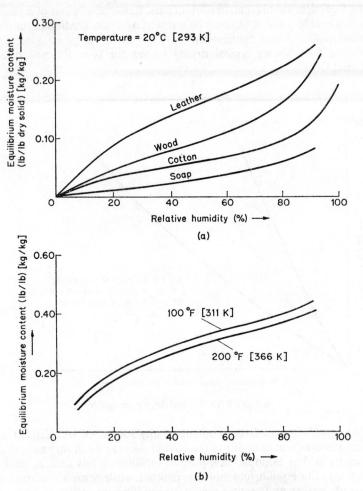

FIGURE 8–17. Equilibrium moisture content and its variation (a) with
temperature and (b) with relative humidity

Selection of suitable drying equipment may be facilitated by knowledge
of previous drying experience of the material. This information may be
found either from published literature or from equipment manufacturers.
In the absence of such detail, a selection procedure must be adopted whereby
dryers suitable for the duty and material are chosen and sized. From suitable cost data, dryers with the highest estimated total capital and operating
costs may be eliminated. Laboratory and pilot plant tests, and performance
and quotations from manufacturers can finally enable an economic selection to be made.

To limit the possible choice, selection between batch and continuous
operation should be made at an early stage. In general a continuous process
is to be preferred, since for a given type and capacity it will require less
labour, fuel and floor space and will discharge a more uniform product

than the corresponding batch dryer.[29] Continuous operation is the automatic choice if the production rate exceeds 100 000 lb/day [1 kg/s] and batch operation if the rate is less than 10 000 lb/day [0.1 kg/s]. Batch drying is also used where cross-contamination must be avoided, such as in the production of pharmaceuticals, dyestuffs and fine chemicals. Efficiency of batch drying increases with the load and decreases with drying time, and the normal efficiency range is 50–60 per cent. The higher efficiencies obtainable in continuous drying arise from the elimination of heat losses at the start and finish of the drying cycle.

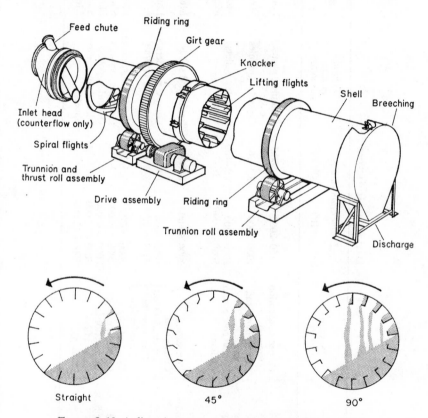

FIGURE 8–18. A direct heat rotary dryer showing the effect of flight geometry on the motion of the solid in the gas stream[29]

The most important types of dryer are included in Table 8–14, which shows a classification according to mode of operation and allows preliminary selection on the basis of residence time. Knowledge of the heat load allows a first estimate of size to be made and use of the cost data in Table 8–15 enables an approximate capital cost to be evaluated.

Details of the construction of the many types of dryer are available elsewhere[14],[29] and the following pages will be concerned with design methods and description which are not so readily available for selected pieces of equipment.

TABLE 8-14. DRYER CHARACTERISTICS[(28),(43)]

Type	System	Evaporation rate (BTU/hr ft²)	[kW/m²]	Typical product suitability	Residence time
Batch dryers					
Ovens	Forced convection vacuum	150–2000	0.5–6.3	Granules, pastes, cake	Long
Agitated pans	Atmospheric and vacuum	1000–5000	3.1–16	Crystals, granules	Medium-long
Fluidized bed	Forced convection	2000–5000	6.3–160	Granules, pellets	Medium-long
Infra-red	Radiant	5000–12 000	16–38	Sheets	Medium
Continuous dryers					
Conveyor band	Convection	2000–10 000	6.3–32	Preformed solids	Medium-long
Rotary	Convection, Direct, indirect, Direct/indirect conduction	1000–6000 BTU/hr ft³	[10–60 kW/m³]	Lumps, powders, pellets, crystals	Long
Film drum	Conduction	3000–6000	9.5–19	Liquids, suspensions	Medium
Fluidized bed	Convection	2000–50 000	6.3–160	Granules, crystals	Short-medium
Pneumatic or flash	Convection	50 000–250 000	160–800	Granules, pellets	Short
Spray	Convection	7000– 33 000	22–105	Fluids, suspensions	Short
Infra-red	Radiant	5000– 12 000	16–38	Sheets	Medium

TABLE 8-15. DRYER COSTS

Type (All in mild steel)	X (A.A.)	X [SI]	Limits of X (A.A)	Limits of X [SI]	Cost equation Y	Cost equation n	$Y = CX^n$ C (A.A.)	$Y = CX^n$ C [SI]	Reference
Tray dryer	ft²	m²	50–1500	5–140	£	0.46	194	576	44
Tunnel dryer	ft²	m²	100–1000	9–90	£	0.83	213	1510	44
*Rotary direct	ft²	m²	300–3000	29–290	£	0.87	54.5	427	46
*Rotary steam tube	ft²	m²	300–5000	29–470	£	0.76	27.1	164	47
Rotary vacuum indirect horizontal	ft²	m²	45–350	4–32	£	0.54	2400	8600	45
Drum dryer—atmospheric	ft²	m²	20–500	2–47	£	0.54	588	2110	46
—vacuum	ft²	m²	20–500	2–47	£	0.54	712	2550	46
Fluidized bed dryer	Bed diameter								
Fluidizing velocity									
1 ft/s [0.3 m/s]	ft	m	3–15	1–45	10³$	0.72	8.8	20.7	35
3 ft/s [0.9 m/s]	ft	m	3–15	1–4.5	10³$	0.72	11.2	26.3	36
5 ft/s [1.5 m/s]	ft	m	3–15	1–4.5	10³$	0.72	15.3	36.0	35
Pneumatic dryer	Evaporative capacity lb/hr	kg/s	1000–20 000	0.12–2.5	$	0.46	2200	137 000	29
					Cost equation				
Spray dryers	Cross section								
Continuous fluidized bed	ft²	m²	40–600	4–56	£ = 118(X + 156) = [1280(X + 13.3)]				43
	ft²	m²	10–120	1–11	£ = 253(X + 14) = [2730(X + 1.3)]				43
Stainless steel pneumatic dryer	ft²	m²	0.4–6	0.04–0.6	£ = 2170(X+1.5) = [23 400(X+0.14)]				43

* for stainless steel multiply mild steel cost by 1.9 and 3.8 respectively.

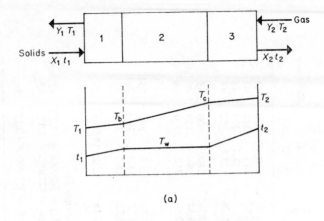

(a)

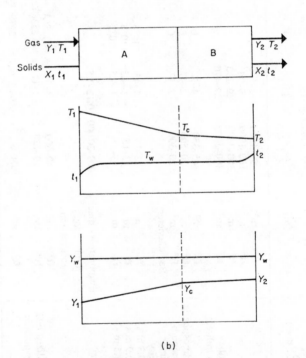

(b)

FIGURE 8–19. (a) Temperature profiles in an ideal countercurrent dryer and (b) Temperature and humidity profiles in an ideal parallel flowdryer

Rotary dryers are the most extensively used items of drying equipment and are available as direct, direct/indirect, and indirected heated types depending upon whether the heating medium is permitted, partially permitted or prevented from making contact with the product. A direct heat rotary dryer is a horizontal cylinder through which a hot gas flows. The cylinder rotates and is usually slightly induced from the horizontal so that the retention time can be varied by adjustment of the angle to give residence times normally between five minutes and two hours [300–7200 s]. From 5 to 15 per cent of the volume of the shell is occupied by the drying solid, and the shell may be fitted with louvred flights to increase the rate of drying and the capacity.[30] Figure 8–18 shows a direct heat rotary dryer and the effect of flight geometry on the passage of the material to be dried.

Flow may be either co-current or countercurrent, the former being used for heat sensitive materials where the hot inlet gas makes contact with the wet feed material. The temperature of the material remains at the wet bulb temperature for as long as moisture is being removed, by which time the dry bulb temperature will have fallen to a point at which it has no harmful effect on the product. Typical temperature profiles are shown in Figure 8–19 for both modes of operation.

The length and diameter of rotary dryers may be estimated from the following procedures.

(i) Countercurrent Flow

Referring to Figure 8–19, the following parameters will either be known or may be assumed:
For the solids,

L = feed rate (lb/dry solid hr) [kg/s],

$X_1 X_2$ = inlet and outlet moisture contents (lb H_2O/lb) [kg/kg],

t_1, t_2 = inlet and outlet temperature (°F) [K],

For the gas,

G = flow rate of dry gas (lb/hr) [kg/s],

G_s = flow rate of dry gas (lb/hr ft²) [kg/s m²],

Y_2 = inlet humidity of gas (lb H_2O/lb gas) [kg/kg],

T_2 = inlet gas temperature (°F) [K].

G_s is chosen to be the highest mass velocity which will not cause excessive dusting or loss of solids by entrainment. The normal range is 200–10 000 lb/hr ft² [0.27–13.5 kg/s m²]. The diameter may then be calculated from

$$\text{diameter } D = (4G/\pi G_s)^{0.5} \qquad (8\text{–}12)$$

(The diameter is normally between 1 and 10 ft [0.3–3.0 m].)

The length, which is normally within the range 4–10D, may be found as follows:

(a) Find Y_1, the exit gas humidity, from an overall moisture balance,

$$Y_1 = Y_2 + (L/G)(X_1 - X_2) \qquad (8\text{–}13)$$

(b) Find the enthalpy of the exit gas H_1 from an overall heat balance,

$$H_1 = H_2 + (L/G)(h_1 - h_2) \qquad (8\text{–}14)$$

where h_1, h_2 = enthalpy of feed at inlet and exit and

$$h = C_d t + (t - 32)X + \Delta H_a \qquad (8\text{–}15)$$
$$[= C_d t + 4.18 \times (t - 273)X + \Delta H_a] \qquad (8\text{–}15a)$$

datum = 0°F [255 K] for dry air and solid; 32°F [273K] for water where C_d = heat capacity of dry solid (BTU/lb °F)[kJ/kg K],

   t = temperature (°F) [K],

   ΔH_a = heat of wetting or absorption (BTU/lb dry solid) [kJ/kg].

(c) Find the exit gas temperature T_1 using either a psychrometric chart (Figure 8–20) or, for air, equations (8–16) and (8–17).

$$H = 0.24T + (1060.8 + 0.45T)Y \qquad (8\text{--}16)$$

$$[= 1.0T + (2460 + 1.88T)Y] \qquad (8\text{--}16a)$$

$$T_1 = (H_1 - 1060.8\,Y_1)/(0.24 + 0.45\,Y_1) \qquad (8\text{--}17)$$

$$[= (H_1 - 2460\,Y_1)/(1.0 + 1.88\,Y_1)] \qquad (8\text{--}17a)$$

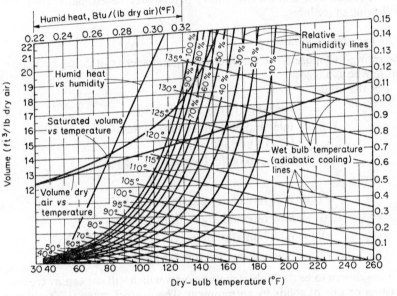

FIGURE 8–20. A psychrometric chart for the determination of the properties of air and water vapour (British Units). For SI use the following factors:

Volume (ft³/lb) × 0.0624 = [m³/kg]

Humid heat (BTU/lb °F) × 4.186 = [kJ/kg K]

Humidity (lb/lb) × 1.0 = [kg/kg]

Temperature (5/9) (°F − 32) + 273 = [K]

(d) Find T_c and T_w by a heat balance over section 3 of Figure 8–19

$$T_c = T_2 - \{(C_d L)/(C_s G)\}(t_2 - T_w) \qquad (8\text{--}18)$$

where C_d = heat capacity of dry solid (BTU/lb °F) [kJ/kg K],

   C_s = humid heat of moist gas (BTU/lb °F) [kJ/kg K],

   T_w = wet bulb air temperature (°F) [K].

Assume a value of T_w and solve the equation for T_c. Use T_c and Y_2 to find T_w on the psychrometric chart and compare with the assumed value of T_w. Repeat until agreement is reached.

(e) Find T_b by a heat balance over section 1 of Figure 8–19,

$$T_b = T_1 + (L/G)\{(C_d + X_1 C_w)/C_s\}(T_w - t_1) \qquad (8\text{–}19)$$

where C_w is the heat capacity of water.

(f) Find the number of transfer units (N_t) for each section from

$$N_t = (T_{in} - T_{out})/\Delta T_{lm} \qquad (8\text{–}20)$$

where ΔT_{lm} is the log mean temperature difference.

(g) Check that the total number of transfer units (N_t) lies in the range 1.5–2.0 for most economical operation.[31] If N_t is outside this range, adjust the gas flow rate and repeat the procedure.

(h) Find the length of a transfer unit for each section from

$$L_t = 0.10 C_s D G_s^{0.84} \qquad (8\text{–}21)$$
$$[= 6.12 C_s D G_s^{0.84}] \qquad (8\text{–}21a)$$

where L_t = length of a transfer unit (ft) [m],
C_s = humid heat of moist gas (BTU/lb °F) [kJ/kg K],
D = estimated diameter (ft) [m],
G = gas flow rate (lb/hr ft^2) [kg/s m^2].

(i) Calculate total length of dryer from $L = N_t L_t$.

(ii) Co-current Flow

The mass transfer unit method will now be summarized to show how the length of the dryer is calculated for co-current flow of solid and gas using the nomenclature of Figure 8–19. The method can be adapted for counter-current flow as indicated:

(a) Establish overall heat and mass balances to determine the end conditions.

(b) Measure the critical moisture (X_c) content experimentally or find from literature.

(c) Find the humidity Y_c from a moisture balance around one section of the dryer.

(d) Find the wet-bulb temperature, T_w, and the corresponding humidity of air saturated at T_w, Y_w from a heat balance at one end in conjunction with a psychrometric chart.

(e) Calculate the number of transfer units for each section from equations (8–22) and (8–23)

$$(N_t)_A = \pm \ln\{(Y_w - Y_c)/(Y_w - Y_1)\} \qquad (8\text{–}22)$$

$$(N_t)_B = \left\{\frac{X_c - X_e}{(Y_w - Y_2)(G/L) \pm (X_2 - X_e)}\right\} \times \ln\left\{\frac{(X_c - X_e)(Y_w - Y_2)}{(X_2 - X_e)(Y_w - Y_c)}\right\} \qquad (8\text{–}23)$$

where X_e = equilibrium moisture content (lbH$_2$O/lb dry solid) [kg/kg]. The positive and negative signs are used for countercurrent and parallel flow conditions respectively.

(f) Calculate the length of a transfer unit for each section from

$$L_t = 48.5 C_s G_s^{0.20}/a \qquad (8\text{–}24)$$
$$[= 4.0 C_s G_s^{0.20}/a] \qquad (8\text{–}24a)$$

where a = effective area for mass transfer/unit volume of dryer (ft^2/ft^3) [m^2/m^3]

(g) Calculate total length from

$$\text{length} = (N_t L_t)_A + (N_t L_t)_B \qquad (8\text{-}25)$$

An experimental method for simulating continuous drying operations of almost any feed in any dryer has been described by Broughton and Mickley.[32] If geometric and hydrodynamic similarity are maintained in scale-up, the method should provide an accurate figure for the total drying time and hence the length of the continuous dryer.

Indirect rotary dryers are used in the following applications:[29]

(a) when higher thermal efficiency is required from the heat in the combustion gases;
(b) when the material to be dried cannot be exposed to the combustion gases;
(c) when steam heating is available at low cost;
(d) when passage of drying gas would cause excessive entrainment losses;
(e) when the liquid to be removed is a valuable solvent, other than water.

In most cases of indirect drying, the drying rate is controlled by the rate of heat supply, i.e.

$$q = UA_h \Delta T \qquad (8\text{-}26)$$

where q = rate of heat transfer (BTU/hr) [W],
U = overall heat transfer coefficient (BTU/hr ft^2 °F) [W/m^2 K],
A_h = surface area for heat transfer (ft^2) [m^2],
ΔT = temperature difference between the heating medium and material at the point where evaporation is taking place (°F) [K].

Values of U or ($U\Delta T$) are difficult to obtain but the values in Table 8–16 provide a guide to the sizing of the heat transfer area.

The situation in which hot moist solids resulting from a process and where the drying operation is to be carried out using cooler air has been considered recently by Oliver.[48]

Air suspended systems. Whilst the rotary dryers which have just been discussed have medium to long residence times, the air suspended systems of flash or pneumatic, spray and fluidized bed dryers have residence times measured in seconds. The combination of the large surface area of small particles and a high degree of agitation leads to a very high rate of heat transfer. For suitable feed applications this group of dryers offers a very attractive means of drying, and all three will be considered in turn.

The rate of drying in air suspended systems is directly proportional to

(a) (particle diameter)$^{1.5}$,
(b) temperature difference between the hot gas and the particles,
(c) (mass velocity of gas)$^{0.5}$,
(d) film coefficient for heat transfer for the drying gas,
(e) pressure.

TABLE 8–16. HEAT TRANSFER DATA FOR DRYERS

Application	Parameter	Value	Reference
Heated surface/fairly dry solid	U	1–2 BTU/hr ft² °F [5.7–11.5 W/m² K]	38
Unagitated dryers	U	1–5 BTU/hr ft² °F [5.7–28.5 W/m² K]	39
Moderate agitation	U	5–15 BTU/hr ft² °F [28.5–85 W/m² K]	39
High agitation	U	15–25 BTU/hr ft² °F [85–140 W/m² K]	39
Caked heating surface	U	1–5 BTU/hr ft² °F [5.7–28.5 W/m² K]	39
Light powdery materials	$(U \Delta T)$	300 BTU/hr ft² [950 W/m²]	31
Coarse granular materials	$(U \Delta T)$	2000 BTU/hr ft² [6300 W/m²]	31

The flash, or pneumatic conveyor, dryer in its simplest form is an air conveyor into which heated air is introduced. A diagram of a typical arrangement is shown in Figure 8–21.

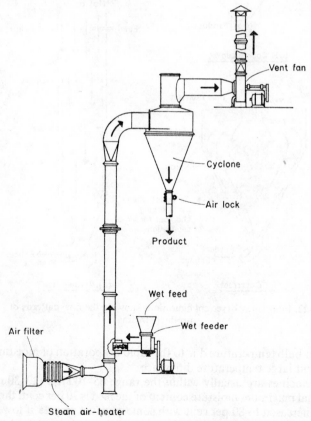

FIGURE 8–21. A typical pneumatic drying system[19]

The residence time normally lies in the range of 1–10 seconds and may often be less than 1 second. This short contact time makes this type of dryer ideal for the rapid drying of heat sensitive materials. High inlet gas temperatures are used, and because of the short residence time the material remains

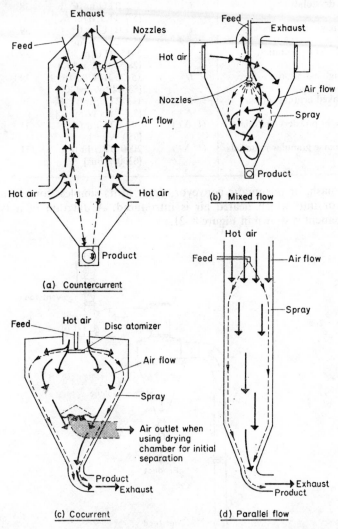

FIGURE 8–22. Four spray dryer configurations showing the flow patterns of the air and product streams

at the wet bulb temperature due to the rapid evaporation of moisture with a consequent large temperature difference.

Gas velocities are usually within the range 65–100 ft/s [20–30 m/s] and the normal maximum moisture content of the feed is 70 per cent, though this may be increased to 80 per cent with some product recycle if low product moisture contents are required. Large particles and pasty feed materials cannot be handled in pneumatic dryers in common with the other 'air-suspended' systems; in particular, abrasive materials cannot be dried and only surface moisture can be removed.

The spray dryer may be considered a modification of the flash dryer in which a liquid or paste feed is introduced as atomized particles of liquid into the gas stream. Spray drying is normally used where the initial moisture content exceeds 80 per cent or in which the solids are more or less dissolved.[29] When water is vaporized it expands about 1500 times, so that the large volume of vapour released from the liquids fed to a spray dryer delays the free fall of the drying particles to such an extent that a tower is unnecessary to achieve the required residence time. Figure 8–22 shows four typical spray dryer configurations. The feed liquid may be atomized to a spray by a rotating disc, a high pressure nozzle or a two-fluid nozzle. Atomization pressures normally lie in the range 300–4000 p.s.i.g. [2000–28 000 kN/m^2]; high pressure nozzles are particularly useful in counter-current spraying. Two-fluid atomization using air or steam between 60 and 100 p.s.i.g. [500–8000 kN/m^2] is used for low capacity operations and for the handling of viscous fluids. Centrifugal atomization ejects the feed from the periphery of a disc at a speed of 250–600 ft/s [75–185 m/s]. Rotational speed of the disc varies between 50 and 340 Hz and by suitable design of the wheel, abrasive materials may be handled without excessive wear.[33],[34]

Table 8–17 below may be used to select between spray, pneumatic and fluidized bed dryers, and the relevant cost correlations are presented in Table 8–15.

TABLE 8–17. SELECTION BETWEEN SPRAY, PNEUMATIC AND FLUIDIZED DRYERS

Particle property	Spray dryer	Pneumatic dryer	Fluidized dryer
Initial moisture greater than 80%	Yes	No	No
Too dry to pump	No	Yes	Yes
Wet enough to pump but moisture less less than 80%	Yes	Yes	No
Solids in dissolved state	Yes	No	No
Partially dry particles are sticky	Yes	No	No
Particles are fragile	Yes	No	Possible
Very small particles	Yes	Yes	No
Residence time (s)	3–10	1–10, often less than 1	Widely variable, greater than 10
Heat sensitive material	Yes	Yes	No
Relative drying speed	Third	First	Second

Fluidized bed drying. Fluidized bed drying may be carried out either as a batchwise or as a continuous process, the latter mode of operation being preferred. Characteristics of the fluidized state which make it useful for drying applications are uniform temperatures coupled with high rates of

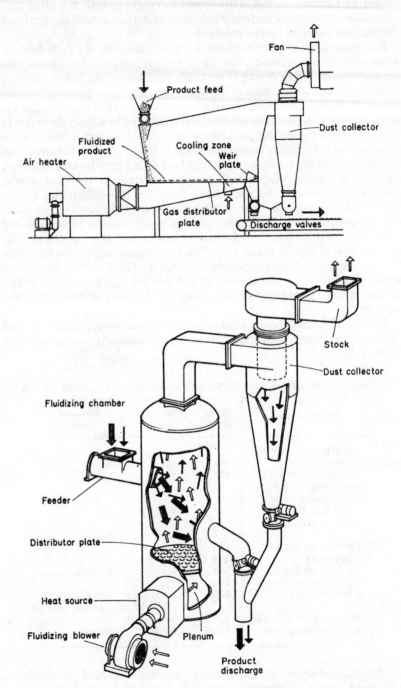

FIGURE 8–23. Two forms of fluidized bed dryers[27],[29]

heat and mass transfer, and although gas velocities may be high, solid particle velocities are low relative to each other and to the wall of the bed. Degradation of the product is lower than in drying applications which rely upon mechanical agitation or spraying of particles through gas streams. Fluidized bed dryers are available in many forms and two types are illustrated in Figure 8–23.

Difficult, sticky feeds cannot be handled in a fluidized bed and neither can a feed with a wide size range distribution. The drying medium may be drawn through the dryer by one fan, which has the advantage of producing

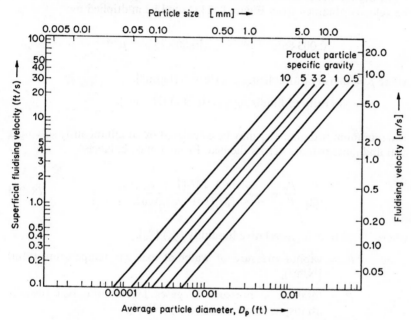

FIGURE 8–24. Superficial operating velocity in fluidized bed dryers[35]

a slight vacuum in the drying chamber so that any leaks are in an inward direction. Fluidized bed dryers are available in sizes from under 1 ft [0.3 m] diameter for laboratory or small-scale applications to over 15 ft [4.5 m] diameter for large industrial users. Normal particle size ranges are from 0.0015 to 0.30 in. [0.04–7.5 mm] diameter, though particles up to 1.5 in. [40 mm] have been handled.[35] Very small particles are better handled in a flash or pneumatic dryer to avoid the use of very large bed diameters necessary to accommodate the low fluidizing velocity.

Description of the fluidized state and properties of the fluidized bed are well covered elsewhere[14],[36],[37] and will not be considered here. However, a design method due to Clark[35] will be presented here as a simple, concise method for the preliminary signing and costing of a fluidized bed dryer.

The minimum bed diameter involves a relationship between the operating velocity, particle characteristics, and the humidity of the drying gas. The hot inlet gas rapidly loses heat and gains moisture as it passes through the bed, eventually leaving the bed at a temperature defined as the bed temperature T_b and a relative humidity R. There is a limit to the value of R which is approximately the relative humidity which would be in equilibrium with the dried product at the temperature of the bed.

The fluidizing velocity may be estimated from equations quoted in the earlier references and the operating velocity taken as twice the minimum fluidizing velocity. Alternatively, either a sample may be used to determine the operating velocity directly by fluidizing on a laboratory scale or, more conveniently, Figure 8–24 may be used.

For drying media other than air at approximately atmospheric pressure, the velocity obtained from Figure 8–24 should be multiplied by

$$0.003\ 68/\rho^{0.29}\mu^{0.43} \qquad [0.009\ 75/\rho^{0.29}\mu^{0.43}]$$

where ρ = density of fluidizing gas (lb/ft^3) [kg/m^3],

μ = viscosity of fluidizing gas (lb/ft s) [N s/m^2].

The minimum bed diameter may be calculated by simultaneously solving a heat and mass balance across the bed. From a mass balance.

$$\frac{R}{100} \cdot \frac{P_v}{P_e} = \frac{w + \{M/(1 + y_m)\}y_m}{w + \{M/(1 + y_m)\}(0.625 + y_m)} \qquad (8\text{--}27)$$

where R = exit gas relative humidity (%) [%],

P_v = vapour pressure of water at exit gas temperature (atm) [N/m^2],

P_e = total static pressure of gases leaving the bed (usually atmospheric pressure) (atm) [N/m^2],

w = evaporative capacity, moisture (lb/hr) [kg/s],

M = flow rate of inlet gas (air) (lb/hr) [kg/s],

y_m = moisture content of entering gas (air) (lb H$_2$O/lb dry gas) [kg/kg],

0.625 = factor of ratio of molecular weights where air and water are used (—),

$M/(1 + y_m)$ = flow rate of dry gas (lb/hr) [kg/s].

The value of P_v may be obtained from Figure 8–25 and that for y_m is the ambient humidity of air when it is indirectly heated. For direct fired heating, Figure 8–26 may be used to obtain values of y_m. This graph is calculated for air heated by the combustion of methane based on ambient conditions of 70°F [293 K] and 0.01 lb water vapour/lb dry air [0.01 kg/kg]. For other ambient conditions, the curve may be transposed either horizontally and/or vertically.

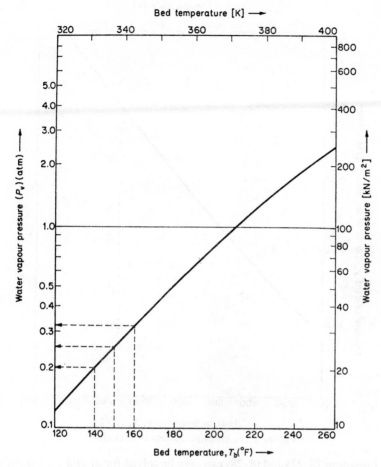

FIGURE 8–25. Water vapour pressure at bed exit[35]

From a heat balance, equation (8–28) is obtained:

$$C_mM(T_m - T_b) = \lambda_b w + C_f F(T_b - T_f) \qquad (8\text{–}28)$$

where C_m = average heat capacity of the heating medium between T_m and T_b (for air equals 0.24 BTU/lb °F) [1.0 kJ/kg K],

C_f = average heat capacity of wet material between T_f and T_b
= $(X_f C_x + C_s)/(X_f + 1)$ (BTU/lb °F) [kJ/kg K],

X_f = moisture content of wet feed on a dry basis (lb/lb) [kg/kg],

C_x = heat capacity of liquid being evaporated (BTU/lb °F) [kJ/kg K],

C_s = heat capacity of dry solid (BTU/hr °F) [kJ/kg K],

λ_b = average latent heat of vaporization of moisture at temperature T_b (BTU/lb) [kJ/kg],

T_m, T_b, T_f = temperatures of entering fluidizing gas, bed and wet feed respectively (°F) [K],

F = wet solid feed rate (lb/hr) [kg/s],

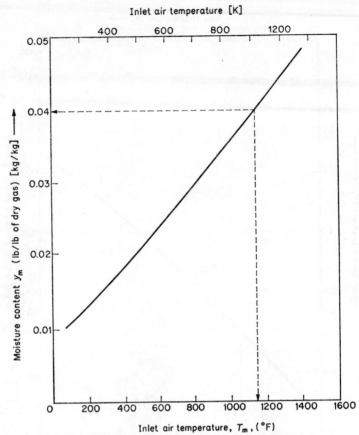

Inlet air temperature [K]

Moisture content y_m (lb/lb of dry gas) [kg/kg]

Inlet air temperature, T_m, (°F)

FIGURE 8–26. Moisture content of inlet air[35]

Equations (8–27) and (8–28) can now be solved for M and T_b. The value of M, T_b and λ_f can then be substituted in equation (8–30) below to calculate the bed diameter,

$$D^2 = (M + 1.58w)(T_b + 460)/111\,000V_f \qquad (8\text{–}30)$$
$$[= (M + 1.58w)T_b/278V_f]$$

The method will be illustrated by means of an example.

Example 8–7

A granular material of specific gravity 5.0 is to be dried using direct air heating at 1000°F [811 K] in a fluidized bed dryer. The particle size is 0.02 in. [0.5 mm] and 10 000 lb/hr [1.26 kg/s] of water is to be removed from a solid feed rate of 100 000 lb/hr [12.6 kg/s] at 70°F [293 K]. What diameter bed will perform this duty?

$\lambda_b = 1000$ BTU/lb [= 2326 kJ/kg], ambient air is at 70°F [293 K] and $y_m = 0.01$, $C_f = 0.40$ BTU/lb °F [= 1.67 kJ/kg K].

From Figure 8–26,

$$y_m = 0.036 \text{ at } 1000°F$$
$$w = 10\,000 \text{ lb/hr} = [1.26 \text{ kg/s}]$$

From equation (8–27), the right-hand side becomes

$$\frac{10\ 000 + \{M/(1 + 0.036)\}0.036}{10\ 000 + \{M/(1 + 0.036)\}(0.625 + 0.036)}$$

$$= (10\ 000 + 0.0347M)/(10\ 000 + 0.638M)$$
$$[= (1.26 + 0.0347M)/(1.26 + 0.638M)]$$

If R is taken as 90 per cent, P_e as atmospheric pressure, the value of M can be found for values of T_b equal to 120, 140, 160°F. At 120, 140, 160°F [321, 333, 344 K],

$$P_v = 0.13, 0.20, 0.32\ \text{atm}$$
$$[= 13, 20, 32\ \text{kN/m}^2]$$

Hence
$$M = 220\ 800, 102\ 000, 47\ 800\ \text{lb/hr}$$
$$[= 27.8, 12.9, 6.02\ \text{kg/s}]$$

From equation (8–28), at $T_b = 120$°F,

$$0.24 \times M(1000 - 120) = 100 \times 10\ 000 + 0.40 \times 100\ 000(120 - 70)$$

or
$$M = 56\ 800\ \text{lb/hr}\ [= 7.16\ \text{kg/s}]$$

At $T_b = 140$°F,

$$M = 62\ 000\ \text{lb/hr}\ [= 7.8\ \text{kg/s}]$$

$T_b = 160$°F,

$$M = 67\ 800\ \text{lb/hr}\ [= 8.54\ \text{kg/s}]$$

Plotting M *vs.* T_b for each equation on the same axis gives a value of $M = 67\ 000$ lb/hr [8.3 kg/s] at $T_b = 152$°F [340 K].

Substituting for T_b in equation (8–30) gives

$$D^2 = (67\ 000 + 1.58 \times 10\ 000)(152 + 460)/111\ 000V_f$$
$$= 455/V_f\ [= 12.6/V_f]$$

From Figure 8–24, the fluidizing velocity is 2.0 ft/s [0.61 m/s]. Hence

$$D^2 = 228\ \text{ft}^2\ [= 21.2\ \text{m}^2]$$
$$D = 15.1\ \text{ft}\ [= 4.60\ \text{m}]$$

This is a very large diameter bed and it will be worthwhile considering whether to increase the fluidizing velocity to reduce the diameter if the subsequent increased rate of elutriation is acceptable. If V_f is increased to 5 ft/s [1.52 m] the diameter is reduced to 9.45 ft [2.88 m] and the capital costs may be calculated from Table 8–15.

Example 8.8

Estimate the costs of the two fluidized bed dryers, one with $V_f = 2.0$ ft/s, [= 0.61 m/s] and $D = 15.1$ ft [= 4.60 m], the other with $V_f = 5.0$ ft/s [= 1.52 m/s] and $D = 9.45$ ft [4.60 m].

The costs may be estimated from Table 8–15. For $V_f = 2.0$ ft/s [= 0.61 m/s] and $X = 15.1$ [= 4.60],

$$\text{cost}\ (10^3\ \$) = 10.0(15.1)^{0.72}$$
$$[= 23.5(4.60)^{0.72}]\quad\text{(by interpolation)}$$
$$\text{cost}\ (\$) = \$70\ 500$$

For $V_f = 5.0$ ft/s [1.52 m/s] and $X = 9.45$ [2.88],

$$\text{cost}\ (10^3\ \$) = 15.3(9.45)^{0.72}$$
$$[= 36(2.88)^{0.72}]$$
$$\text{cost}\ (\$) = \$77\ 000$$

This example shows that although the diameter of the bed has been reduced by employing a higher fluidizing velocity, the cost of the resulting smaller unit shows an increase over the original design. This is because the cost of the equipment needed to produce the higher gas velocity has increased at a rate greater than the reduction of cost due to the use of a smaller diameter bed. This is a good example of where the cost may be optimized between the gas velocity and the bed diameter.

8.5 Nomenclature

A	cross-section of settling tank	(ft^2)	$[\text{m}^2]$
A_h	area for heat transfer	(ft^2)	$[\text{m}^2]$
C_d	heat capacity of dry solid	(BTU/lb °F)	$[\text{kJ/kg K}]$
C_s	humid heat of moist gas	(BTU/lb °F)	$[\text{kJ/kg K}]$
d	particle diameter	$(\text{cm})^*$	$(-)$	$[-]$
D	diameter		(ft)	$[\text{m}]$
e	voidage		$(-)$	$[-]$
F	solids feed rate		(lb/hr)	$[\text{kg/s}]$
F'	ratio of liquid to solid in feed		(lb/lb)	$[\text{kg/kg}]$
g	acceleration due to gravity	$(\text{cm/s}^2)^*$	(ft/s^2)	$[\text{m/s}^2]$
G	flow rate of gas		(lb/hr)	$[\text{kg/s}]$
G_s	flow rate of gas		(lb/hr ft^2)	$[\text{kg/s m}^2]$
h	enthalpy of solid		(BTU/lb)	$[\text{kJ/kg}]$
h	depth		(ft)	$[\text{m}]$
H	enthalpy of gas		(BTU/lb)	$[\text{kJ/kg}]$
L	feed rate		(lb/hr)	$[\text{kg/s}]$
L_t	length of a transfer unit		(ft)	$[\text{m}]$
M	flow rate of inlet gas		(lb/hr)	$[\text{kg/s}]$
N_t	number of transfer units		$(-)$	$[-]$
P_e	pressure of gases leaving bed		(atm)	$[\text{N/m}^2]$
P_v	water vapour pressure		(atm)	$[\text{N/m}^2]$
q	rate of heat transfer		(BTU/hr)	$[\text{w}]$
Q	overflow rate (centrifuges)	$(\text{cm}^3/\text{s})^*$	$(-)$	$[-]$
R	relative humidity		$(-)$	$[-]$
t	solids temperature		$(°\text{F})$	$[\text{K}]$
t_r	retention time		(hr)	$[\text{s}]$
T	gas temperature		$(°\text{F})$	$[\text{K}]$
T_w	wet bulb temperature		$(°\text{F})$	$[\text{K}]$
u_c	settling velocity of suspension	$(\text{cm/s})^*$		
U	overall heat transfer coefficient		$(\text{BTU/hr ft}^2 \, °\text{F})$	$[\text{W/m}^2 \text{ K}]$
U'	ratio of liquid to solid in underflow		(lb/lb)	$[\text{kg/kg}]$
V_f	fluidizing velocity		(ft/s)	$[\text{m/s}]$
V_g	setting velocity		(ft/s)	$[\text{m/s}]$
w	evaporative capacity		(lb/hr)	$[\text{kg/s}]$
W	mass feed rate of solid		(lb/hr)	$[\text{kg/s}]$
X	factor in cost equations		various units	
X	moisture content		$(\text{lb H}_2\text{O/lb})$	$[\text{kg/kg}]$
X'	mass ratio of liquid to solid in thickening section	(lb/lb)	$[\text{kg/kg}]$
Y	humidity of gas		$(\text{lb H}_2\text{O/lb})$	$[\text{kg/kg}]$
ρ	density	$(\text{gm/cm}^3)^*$	(lb/ft^3)	$[\text{kg/m}^3]$
μ	viscosity	$(\text{centipoise})^*$ (or poise)	(lb/ft s)	$[\text{N s/m}^2]$
η	efficiency		$(-)$	$[-]$
Σ	sigma value	$(\text{cm}^2)^*$	$(-)$	$[-]$
λ_f	latent heat of vaporization		(BTU/lb)	$[\text{kJ/kg}]$
ΔT_{lm}	log mean temperature difference		$(°\text{F})$	$[\text{K}]$

 * C.G.S. system of units—applies only to calculation of sigma values and sedimentation rates.

Subscripts

c   suspension
s   solid
b   bed

REFERENCES

1. Matthews, C. W. *Chem. Eng.* 15 Feb. 1971, **78**, 99.
2. Perry, J. H. *Chemical Engineer's Handbook*, 4th ed. New York: McGraw-Hill Book Co., 1963.
3. Roberts, E. J., Stavenger, P., Bowersox, J. P., Walton, A. K., Mehta, M. *Chem. Eng.* 15 Feb. 1971, **78**, 89.
4. Gaudin, A. M. *Flotation* 2nd ed. New York: McGraw-Hill Book Co., 1957.
5. Fuerstenau, D. W. Ed. *Froth Flotation*, 50th Anniversary vol. New York: Am. Inst. of Mining, Met. and Pet. Engineers, 1962.
6. Sargent, G. D. *Chem. Eng.* 15 Feb. 1971, **78**, 11.
7. Kane, J. H. *Heating and Ventilating*, 1954, **51**, 10, 77.
8. Davies, E. *Tr. Inst. Chem. Eng.* 1965, **43**, T256.
9. Dahlstram, D. A. and Carnell, C. F. *Chem. Eng.* 15 Feb. 1971, **78**, 63.
10. Porter, H. F., Flood, J. E. and Rennie, R. W. *Chem. Eng.* 15 Feb. 1971, **78**, 39.
11. Mais, L. G. *Chem. Eng.* 15 Feb. 1971, **78**, 49.
12. Steinour, H. H. *Ind. Eng. Chem.* 1944, **36**, 618, 840, 901.
13. Hawksley, P. G. W. 'The Effect of Concentration on the Setting of Suspensions and Flow through Porous Media', *Inst. of Phys. Symp.* 1950, 114.
14. Coulson, J. M. and Richardson, J. F. *Chemical Engineering*, Vol. II, 2nd ed. London: Pergamon Press, 1968.
15. Chemical Processing Supplement, *Processing Equipment Selector*, 1969, **15**, No. 11.
16. Forbes, F. *Chem. Proc. Eng.* 1970, **52**, 12, 49.
17. Barber, T. P. *Chem. Proc. Eng.* 1970, **52**, 12, 53.
18. Purchas, D. B. *Chem. Proc.* 1971, **17**, 1, 31, and 1971, **17**, 2, 55.
19. Morris, B. G. *Brit. Chem. Eng.* 1966, **11**, 5, 347.
20. Holley, F. V. *Brit. Chem. Eng.* 1969, **14**, 11, 1540.
21. Sutherland, K. S. *Chem. Proc.* 1970, **16**, 5, 10.
22. Morns, B. G. *Brit. Chem. Eng.* 1966, **11**, 8, 846.
23. Ambler, C. M. *Chem. Eng. Prog.*, 1952, **48**, 150.
24. Trowbridge, M. E. O'K. *The Chem. Eng.* Aug. 1962, No. 162, A73.
25. Trowbridge, M. E. O'K. and Bradley, D. Paper presented at the 4th Congress of the European Federation of Chemical Engineering, London, 1966.
26. Bradley, D. *Chem. Proc. Eng.* 1965, **46**, 11, 595.
27. Rosin, S. N. *Chem. Proc.* 1970, **16**, 7, 9.
28. Williams-Gardner, A. *Chem. Proc. Eng.* 1965, **46**, 11, 609.
29. Sloan, C. E. *Chem. Eng.*, 19 June 1967, **74**, 169.
30. Bowmer, G. C. *Chem. Proc. Eng.* 1969, **50**, 10, 77.
31. Marshall, W. R. Jnr. and Friedman, S. J. Section in *Chemical Engineers Handbook*. Ed. J. H. Perry. New York: McGraw-Hill Book Co. 1963.
32. Broughton, D. B. and Mickley H. S. *Chem. Eng. Prog.* 1953, **49**, 6, 319.
33. Master, K. *Chem. Proc. Eng.*; 1969, **50**, 10, 91.
34. British Patent No. 1136952, 1968.
35. Clark, W. E. *Chem. Eng.* 1967, **74**, 6, 177.
36. Davidson, J. F. and Harrison, D. *Fluidised Particles*. Cambridge: Cambridge University Press, 1963.
37. Zenz, F. A. and Othmer, D. F. *Fluidisation and Fluid Particle Systems*. New York: Reinhold, 1960.
38. Fischer, J. J. *Ind. Eng. Chem.* 1963, **55**, 2, 18.
39. Lapple, W. C., Clark, W. E. and Dyhal, E. C. *Chem. Eng.* 1955, **62**, 11, 177.
40. Bradley, D. *International Series of Monographs in Chemical Engineering*, Vol. 4. London: Pergamon Press, 1965.
41. Guthrie, K. M. *Chem. Eng.* 1969, **76**, 6, 114.
42. Imperazo, N. F. *Chem. Eng.* 1968, **75**, 22, 154.

43. Noden, D. *Chem. Proc. Eng.* 1969, **50**, 10, 67.
44. Anon. *Ind. Chemist.* 1964, **40**, 2, 90.
45. Mills, R. E. *Chem. Eng.* Mar. 1964, **71**, 16, 133.
46. Anon. *Ind. Chemist.* 1963, **39**, 12, 649.
47. Anon. *Ind. Chemist.* 1964, **40**, 1, 34.
48. Oliver, E. D. *Chem. Eng.* 14 July 1969, **76**, 132.

Chapter 9: Ancillary Equipment

9.1 Introduction

Earlier chapters have covered process design aspects of heat exchange equipment and towers for mass transfer operations. In the preceding chapter, emphasis changed from design to selection and specification of equipment for physical separations. The same theme is continued in this chapter, where discussion of other essential components of the chemical plant is included. Topics covered include pumping, crystallization, grinding and product storage, and whilst the total number of operations covered in these two chapters is by no means comprehensive, it is hoped that the material presented will form a concise guide to the selection and costing of equipment most frequently encountered in process design.

9.2 Pumping

There are six methods by which any fluid can be made to flow[1] and these will be considered briefly in turn.

(i) *By centrifugal force*. Kinetic energy, produced by the action of centrifugal force, is partially converted to pressure by reduction of the fluid velocity. The basic function of all centrifugal equipment is the same, though physical appearances may differ widely. High throughputs may be handled with a pulse-free output at high efficiency and these factors generally make this type of equipment the first choice in preliminary selection.

(ii) *By displacement*. Discharge of fluid is produced by partially or completely displacing its internal volume by either mechanical means or by means of a second fluid. Examples in this category include reciprocating pistons and diaphragms, rotary vane and gear types, fluid piston compressors and air lifts. High pressure discharges are obtainable, though pulsating flow may be a problem; throughputs are limited by physical size of the equipment, but efficiencies can be high.

(iii) *Mechanical impulse*. This is usually combined with one of the other methods as, for example, in axial flow.

(iv) *Transference of momentum*. Acceleration of one fluid in order to transfer its momentum to another is commonly used for handling corrosive materials or for evacuation problems. The devices are usually inefficient but have low capital and operating costs which make them attractive.

(v) *Electromagnetic force*. This is used where the fluid to be moved is a good electrical conductor. Flow is produced by the application of a magnetic field.

(vi) *By gravity*. This principle needs no further description.

In this section, pumping will be considered under three headings: pumping of liquids, compressors, and the production of vacuum. A detailed description of types of pumps will not be included as they are available elsewhere.[1],[2] Emphasis will be placed on the selection of equipment for specific duties, and methods of obtaining a first estimate of the costs involved will be included where such data is available.

9.2.1 Pumping of Liquids

The great variety of liquids handled by the chemical industry, together with the physical condition of the fluids, has given rise to a large number of different types of pumps to meet the problems involved. Liquids may range from liquefied gas at temperatures as low as $-200°C$ [73 K] to molten salts at 750°C [923 K]. Viscosities may range from very low values with the liquefied gases, to values as high as 10 000 cP [$10 N s/m^2$]. Pressures developed by pumps may range from a few inches of liquid to 10 000 lb/in.^{2} [69 MN/m^2]. Frequently the fluid is contaminated with a small amount of solid material and it may be required to pump a slurry of crystals without breakage of the solids. These are some of the problems involved, and as a first step in a selection procedure the following factors may be listed for consideration:[2]

(a) The quantity of liquid to be handled. This will affect the size and the number of pumps required for the duty.

(b) The head against which the liquid is to be pumped. This is made up of the pressure and vertical height of the downstream receiver and by the friction losses in the pipeline.

(c) The nature of the liquid itself will determine the material of construction from a corrosion viewpoint, and its physical properties will have an effect on the type of equipment which can be used.

(d) The type of power supply available can affect the type of pump. Conversely, certain types of pump, especially those capable of high pressure duty, demand a particular type of drive.

(e) The mode of operation can have an effect on the pump type, as intermittent use can cause corrosion problems which would normally be neglected in the case of continuous operation.

TABLE 9–1. RANGE OF OPERATION OF PUMPS[3]

Type of pump	Normal capacity		Normal maximum head	
	(gal/min)	[m³/s]	(ft of water)	[m of water]
Centrifugal	1–3000	0.76×10^{-4}–0.23	300 (3000 if multistage)	90 (900 if multistage)
Vertical centrifugal	5–3000	0.38×10^{-3}–0.23	300	90
Reciprocating	2–2000	0.15×10^{-3}–0.15	700	200
Diaphragm	0.25–200	0.19×10^{-4}–0.015	230	70
Rotary (gear)	0.25–2000	0.19×10^{-4}–0.15	700	200
Screw	0.5–280	0.36×10^{-4}–0.021	1100	340
Sliding vane	1–250	0.76×10^{-4}–0.019	230	70
Peripheral	1–40	0.76×10^{-4}–0.003	350	105

Table 9-1[3] enables a pump selection to be made on the basis of capacity and the head to be overcome. The method of determining the total head is fully discussed by Perry[1] and is readily calculated by procedures described in that text. The centrifugal pump should always be considered first on the grounds of wide operating ranges, uniform flow and low capital cost. Again, Perry discusses their characteristics in great detail and no further consideration is warranted in this book. Table 9-2[4] is included here as a guide to the use of centrifugal pumps with viscous liquids, but for higher viscosities alternative types of pump should be sought.

TABLE 9-2. VISCOSITY EFFECT ON CENTRIFUGAL PUMP SELECTION[4]

Maximum viscosity		Minimum outlet bore of pump	
(poise)	[N s/m²]	(in.)	[mm]
0.3	0.03	0.75	19
0.5	0.05	1.0	25
0.8	0.08	1.5	38
1.0	0.10	2.0	50
2.0	0.20	3.0	75
3.0	0.30	5.0	125
4.0	0.40	6.0	150
5.0	0.50	8.0	200
6.0	0.60	10.0	250
8.0	0.80	12.0	300
9.0	0.90	14.0	350
20·0	2.00	16.0	400

To determine the theoretical work required of a pump, the hydraulic horse power is calculated from a knowledge of the head, capacity and specific gravity of the fluid, and may be obtained from

$$hp = HSQ/3960 \qquad (9\text{-}1)$$

$$[KW = 8.13 \, HsQ] \qquad (9\text{-}2)$$

where H = total head (ft) [m],
   s = specific gravity (—) [—],
   Q = volumetric flow (gal/min) [m³/s].

The actual power requirement is obtained by dividing the value obtained from equations (9-1) or (9-2) by the efficiency of the pump. If possible the true efficiency should be obtained from the pump manufacturer, but for a preliminary estimate the values from Table 9-3 may be used.[3]

TABLE 9-3. APPROXIMATE PUMP EFFICIENCIES[3]

Type of pump	Efficiency (%)
Positive displacement	90
Peripheral	80
Large capacity centrifugal	65
Small capacity centrifugal	30

Cost Data

Cost data for pumps other than the centrifugal type are difficult to correlate with any degree of accuracy and individual costs are best obtained from manufacturers. The Institution of Chemical Engineers[5] have correlated costs for centrifugal pumps over the size range 1–1000 hp [0.75–750 kW]. The data is presented as the cost of the pump unit, the cost of various drives, and the cost of fluid couplings.

For single-stage centrifugal pumps excluding drive, the costs may be obtained from equations (9–3) or (9–4) and Table 9–4.

$$\text{Cost £/bhp} = 210 \, (\text{bhp})^{-0.713} \qquad (9\text{–}3)$$

$$[\text{Cost £/kW} = 239 \, (\text{kW})^{-0.713}] \qquad (9\text{–}4)$$

for 1–400 hp [0.75–300 kW].

TABLE 9–4. COST FACTORS FOR SINGLE-STAGE CENTRIFUGAL PUMPS[5]

	Multiplying factor for equations (9–3) and (9–4)
Material	
Mild steel	1.0
Silicon iron	1.25
18/8 Stainless steel 304	2.0
18/8 Stainless steel 316	2.3
Monel	2.4
Nickel	4.0
Hastelloy	6.0
Rubber lined	1.4
Lead	1.5
Stoneware	1.9
Drive type	
Horizontal	1.0
Vertical, close coupled	1.25
Vertical submerged	1.4

For multistage centrifugal pumps, the cost may be obtained from equations (9–5) or (9–6) and Table 9–5.

$$\text{Cost £} = 190(\text{bhp})^{0.413} \qquad (9\text{–}5)$$

$$[\text{Cost £} = 214(\text{kW})^{0.413}] \qquad (9\text{–}6)$$

for 1–1000 hp [0.75–750 kW].

The approximate cost of motors and turbines in the range 1–1000 hp [0.75–750 kW] may be obtained from

$$\text{cost £} = 10.8(\text{hp})^{0.885} \qquad (9\text{–}7)$$

$$[\text{cost £} = 14(\text{kW})]^{0.885} \qquad (9\text{–}8)$$

For steam turbines, including condensers in the range 1000–10 000 hp [750–7500 kW] use

$$\text{cost £} = 295(\text{hp})^{0.63} \qquad (9\text{–}9)$$

$$[\text{cost £} = 355(\text{kW})^{0.63}] \qquad (9\text{–}10)$$

TABLE 9-5. COST FACTORS FOR MULTISTAGE CENTRIFUGAL PUMPS[5]

		Multiplying factor for equations
(a) *Material*		
Casing	Impeller	
Mild steel	Mild steel	1.0
18/8 St.St.	18/8	1.7
13% Cr.St.St.	13% Cr.St.St.	1.25
Mild steel	13% Cr.St.St.	1.1
(b) *Number of Stages*		
1		1.0
2		1.3
3		1.4
4		1.5
5		1.7
6		1.9

For fluid coupling costs for drives in the range 100–1000 hp [75–750 kW] use

$$\text{cost } £ = 360(hp)^{0.53} \qquad (9\text{–}11)$$
$$[\text{cost } £ = 398(kW)^{0.53}] \qquad (9\text{–}12)$$

In further information beyond the scope of the quoted sources,[1],[2] reference may be made to a series of papers by Thurlow on the selection,[6] installation and operation,[7] and on maintenance and economics,[8] to an article by Rost and Visich[9] on all aspects of pump selection, and to the book of Holland and Chapman on the pumping of liquids.[9a] To illustrate the use of the cost data above, an example will be included at this stage.

Example 9–1

Estimate the cost of a 750 hp [559 kW], 3-stage centrifugal pump using 18/8 stainless steel for both impeller and casing, together with the cost of the appropriate motor.

From Table 9–5, the multiplying factors for the material of construction and for the number of stages are 1.7 and 1.4 respectively. From equation (9–5) or (9–6), the cost is given by

$$\text{cost } £ = 190(750)^{0.413} \times 1.7 \times 1.4 \qquad (9\text{–}5)$$
$$[= 214(559)^{0.413} \times 1.7 \times 1.4] \qquad (9\text{–}6)$$
$$= £6980$$

The cost of the motor is obtained from equations (9–7) or (9–8),

$$\text{cost } £ = 10.8(750)^{0.885} \qquad (9\text{–}7)$$
$$= [14(559)^{0.885}] \qquad (9\text{–}8)$$
$$= £2930$$

9.2.2 Compressors

Compressors are one of the most expensive components in a process plant[10] and can be worth nearly £200 000 excluding drive, foundations, and erection costs. Their function in a process is frequently critical and the range of choice of equipment is often wide so that selection may become a difficult task. In order to attempt to simplify the selection problem, the three main considerations of duty, operating characteristics, and cost, will be

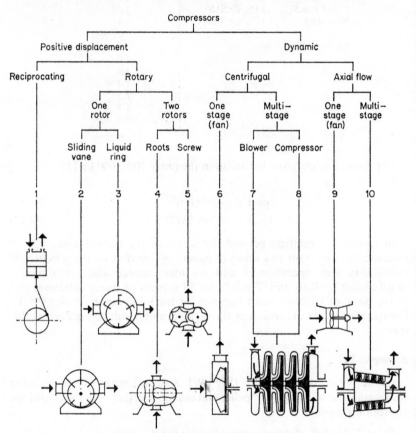

FIGURE 9–1. Classification of compressor types[11]

discussed in turn. Firstly, although it is not necessary to describe in any detail the various compressors which are available, Figure 9–1 shows a concise classification of ten different types. For further description of these compressors a paper by Begg[11] and various textbooks[1],[2] may be considered.

In order to make a preliminary selection of the type of compressor for a given duty, Table 9–6 may be consulted, where the normal maximum volume and pressure characteristics of each of the ten compressors classified earlier, are given. Whilst useful, this table takes no account of occasions when a combination of compressors may be worthy of consideration. With

TABLE 9–6. SIZE AND CAPACITY OF GAS COMPRESSORS

Type of compressor	Normal maximum speed (r.p.m.)	Normal maximum volume		Normal maximum pressure			
				1 cylinder		n cylinders	
		(ft³/min)	[m³/s]	(lb/in.² g)	[kN/m²]	(lb/in.² g)	[kN/m²]
Positive displacement							
1. Reciprocating [12], [13]	300	50 000	24	50	450	75 000	518 000
2. Sliding vane	300	2000	0.95	50	450	120	950
3. Liquid ring	200	1500	0.17	10	170	25	275
4. Roots	250	2500	1.2	5	135	25	275
5. Screw [14]	10 000	7500	3.5	50	450	250	1850
Dynamic							
6. Centrifugal fan [15]	1000	100 000	47	1.5	110	3	122
7. Turbo blower	3000	5000	2.4	5	135	25	275
8. Turbo compressor	10 000	80 000	38	50	450	1500	10 500
9. Axial flow fan	1000	100 000	47	0.5	105	3	122
10. Axial flow blower	3000	100 000	47	50	450	150	1050

high costs of drives, it may be worth connecting four centrifugal compressors through a gear train to a common drive, the whole to replace, for example, an axial compressor. In achieving very high pressures, a rotary screw compressor may be used to reach an intermediate pressure, after which the gas is piped to a reciprocating unit to complete the process.

The operating characteristics of compressors typical of the centrifugal, axial and positive displacement types are shown in Figure 9–2.

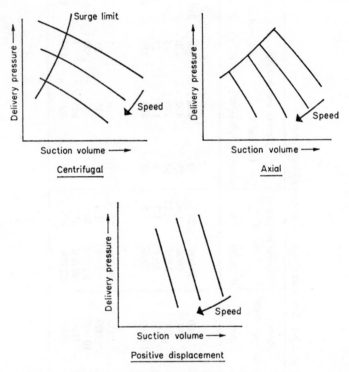

FIGURE 9–2. Comparative characteristics of compressors

These characteristics show the sensitivity of the compressor to variation of flow conditions, and consequently a positive displacement type would be totally unsuitable for conditions where wide variations in duty are called for. The efficiency of a centrifugal compressor is lowest of the three types, providing a penalty for its flexibility. Efficiencies some 5–20 per cent lower than the reciprocating machine have been quoted,[16] whilst the axial compressor efficiency exceeds that of the reciprocating type by 5–10 per cent.

Cost correlations for compressors represent only a part of the total cost of a complete installation. Ancillary equipment, in the form of motor turbine drive, coolers, foundations and other auxiliaries, will add between 20 and 40 per cent and 15 and 60 per cent to the cost of reciprocating and centrifugal compressors respectively.[10]

In general, for low pressures and large flows the purchase cost of a reciprocating compressor may be twice that of a centrifugal machine of the same

TABLE 9–7. COMPRESSOR COST DATA[5]

Type	Range of operation (lb/in.² g)	(kN/m²)	(ft³/min)	(m³/s)	Costs (£/bhp)	(£/kW)	(£/(ft³/min))	(£/(m³/s))
Rolling drum rotary	0–15	0–200	80–2000	0.04–0.95	13	17	4.3	9100
Sliding vane rotary	15–40	200–380	20–1000	0.01–0.47	11	15	2.2	4560
Reciprocating								
Single-stage vertical	15–150	200–1130	0–150	0–0.07	19	25	6.5	13 800
Two-stage vertical	40–150	380–1130	45–400	0.02–0.20	12	16	3.2	6800
	150–450	1130–3200	0–300	0–0.14	14	19	6.5	13 800
	450–1000	3200–7000	0–150	0–0.07	29	39	17	36 000
Three-stage vertical	450–600	3200–4200	150–300	0.07–0.14	16	22	6.5	13 800
	1000–4500	7000–36 000	0–50	0–0.02	71	95	6.3	133 000
Four-stage vertical	600–6300	3200–43 500	0–200	0–0.01	89	120	77	164 000
Centrifugal	0–350	0–2500	2000–100 000	0.95–47	7.6	10	0.65	1380

capacity, while if the pressure is high and the flow low, costs will be almost equal.[16] Using the same approximations, rotary screw and axial compressors cost approximately the same as the centrifugal unit, though when particularly suited to their application, their costs may be considerably lower.

Specific cost data has recently been published by the Institution of Chemical Engineers[5] and their figures, updated to June 1970, are shown in Table 9–7.

The base price of reciprocating and centrifugal compressors, according to the power of the unit, has been published by Bresler.[16] The limit of the base price is a pressure of 1000 lb/in.^{2} [7000 kN/m^2], and above this operating pressure the cost/unit power should be increased by the percentage indicated. The relevant equations are presented in Table 9–8.

TABLE 9–8. COST DATA FOR CENTRIFUGAL AND RECIPROCATING COMPRESSORS[16]

| | Cost equations | |
	(A.A.)	[SI]
Base price		
Centrifugal	$/hp = 1950(hp)$^{-0.50}$	$/kW = 2260(kW)$^{-0.50}$
Reciprocating	$/hp = 772 (hp)$^{-0.50}$	$/kW = 950(kW)$^{-0.29}$
Increase in price for		
pressures > 1000 lb/in.^{2}		
Centrifugal	% increase = 0.24$(P)^{0.68}$	0.064$(P)^{0.68}$
Reciprocating	% increase = 0.0107$(P)^{0.99}$	0.0015$(P)^{0.99}$

Example 9–2

Estimate the cost of a reciprocating compressor rated at 4500 hp [3350 kW] capable of a delivery pressure of 6000 lb/in.^{2} [41 400 kN/m^2].

From Table 9–8, the base cost is given by

$$\$ \, hp = 772(4500)^{-0.29} = 67.2$$

$$[\$/kW = 950(3350)^{-0.29} = 90.2]$$

$$\text{Base cost} = 67.2 \times 4500$$

$$[= 90.2 \times 3350]$$

$$= \$302 \, 000$$

To allow for the higher pressure, calculate the percentage increase over the base cost.

$$\% \text{ increase} = 0.0107(6000)^{0.99}$$

$$[= 0.0015(P)]^{0.99}$$

$$= 59\%$$

$$\text{Hence total cost} = 1.59 \times 302 \, 000$$

$$= \$480 \, 000$$

9.2.3 Production of Vacuum

Equipment for the production of low pressure within a vessel is dependent upon the capacity of the system and the absolute pressure required. Reciprocating pumps are readily available in either single or multiple stages, but their efficiency falls as absolute pressures of approximately 1 mm Hg are reached.

The basic functions of a vacuum pump are to remove vapour from the system until the desired low pressure has been reached and then to maintain this pressure by the removal of non-condensable vapours or gases resulting

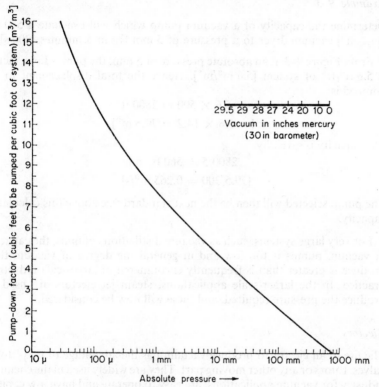

FIGURE 9–3. Vacuum pump-down chart[17]

from decomposition, vaporization, reaction, leakage of air into the vacuum and other sources. The size of a vacuum pump depends primarily upon five factors:

   (i) absolute pressure required,
  (ii) the size of the system,
 (iii) the nature and quantity of vapours and gases to be handled,
 (iv) pump-down time,
  (v) the frequency with which the system is pumped down to operating vacuum.

Manufacturers of vacuum pumps published extensive literature on their products and one example of the range available is provided by the Microvac pumps of Pennwalt Limited.[17] The capacity of this range varies from 30 to 728 ft³/min [0.014–0.34 m³/s] using motors from 1.5 to 30 hp [1.1–22.4 kW], and they are capable of producing vacuum down to a pressure of 5 microns Hg. Efficiencies of 80 per cent or higher are maintained to a pressure as low as 0.1 mm Hg.

As vacuum pumps are rated by volumetric displacement, Figure 9–3 is included to aid pump specification, and an example is included to illustrate the method of use of the chart.

Example 9–3

Determine the capacity of a vacuum pump which will evacuate a 500 ft³ [14.2 m³] vacuum dryer to a pressure of 5 mm Hg in 5 minutes [300 s].

From Figure 9–3 at an absolute pressure of 5 mm, the pump-down factor is 5.6 ft³/ft³ of system [5.6 m³/m³]. Hence the total displacement to be pumped is

$$5.6 \times 500 = 2800 \text{ ft}^3$$
$$[= 5.6 \times 14.2 = 79.5 \text{ m}^3]$$

The capacity is given by

$$2800/5 = 560 \text{ ft}^3/\text{min}$$
$$[79.5/300 = 0.265 \text{ m}^3/\text{s}]$$

The pump selected will then be the next standard size above this calculated capacity.

For very large systems, such as vacuum distillation columns, the capacity of vacuum pumps is too low and in general the degree of vacuum they produce is greater than is frequently encountered in large-scale industrial practice. In the larger-scale applications, steam jet ejectors are used to produce the pressure required, and these will now be considered.

Ejectors

The use of an ejector to produce vacuum has the advantage of no pistons, valves, rotors or any other moving part. They are widely used in the chemical industry for vacuum applications, are easy to operate and have low capital, installation and maintenance costs. They are capable of handling a very

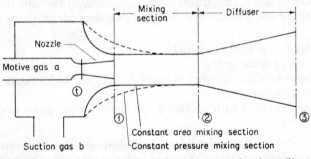

FIGURE 9–4. Diagrammatic section of a stream jet ejector[1]

wide range of gases or vapours as they may be constructed from a large number of corrosion-resistant materials.

The ejector consists essentially of a steam nozzle that discharges a high velocity jet across a suction chamber connected to the equipment to be evacuated. The gas or vapour is entrained by the steam and carried into a venturi-shaped diffuser which converts the velocity energy of the steam into pressure energy. Figure 9–4 shows a typical steam jet ejector.[1]

Condensers may be used in conjunction with ejectors to reduce the vapour load in subsequent ejectors or to reclaim the condensate. Two or more ejectors may be connected in series or stages in a variety of arrangements designed to produce a higher vacuum or, when connected in parallel, to handle a greater throughput. Figure 9–5[1] shows several typical ejector arrangements, and Table 9–9 may be used to select a suitable system depending upon the vacuum required.

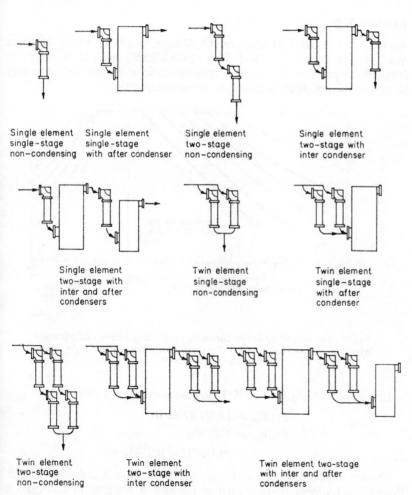

FIGURE 9–5. Commonly used ejector condenser arrangements[1]

TABLE 9–9. STEAM JET EJECTOR SELECTION[(1)]

| Design suction pressure (absolute) | | Arrangement of ejector |
(in. Hg)	[mm Hg]	
30–3.0	760–76	Single-stage
5.0–0.5	125–12.5	Two-stage
—	50–1	Three-stage
—	5–0.05	Four- or five-stage

This table is based on available steam at 100 lb/in.^{2} a [689 kN/m^2] with cooling water at 85°F [303 K] and a maximum discharge back pressure of 1 lb/in.^{2} g [108 kN/m^2].

Selection of the optimum design of a single-stage ejector is best illustrated by means of an example.

Example 9–4

Air is to be evacuated to pressure of 1.47 lb/in.^{2} a [10.1 kN/m^2]. Steam is available at a pressure of 100 lb/in.^{2} a [689 kN/m^2] and will discharge to atmospheric pressure. Estimate the optimum ratio of the areas of the mixing section to the nozzle section and the entrainment ratio.

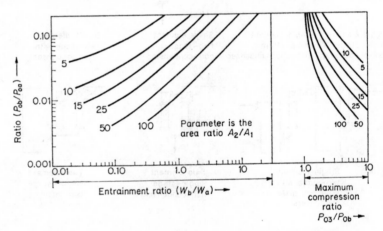

FIGURE 9–6. Simplified design curves for single stage ejectors (applicable where $M_b/M_a = T_{oa}/T_{ob} = 1.0$. For other conditions, correct w_b/w_a with equation (9–13)).[(18)]

Use is made of Figure 9–6.[(18)] Using the notation of Figure 9–4.

$$P_{ot}/P_{ob} = 14.7/1.47 = 10$$
$$P_{ob}/P_{oa} = 1.47/100$$
$$[= 10.1/689] = 0.0147.$$

From the right-hand side of Figure 9–6, the optimum area ratio is 12.5. Proceeding horizontally to the left of Figure 9–6, the entrainment ratio is approximately 0.05 lb air/lb steam [0.05 kg/kg].

Knowing the temperatures of the air and steam, the entrainment ratio is corrected by equation (9–13).

$$(W_b/W_a)' = (W_b/W_a)\{(T_{oa}M_b)/(T_{ob}M_a)\}^{0.5} \qquad (9\text{–}13)$$

where $(W_b/W_a)'$ = corrected entrainment ratio,
$\quad T_{oa}, T_{ob}$ = absolute temperature of a and b,
$\quad M_a, M_b$ = molecular weight of a and b.

Thus Figure 9–6 may be used for any pair of fluids using equation (9–13). If in this example $T_{oa} = [437\ K]$ and $T_{ob} = [300\ K]$, $M_a = 18$, $M_b = 29$, then

$$(W_b/W_a)' = 0.05(437 \times 29/300 \times 18)^{0.5}$$
$$= 0.0765\ \text{lb air/lb steam} \ [0.0765\ \text{kg/kg}]$$

This procedure is for preliminary sizing, and final specifications should always be produced in conjunction with the equipment manufacturer.

9.3 Crystallization

Crystallization has been defined[19] as the formation of a systematically organized solid phase from a solution, melt or vapour. Its principal objective is to form a solid phase from a solution for convenience in later handling. There are, however, other ways of attaining this objective, so that crystallization must offer other clear and decisive advantages.

The most important consideration is purification, either of the crystals themselves or sometimes of the mother liquor, when the crystals may be considered the impurity. An example of the latter is the removal of sodium chloride from caustic soda in the electrolysis of brine.

Crystallization offers a product in the form of a solid whose shape and size may be specified. The specifications may range from small particles, as opposed to a solid mass, to very close tolerances on the uniformity and dimensions of the product.

Whilst crystallization is a basic and distinct unit operation, it cannot be grouped with any of the other normal unit operation categories. It is a diffusional process but the mathematical methods associated with other diffusional operations cannot be applied. It is not a separation operation (that is the step which normally follows crystallization) and it is only concerned with new phase formation. The process is thus chemical, kinetic and diffusional, and while it is beyond the scope of this text to consider theory in detail, a summary of the most important factors will be discussed in order to understand more fully the process of crystallization equipment selection.

9.3.1 Crystallization Theory

Details of crystallization theory are available in several books and reviews[2],[20]–[22] but the basic theory of Ostwald[23] still forms the fundamentals of the subject. Every system has an equilibrium stability relationship that denotes the amount of material that can be dissolved in a solution at a given temperature. Figure 9–7 illustrates such a solubility curve of a typical inorganic salt of normal solubility.

Ostwald postulated that it was possible under certain conditions to have a 'metastable supersaturation' range in which crystallization may occur on existing seed crystals without spontaneous nucleation in the bulk of the solution. However, after a higher 'critical' supersaturation has been reacted, the solution will nucleate spontaneously and return to its saturation concentration. These regions, Ostwald called the 'metastable' and 'labile' zones, and they are indicated on Figure 9–7.

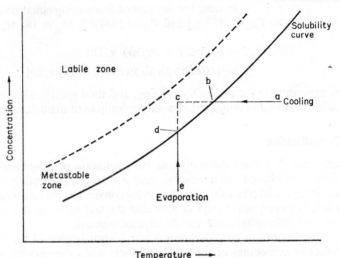

FIGURE 9–7. A typical solubility curve showing the zones postulated by Ostwald[23]

This theory, whilst satisfactory, has been shown to be incomplete by later workers.[24]–[28] There is now agreement that even if the degree of supersaturation is small, spontaneous nucleation will occur if sufficient time is allowed to elapse and the level of supersaturation which can exist for any given time is inversely proportional to the mechanical stimulus or shock in the solution. Thus the degree of agitation in a crystallizer is a vital factor in the selection of such a unit since, together with the above, the quantity of seed crystals present, their size and the speed of agitation in the system all influence the rate of formation of new nuclei.[28] The influence of the degree of agitation present in several different types of crystallizer is shown in Figure 9–8.

The rate at which new crystals are born may be taken as the sum of the rates of nucleation, seed addition and fines formation by attrition, less the rate of fines removal. With the exception of the rate of nucleation, which may be controlled at a low level by reduction of the level of super-saturation, all the other factors lie in the hands of the crystallizer designer, and it is the nature and size of the required product which will determine how these factors are applied in practice to the design and selection of crystallizing equipment.

Reference to Figure 9–7 shows that there are two ways of producing supersaturation—by cooling, or by increasing the concentration of a solution by evaporation, thereby proceeding along lines abc and edc respectively.

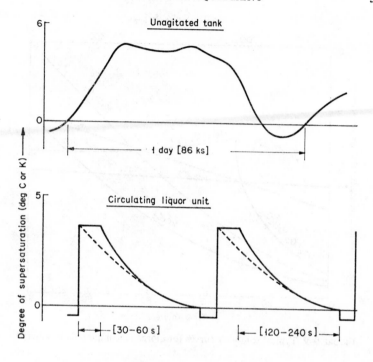

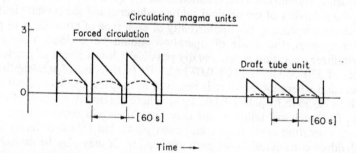

FIGURE 9–8. Supersaturation-time characteristics for different types of crystallizer

It is this simple fact which forms the basis of a selection procedure. Figure 9–9 shows the saturation solubility curves for a number of inorganic salts in water. Examination of the data of Figure 9–9 shows that the curves form three distinct types—steep or normal, flat and inverted solubility curves. Examples of each are $Na_2CO_3.10H_2O$, $NaCl$, and $Na_2CO_3.H_2O$ respectively. Where the solubility curve is steep, supersaturation is most easily and economically effected by cooling with indirect heat exchange. Where the curve is flat or inverted, evaporation of solvent is the means used to produce supersaturation, while for the intermediate cases vacuum cooling by flash evaporation is used. One other type of crystallizer, often omitted from this type of classification, is the salting-out crystallizer. A third component,

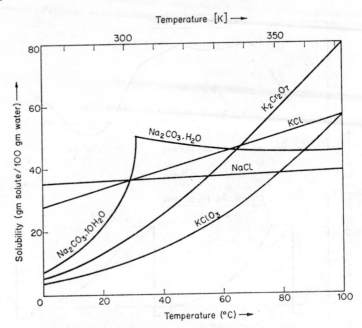

FIGURE 9–9. Typical solubility curves for inorganic compounds in water[1]

frequently sodium chloride, is added to the system with the effect of depressing the solubility of the original solute and increasing the crystal yield.

Before considering the applicability of each type of crystallizer to particular systems, the mode of operation should be decided. Continuous crystallizer operation is not usually possible below a range of production rates of 330–550 lb/hr [0.04–0.07 kg/s].[29] In the case of semicontinuous operation where the product is removed in batches, the lower limit range falls to 22–55 lb/hr [0.003–0.007 kg/s]. There are no lower limits of operation for batch crystallizers and they may in certain circumstances show lower operating costs than continuous plant, but this cost factor needs individual consideration for each application. It may also be desirable to run two similar continuous installations side by side, as crystallizers require more frequent shut-downs than other types of plant and continuity of production is thus assured.

In his book *Industrial Crystallisation*, Bamforth[30] has devoted four chapters to the description of industrial crystallizers. In the following, mention of each type will be made with emphasis placed on applications and limitations rather than detailed description.

9.3.2 Tank Crystallizers

Simplest of all crystallizers is a tank or trough used principally for cooling hot saturated solutions with steep solubility curves. Operation is batchwise and the process may cover a period of days, the crystals produced may lie in a wide size range and, as in the case of copper sulphate, may be very large.

Tanks are useful for the production of small batches of crystals of fine chemicals and pharmaceuticals where the cost of the product is high enough to absorb the cost of further treatment of the crystals produced. The low heat transfer rate may be improved by either the addition of cooling coils or by agitation of the solution, but in general, unless the batches are small, this type of crystallizer is not normally the most economic solution to a crystallization problem.

9.3.3 Scraped Surface Crystallizers

The disadvantage of the addition of cooling coils to a tank crystallizer is that crystals are deposited on the cooling surface with the effect of

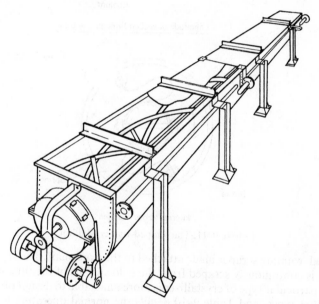

FIGURE 9–10. The Svenson–Walker crystallizer[2]

drastically reducing the heat transfer coefficient between the surface and solution. In the Swenson–Walker crystallizer shown in Figure 9–10, cooling takes place through an external water jacket. A spiral scraper keeps the cooling surface free of adherent crystals and the crystallizer can produce uniform crystals, though a certain amount of breakage from the scraper is inevitable. The Swenson–Walker crystallizer has found particular application in the production of potassium chloride, but it is not suited to crystal production from many organics, waxes and slurries. In these cases, and where high temperature differences are to be employed, scraped wall tubes are necessary. Figure 9–11 shows a longitudinal section and an internal cross-section of such a crystallizer.

The crystallizer consists of one or more sections of large diameter tube with a concentric cooling jacket through which the coolant flows. The process fluid flows through the inner tube which is equipped with a spring

Crystallization tube

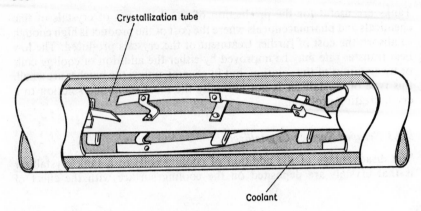

Coolant

Longitudinal section through one tube

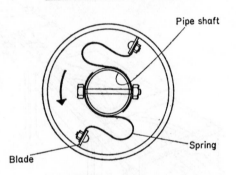

Pipe shaft

Spring

Blade

Internal cross-section

FIGURE 9–11. The scraped-wall crystallizer

mounted, rotating scraper blade attached to the axial shaft. The crystallized deposit is continuously scraped from the cold surface of the tube wall.

This particular type of crystallizer is more amenable to design procedures than other types, and Table 9–10 details the normal operating limits and design parameters which may be considered typical.

TABLE 9–10. OPERATING RANGES FOR SCRAPED SURFACE CRYSTALLIZERS[31]

Parameter	(A.A.)	[SI]
Normal diameter	Usually 6 in., 8 in. and 12 in. are used	0.15, 0.2, 0.3 m
Surface area of installation	10–5000 ft²	1–500 m²
Speed of rotation of scraper	10–50 r.p.m.	0.1–0.8 Hz
Normal temperature range of operation	5–100 °F	258–311 K
Throughput	1–1000 gal/min.	80 cm³/s–0.08 m³/s
Production rate	50 tons/yr	0.016 kg/s
Heat transfer coefficient range	10–150 BTU/hr ft² °F	56–850 W/m² K
Crystal size	80 μm average	80 μm

It is usual to operate the scraped surface crystallizer with several shells in series, with the process liquor entering at one end and the cooled product slurry leaving the other. Solid contents greater than 50 per cent by weight have been handled[31] but a thinner product is to be recommended. One typical application for this type of equipment is the dewaxing of lubricating oils where tube stocks are chilled with selective solvents and waxes crystallized out.[32],[33]

9.3.4 Vacuum Crystallization

An alternative method to cooling by water or air is by evaporation under reduced pressure, where the evaporation of the solvent is used as a means of removing heat to reduce the temperature of the solution. In continuous operation, a large bulk of material is maintained at a fixed temperature and hot liquid is fed continuously into the plant with continuous heat removal by flash evaporation from the surface. Operating batchwise, the vacuum is progressively raised until the required degree of cooling has been achieved. In both cases the vapour is then condensed and vacuum maintained by compressing any non-condensable vapours to atmospheric pressure.

Vacuum cooling and crystallizing offers certain advantages over the other methods discussed:[34]

(a) There is no heat transfer across a cold surface so that no fouling can occur.

(b) With low operating temperatures and no heat transfer surfaces, corrosion problems are reduced. Use may be made of plastic and other lining materials in corrosive applications to reduce plant costs.

(c) Very low crystallizing temperatures may be obtained by compressing the solvent vapour given off to a higher pressure where condensation can be achieved using normal cooling water temperatures. The alternative to this procedure is to use chilled water, brine or a refrigerant in an indirectly cooled plant, which would lead to much higher capital and operating costs.

(d) For the above reasons, capital costs are low.

(e) Control of the system is simple.

One simple form of vacuum crystallizer is a baffled vessel which is agitated by means of a stirrer and connected to a vacuum supply. The stirrer and baffle are designed to circulate both crystals and liquor to the liquor surface, where supersaturation exists due to the flash evaporation. Another common form is one of the variations of the Oslo crystallizer which is described below.

9.3.5 The Oslo or Krystal Crystallizers

The word 'Oslo' is now used as a generic term for a system invented by Isaachsen and Jeremiassen[35] and subsequently developed by the latter and his company, A/S Krystal of Oslo, Norway. (This Company was subsequently bought by the Power Gas Corporation in 1961.) The process is unique in that the same basic design and operating principles are used whatever the method of achieving supersaturation—indirect cooling,

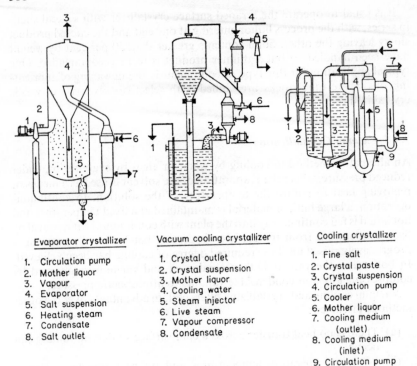

Evaporator crystallizer	Vacuum cooling crystallizer	Cooling crystallizer
1. Circulation pump	1. Crystal outlet	1. Fine salt
2. Mother liquor	2. Crystal suspension	2. Crystal paste
3. Vapour	3. Mother liquor	3. Crystal suspension
4. Evaporator	4. Cooling water	4. Circulation pump
5. Salt suspension	5. Steam injector	5. Cooler
6. Heating steam	6. Live steam	6. Mother liquor
7. Condensate	7. Vapour compressor	7. Cooling medium
8. Salt outlet	8. Condensate	(outlet)
		8. Cooling medium
		(inlet)
		9. Circulation pump

FIGURE 9–12. Three forms of the Oslo crystallizer

vacuum cooling or evaporation. These three modes of operation are shown in Figure 9–12.[36]

The equipment consists of a container for the suspension of the crystals where growth can take place, a vessel or means of producing supersaturation, and a means of circulating saturated mother liquor from the suspension container, mixing it with fresh feed liquor, passing through the supersaturation zone, and finally passing the solution through the suspension vessel.

The Oslo crystallizer is normally operated as a 'circulating liquor' type where the crystals are retained in the suspension vessel and the liquor alone circulated. The crystals are maintained in suspension by the upward flowing liquor, each face of the crystal being continuously exposed to fresh supersaturated solution. This is the ideal condition for the growth of large, regular crystals, and by controlling the upward liquor flow, control over the output crystal size can be achieved.

The feed to the crystallizer forms only a small proportion of the circulating liquid and nucleation does not normally take place in the heat transfer equipment. Some nuclei form in the crystal bed and others are formed by attrition and they will circulate with the liquor until they grow to a sufficient size to be retained in the bed. If large crystals are required, those nuclei circulating externally may be removed by the addition of a decanter to the system.

9.3.6 Circulating Magma Crystallizers

This classification includes all crystallizers in which growing crystals are intentionally circulated to the zone where the supersaturation is being produced. As shown in Figure 9–8 the level of supersaturation with this type of crystallizer has been reduced to lower correspondingly the rate of nuclei formation. In this way, with the residence time properly adjusted, product

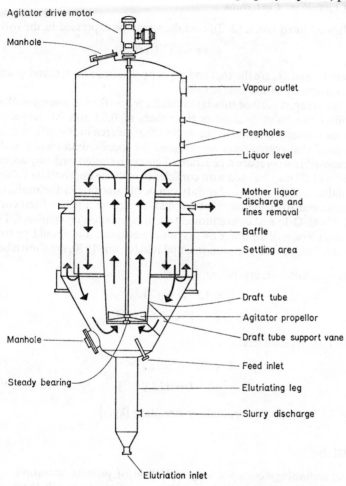

FIGURE 9–13. Circulating magma type, draft-tube baffle crystallizer

crystals within a narrow size range can be produced.[37] Figure 9–13 shows a draft-tube baffle crystallizer of the circulating magma type which has been developed by the Swenson Evaporator Co. of Illinois.

In the draft tube baffle crystallizer a large internal circulation rate is maintained with an agitation rate of approximately 125 rpm [2 Hz]. Use of the vertical draft tube surrounding the propeller ensures circulation of both liquid and crystals from the bottom to the top of the vessel, and the partitioned settling area allows regulation of the magma density and control of nuclei removal. The problems of crystallization, with particular reference to nucleation and the question of fines removal, have been discussed by Randolph.[38] The low level of supersaturation and the rapid circulation of growing crystals to the boiling surface minimize nuclei formation and enable production of large crystals within narrow size specification limits.

9.3.7 Crystallizer Cost Data

As indicated in equation (2–2), cost data may be expressed in the form

$$C = C_b(Q/Q_b)^n \tag{2–2}$$

Where C_b and Q_b are the cost and capacity of a basic unit, C and Q are the cost and capacity of the selected size.

For the scraped surface tubular crystallizer and for the Swenson–Walker crystallizer the index n assumes the values of 0.83 and 0.6 respectively, where the capacity is taken as the heat exchange area in ft² [m²]. It must be appreciated that crystallizers are designed for specific duties and it is therefore impossible to prepare a generalized cost correlation with any accuracy. Equation (2–2) may be used with confidence with the appropriate value of n if a reliable cost for a particular duty can be obtained from a manufacturer.

Garrett and Rosenbaum[19] attempted to prepare a generalized correlation for both Oslo and conventional forced circulation crystallizers. This is very much a case of an order-of-magnitude estimate and should be treated with care. The respective equations, updated to June 1970, are shown below.

Forced circulation crystallizers:

$$\$ \times 10^3 = 6.37X^{0.57} \tag{9–14}$$

$$[= 79.6X^{0.57}] \tag{9–15}$$

Oslo type crystallizers:

$$\$ \times 10^3 = 6.22X^{0.65} \tag{9–16}$$

$$[= 112X^{0.65}] \tag{9–17}$$

where X = capacity of crystallizer (ton/day) [kg/s].

9.4 Mixing

Mixing technology covers a very wide field of process operations—from the simple stirring of a single liquid to situations in which three phases and many components may be present. Conditions of temperature, viscosity, pressure and flow rate, in both batch and continuous processes, make the logical selection of types of mixing equipment a difficult task.

The types of equipment and the problems involved may most conveniently be divided into two classes—solids mixing and the mixing of liquids, both of which can be considered as either batch or continuous processes. Whilst mixing systems are more commonly batch in operation, the substitution of a continuous process may offer several advantages which have been discussed in detail elsewhere,[39] and are summarized below.

The advantages of continuous mixing are:

(i) Labour requirements are minimized.
(ii) The capacity of continuous mixing units is greater than that of batch mixers. Whilst the power consumption per unit weight is not reduced, the large power units necessary to mix large batches are eliminated.
(iii) A continuous system exhibits low hold-up so that the undesirable effects of over-mixing, which is considered later, may be avoided.
(iv) In the case of a solids mixer, segregation can be minimized and intensive mixing of a minor ingredient achieved, providing that care is taken in the design of subsequent handling stages.
(v) The continuous system is readily adaptable to automatic control.
(vi) Because the equivalent capacity continuous mixer is smaller than that of its batch counterpart, the capital cost is lower. Together with the lower labour requirements already mentioned, this is a significant factor in the selection of a system.

The disadvantages of continuous mixing may be listed in a similar manner:

(i) The mixing process depends upon the accurate functioning of associated parts of equipment and is susceptible to process fluctuations. The mixing system is usually designed for a specific duty and the flexibility of a batch mixer when dealing with varying through-puts cannot be matched.
(ii) The continuous mixing process requires calibration and checking, and in the case of solids mixing it is particularly difficult to measure and record feed rates.
(iii) Also in the case of solids mixing, continuous control of component specification is difficult, in that a high frequency of sampling is necessary and this may increase running costs considerably.

The advantages of continuous operation for liquid mixing and liquid–solid mixing clearly outweigh the disadvantages, but individual systems for the mixing of solids must be considered on their own merits before a choice can be made.

9.4.1 Solids Mixing

In 1964, the Institution of Chemical Engineers sponsored a report on the 'State of the Art' and industrial problems in the mixing of difficult materials.[40] This report was later discussed with reference to the particular problems of industrial users, and the hope was expressed that further laboratory scale research would lead towards the design of large-scale mixers becoming more of a science than an art.[41] Until this time arrives, selection based upon experience and empiricism must remain.

For the selection of a batch mixer, Figure 9–14, which was originally presented by Miles and Schofield, cannot be bettered.[39] Emphasis has been placed upon the problem of separation and segregation of the

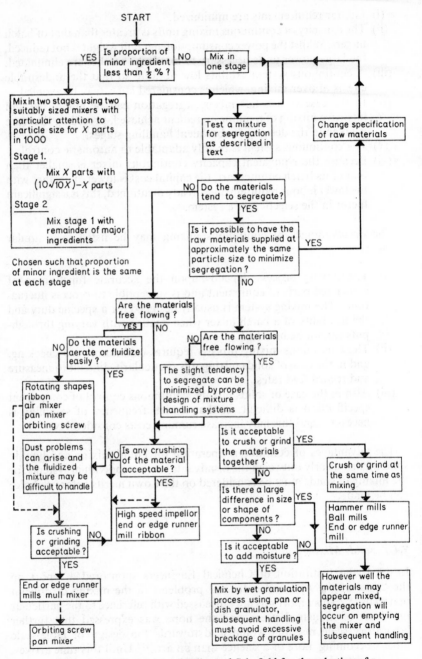

FIGURE 9–14. The scheme of Miles and Schofield for the selection of a solids mixer[39]

components of the mix, this being of prime importance in the majority of applications. Consideration must also be given to the abrasiveness of the mix, attrition, formation of dust and electrostatic change development. These factors, if they were to be incorporated into Figure 9–14, would make the resultant diagram so complex as to be worthless. If the chart is to be used for preliminary selection, full details of the physical properties of the components should subsequently be passed on in the first enquiry to the equipment manufacturer, who can then advise his client with the maximum certainty on the best type of a specific piece of equipment. In addition to those mixers mentioned in Figure 9–14, attention should also be paid to the air vortex mixing of solids which has recently been discussed by Griffin.[42] Air is supplied in a series of blasts through nozzles in a mixing head such that the material to be mixed is fluidized in an upward and spiral motion. Blasts of one second duration are commonly used and mixing may be completed in 30 s.

A typical 'Airmix' unit has been used to successfully disperse a minor component of 0.001 per cent and the range of density and size which can normally be mixed is 6:1. Production units range in size from 6 to 800 ft^3 [0.17–23 m^3], the most popular size range being 10–60 ft^3 [0.3–1.7 m^3]. Costing data is included later in this chapter.

A guide to the published literature on the mixing of powders, in terms of the particle properties and the equipment in which experimental work has been carried out, has been presented by Lloyd and Young.[43] This provides information on the success authors have had with the mixing of many different materials, and should act as a useful selection tool.

Not all the mixers previously mentioned are suitable for continuous operation. Most commonly used continuous types are the ribbon bladed and tumbler mixers, the former being made longer and narrower, and the latter modified to produce a given hold-up of material and to promote a general movement of matter in one direction. A review of mixing equipment has recently been published[44] in which equipment and recent innovations in the fields of solids mixing, liquid–solids mixing, mixer–processors, and portable mixers are described together with details of capacity and manufacturer.

To conclude this section on solids mixing, one piece of equipment suitable for the mixing of multicomponent mixes and for homogenizing single heterogeneous materials will be described. In order to achieve intimate mixing of two or more components, it is an attractive concept to visualize their elongation into continuous streams before progressively mixing from their point of confluence. It is this type of action which has been described by Jones[45] in discussing the 'Intramix'.[46] This is a continuous device which is composed of an externally threaded rotor moving in an internally and oppositely threaded stator. Both rotor and stator are complementarily threaded so that the rotor thread is parallel at the material feed zone, tapering in the mixing zone, and having no thread at the exit. The feed materials are subjected to an intense mixing–shearing action as they proceed towards the exit. This type of mixer has found application in the mixing of cement and other components to form building blocks, in the mixing of compound fertilizers and in the production of smooth consistency clays.

9.4.2 Liquid Mixing

Descriptions of the different types of liquid mixers abound in the literature, with nearly 140 pages in *Mixing in the Chemical Industry*.[47] In spite of the mass of descriptive matter, four main criteria emerge from consideration of selection factors:

   (i) Is the process batch or continuous? If there is a choice, earlier comments may aid selection.

  (ii) What is the liquid viscosity?

 (iii) What is the vessel volume or the desired residence time?

 (iv) Is the flow laminar or turbulent?

In the selection of suitable mixing equipment, consideration should always be given to simple pipeline equipment for conditions where short

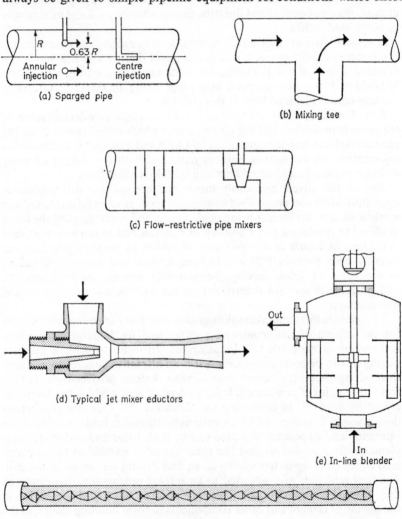

(a) Sparged pipe

(b) Mixing tee

(c) Flow–restrictive pipe mixers

(d) Typical jet mixer eductors

(e) In-line blender

(f) The 'Static' mixer

FIGURE 9–15. 'In-line' mixing equipment[50]

residence times are required. The most commonly used types of 'in-line' mixers are shown in Figure 9–15 and are self explanatory. Conditions must be turbulent, and for conditions of equal viscosity and density complete mixing can be achieved in 10, 50 or 80 pipe diameters for the mixing tee, annular sparger, and centre injection respectively.[48],[49] A viscosity ratio of greater than 10:1 will need longer mixing lengths, as will the mixing tee where the mass flow through the tee is less than three times the flow through the run. In cases where viscosities differ, the more viscous liquid should be sparged into the less viscous.[50]

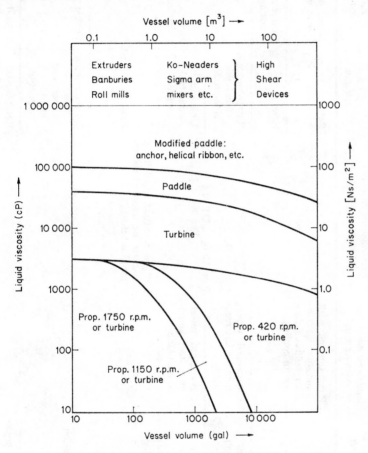

FIGURE 9–16 Agitator selection guide[50]

The relationship between vessel volume, viscosity and mixing equipment is illustrated in Figure 9–16 to enable the selection of the most suitable agitator. The mixing of liquids in tanks, different types of agitators, and power requirements are discussed in detail by Holland[51] and will not be considered further in this text.

Table 9–11, which is adapted from a paper by Penney,[50] has been included to cover the selection of mixing equipment for most common applications. Alternatives to the simple agitated vessel are recommended where

TABLE 9–11. SELECTION GUIDE TO MIXING EQUIPMENT[50]

Process result	Recommended equipment other than agitated vessel	Recommended impeller for agitated vessel*	Measure of effective agitation	Effect of excessive agitation on process
Heating	Pump-through shell and tube exchanger, reflux condenser for cooling		Rate	None, except backmixing is increased
Cooling			Rate	Heat removal decreases; backmixing increases
Chemical reaction	Gas-sparged vessel; pipe for last reactions	P or AFT$\frac{1}{4}$ < DT < $\frac{1}{2}$ $\frac{1}{8}$ < W/ < $\frac{1}{2}$	Rate; absence of undesirable side-products	
Blending	Gas-sparged vessel; pipe; tee; pump-around loop		Uniformity; rapid tank turnover	None, except back-mixing increases
Dissolving	For gas–liquid and liquid–liquid systems use impellers recommended below			
Melting			Rate	
Leaching				
Solid suspension			Bottom motion; off-bottom suspension; uniform suspension	Friable particles are broken
Crystallization			Rate; crystal size and uniformity of size	Crystals are broken and more nuclei may form

(Solid–liquid systems)

System	Operation	Equipment	Impeller type and ratios*	Purpose	Disadvantage
	Slurrying	Pipe; restriction in pipe; colloid mill; homogenizer	P or AFT; $\frac{1}{7} < D/T < \frac{1}{2}$; FBT; $\frac{1}{10} < W/D < \frac{1}{5}$	Agglomerate break-up; particle wetting	Friable particles are broken
	Liquid suspension		P or AFT $\frac{1}{4} < D/T < \frac{1}{2}$; $\frac{1}{8} < W/D < \frac{1}{5}$	Uniform suspension; ease of breaking dispersion	Difficult to break emulsions are formed, and backmixing increases
Liquid–liquid systems	Extraction	Pipe; restriction in pipe.	FBT DT; $\frac{1}{3} < D/T < \frac{1}{2}$; $\frac{1}{8} < W/D < \frac{1}{5}$	Rate; efficiency ease of breaking dispersion	
	Emulsification	Colloid mill, homogenizer	FBT SD; $\frac{1}{7} < D/T < \frac{1}{3}$; $\frac{1}{10} < W/D < \frac{1}{6}$	Rate; particle size and uniformity of size	Too-small particle
	Gas dispersion		DT; $\frac{1}{8} < D/T < \frac{1}{3}$; $\frac{1}{8} < W/D < \frac{1}{5}$	Efficient vapour-liquid contracting	Difficult-to-break foams are formed, and backmixing increases
Gas–liquid systems	Absorption	Pipe; restriction in pipe		Rate; efficiency; ease of foam separation	
	Foaming	Pipe; restriction in pipe; gas flow through wetted screen	FBT or DT; $\frac{2}{3} < D/T < \frac{1}{3}$ Whip or egg-beater type	Rate; cell size and uniformity of size	Too small cells

* P = Propeller; AFT = Axial-flow turbine; FBT = Flat-blade turbine; DT = Disc-type turbine; SD = Serrated disc; D = Impeller diameter; T = Tank diameter; W = width of impeller blade in axial direction

applicable, and for cases where agitation is required, suitable impeller dimensions are indicated.

Continuous operation fluid mixing equipment has been described by Jones,[52] with reference to mixing equipment manufactured by Lightnin' Mixers Ltd. This company has devised a column applicable to a number of processes which enables a continuous operation to be completed in one vessel instead of the more conventional series of stirred tanks. The five main examples of this type of plant are shown in Figure 9–17. The equipment

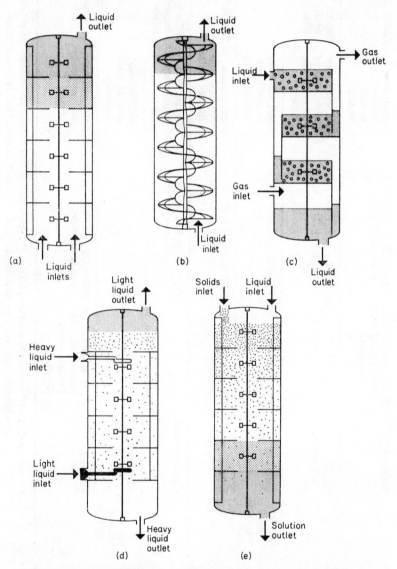

FIGURE 9–17. Fluid mixing equipment manufactured by Lightnin' Mixers Ltd, showing (a) general reactor column, (b) the high-viscosity column, (c) the gas absorber, (d) liquid–liquid extractor, (e) continuous dissolver

consists of one column divided into a number of stages with one mixer and shaft system. The whole range is flexible in that each piece may be fitted with heat transfer coils, jackets or multiple inlets and may be constructed from a wide range of materials. The general reactor, the liquid–liquid extractor and the continuous dissolver all act on the same principle, and flow can be either countercurrent or co-current. The high viscosity reactor is capable of dealing with viscosities in the range 20 000 to 1 000 000 cP [20–1000 Ns/m²] with guaranteed residence time and heat transfer characteristics. The gas absorber differs from the other types in that the liquid phase is staged through the column, and with the stirring at each stage a high efficiency column results.

9.4.3 Costing Data

The very wide range of mixing equipment available makes it impossible to produce any simple all-embracing cost correlation. The best attempt, made by Drew and Ginder[53] covers only the costs of agitators for use in mixing vessels. Their figures, updated to 1970 prices, are shown below in Table 9–12.

TABLE 9–12. AGITATOR COST DATA.[53] $ (1970) $= CX^n$

Type of agitator (all with explosion proof motor and made of mild steel*)	Range of X		n	C	
	(A.A.) (hp)	[SI] [kW]		(A.A.)	[SI]
Portable, propeller type	0.25–2	0.2–1.5	0.58	312	371
direct drive	0.25–1	0.2–0.75	0.44	431	493
geared drive	1–3	0.75–1.5	0.34	431	479
Turbine, geared drive mechanical seal					
constant speed	1–2	0.75–1.5	0.10	1470	1570
	2–3	1.5–2.3	0.72	1570	1950
variable speed	1–2	0.75–1.5	0.10	1580	1630
	2–3	1.5–2.3	0.49	1690	1960
Anchor, geared drive stuffing box seal					
constant speed	1–3	0.75–2.3	0.49	1100	1280
variable speed	1–3	0.75–2.3	0.41	1420	1610

```
* Multiplying Factors for other materials:
    Stainless Steel Type 304   = 1.13      Monel                        = 1.41
    Stainless Steel Type 316   = 1.18      Rubber covered mild steel = 1.53
```

For any other type of mixing equipment reference should be made to the manufacturer of the particular item for an accurate cost estimate.

9.5 Grinding

The problem of the size reduction of solids ranges from coarse crushing to very fine grinding. It is a particularly important unit operation which unfortunately is not amenable to theoretical treatment and analysis. It may be the first operation in a chemical process or the last step before product packaging. Examples may be taken as the preliminary grinding of phosphate rock in phosphoric acid production, and the final grinding of clinker in the manufacture of cement.

The specification of crushing or grinding equipment is not a simple task since size reduction is primarily an empirical subject. The more available knowledge on the feed material, the simpler will be the selection of equipment. These are twelve characteristics which should be considered[54] and these may be listed as hardness, toughness, abrasiveness, stickiness, softening or melting temperature, structure, specific gravity, free moisture content, chemical stability, homogeneity, physiological effect, and purity.[55] It is important to remember that no two materials are exactly alike, just as no two product specifications are identical. It is therefore necessary, before purchasing grinding equipment, to supply the manufacturer not only with a representative sample of the material but also with the finished product specifications.

The choice of final product size is important from cost considerations. Bond[56],[57] produced an expression in an attempt to improve upon the earlier equations of Rittinger[58] and Kick[59] for the prediction of power consumption in a size reduction process. This equation is

$$W = 10W_i(1/\sqrt{P} - 1/\sqrt{F}) \tag{9-18}$$

where W = work input (kWh/ton),
W_i = work index (kWh/ton),
P = product size (micron),
F = feed size (micron).

The work index for approximately eighty materials has been published by Bond,[60] and its values range from 1.43 for calcined clay to 134.5 for mica with an average value of 13.81. The particle sizes used in equation (9–18) are those sizes in microns which 80 per cent of the product and feed passes.

Example 9–5

Compare the power requirements to grind a material with a work index of 20.0 from 10 to 1 mm, 1 mm to 100 μ, and 100 μ to 10 μ.
From equation (9–18), for the size range 10–1 mm,

$$W = 10 \times 20 \, (1/\sqrt{1000} - 1/\sqrt{10\,000})$$
$$= 4.32 \text{ kWh/ton}$$

For the size range 1 mm–100 μ,

$$W = 10 \times 20 \, (1/\sqrt{100} - 1/\sqrt{1000})$$
$$= 13.7 \text{ kWh/ton}$$

For the size range 100 μ to 10 μ,

$$W = 10 \times 20 \, (1/\sqrt{10} - 1/\sqrt{100})$$
$$= 43.2 \text{ kWh/ton}$$

It will be seen from Example 9–4 that more power is required per unit weight to grind from medium to fine than from coarse to medium, and even more is required to grind from fine to very fine. Hence the product size should be selected as the largest that any subsequent processing operations or usage will permit.

The published literature contains many descriptions of available grinding equipment. Most recent are those of Harley[61] and Hiorns.[62] Crushers and mills may be classified into six groups—crushers, ball mills, roller mills, high speed mills, fluid energy mills, and non-mechanical systems. Table 9–13 shows the classification of Hiorns, and is useful for selection of the most suitable type of crushing and grinding equipment.

Generalized cost data on grinding equipment is difficult to obtain. The only information published has been by Mills[63] and Guthrie,[64] both of whom were primarily concerned with crushers. Their data has been updated and is presented in Table 9–14.

Sizing of crushing and grinding plant can also present problems in that the equipment will have a capacity in terms of weight processed per unit time for a material of a particular hardness. Manufacturers will therefore have to be consulted to obtain a reliable capacity for the particular material in the selected machine, thus emphasizing again the importance of the knowledge of the properties of the feed.

Consideration should also be given at the selection stage to the provision of the necessary solids handling equipment, whether the process be continuous or batchwise. To improve the quality of the product and the capacity of the plant, it may be desirable to install a classification system whereby fines removal and oversize recirculation may be effected where these features are not included in the size reduction equipment.

With grinding, probably more than with most equipment, close co-operation between the process designer and the manufacturer is essential if the most suitable and economic selection of plant for the particular duty is to result.

9.6 Storage and Storage Vessels

Storage of solids liquids and gaseous products at intervals between production, transportation, refining, blending and marketing, ensures sufficient balance of stock to allow continuous operation. In general, large vessels are employed which can be either installed as shopbuilt units or, in the case of very large units, manufactured on site. Mild steel is most commonly used as a material of construction, but storage vessels are made from a wide range of materials. The technique of lining mild steel vessels with a suitable inert material such as plastic, rubber or glass is now commonplace, where previously an expensive corrosion resistant metal would normally have been employed.

Materials which are normally stored in bulk may be classified into four groups:

(i) non-volatile liquids,
(ii) volatile liquids,
(iii) gases,
(iv) solids.

Each of these groups will be considered in turn.

TABLE 9–13. CRUSHING AND

Type	Variants	Wet (W) or Dry (D)	Typical size Feed
Jaw crusher	Position of moving jaw pivot	D	Several feet to $\frac{1}{2}$ in. [12 mm]
Rotary crusher	Gyratory cone Gyratory jaw	D	Several feet to 1–2 in. [25–50 mm]
Rolls	Smooth, fluted, spikes, single, double, multiple systems breakers	W, D	About 3 in. [80 mm] but depends on gap
Autogenous	Semi-autogenous	W, D	Several feet to several inches
Pan	Edge runner	W, D	About 3 in. [80 mm]
Tumbling	Ball, pebble, rod with length and diameter variations (sometimes with integral classifiers); compartment mills; sub- and super-critical	W, D	Depends on ball size and feed strength; not more than 1–2 in. [25–50 mm]
Vertical spindle	Ring-ball Ring-roll Bowl	D	1–2 in. [25–50 mm]
Rotary impact	Fixed and swing hammer. Paddle (with or without integral screens); cage disintegrators; pin-disc shredders; rotary knives; fan beaters	D	Several feet to $\frac{1}{16}$ in. [2 mm]
Vibration	Horizontal and vertical types	D, W	$\frac{1}{8}$ in. [3 mm]
Fluid energy	Opposed jet Spiral jet Colloid	D W	$\frac{1}{2}$ in.–100 mesh [12 mm–150 μ]
Non-mechanical	Electro-hydraulic; thermal shock	W	Large blocks to inch sizes

GRINDING EQUIPMENT[62]

range

Product	Material type	Notes
Down to $\frac{1}{8}$ in. [3 mm]	All hardness ranges	Relies on direct compression; not suitable for tough or plastic material.
Down to $\frac{1}{8}$ in. [3 mm]	All hardness ranges	Variable speed; feed limitations generally those of jaw crushers.
Micron sizes	All materials except plastic, readily aggregating or rough	Variable (including differential) speed and roll surfaces give wide applicability. Angle of nip criterion must be satisfied, first cost tends to be high. Uniform feed along roll length essential to prevent uneven wear. Short residence time.
Down to about 100 mesh [150 μ]	Best suited to dense brittle or degradable materials	Can be used as combined mill–drier with hot air. Semi-autogenous have some grinding media added.
About [50 μ] in closed circuit or batch	Generally used for clays and soft minerals	Material hardness limited by wear effects, and size by nip angle.
About [10 μ]	Suitable for most materials that will fragment under compression	Used in batch or open or closed circuit. Dry milling hot or cold air swept, or overflow type. Efficiency falls off as product particle size decreased. Production of ultrafine sizes aided by wet operation and surface active agents. Long residence time.
About [50–100 μ]	Used particularly for coal and grinding feed materials for cement production	Air swept; efficient drier. May be used with integral classifiers. Intermediate residence time.
To [100 μ] or below depending on hardness of feed	Wide range including brittle and non-brittle materials, organics, etc. Used for refuse disintegration	Some suitable for moist sticky materials. Not suitable for very hard materials because of high wear rates, except at low speeds, and hence best for relatively soft materials. Short residence time; not good mixer.
[100 to 10 μ] and below	All ranges of hardness; not suitable for tough plastic or rubber-like materials	Can readily be adapted to a cooled unit for temperature sensitive feeds in small sizes. Transmission of grinding forces poor with viscous fluids. Rate of loading small.
[100 to 10 μ]	Any; operating fluid can be chosen to suit material	Economics depends on availability of high-pressure fluid. Colloid mills used mainly for aggregate dispersion. Narrow size distribution; useful for temperature sensitive materials. Residence time of the order of 100 s.
Large pieces down to [100 μ]	Any	Efficiency generally inferior to mechanical methods.
	Specific thermomechanical properties required	Electrohydraulic can give narrow size range and favours formation of equi-dimensional fragments.

TABLE 9-14. CRUSHING EQUIPMENT COST DATA; COST EQUATION $Y = Cf(X)^n$

Type	X (A.A.)	X [SI]	Range of X (A.A.)	Range of X [SI]	Y(1970)	n	Cost equations $Y = C \cdot f(X)$ (A.A.)	[SI]	Reference
Roll crushers	ton/hr	kg/s	15–150	4–40	$	1.36	55.5(X + 20)	310(X + 5.65)	63
Cone crushers	ton/hr	kg/s	—	—	$	0.85	840X	2450X	64
Gyratory crusher	ton/hr	kg/s	100–1000	30–300	10^3 $	0.27	24.4(X − 90)	35.6(X − 25.4)	63
Jaw crusher	ton/hr	kg/s	—	—	$	2.20	62X	1000X	64
	ton/hr	kg/s	—	—	$	1.20	95X	433X	64
Ball mills	ton/hr	kg/s	10–800	3–250	$	0.70	1950X	4720X	63

9.6.1 Liquid Storage

(i) Non-volatile Liquids

These present the easiest storage problem and are usually stored in standard cylindrical tanks. Tank sizes will normally lie in the range of up to 200 ft [60 m] diameter and 100 ft [30 m] in height. Construction is in the form of a butt-welded vertical cylinder and the roof is usually supported on a roof truss but not attached to it. Thus the tank may 'breathe' with changes in temperature and atmospheric pressure.

(ii) Volatile Liquids

The storage of volatile liquids involves the design of tanks which are able to conserve the contents. There are three main ways of achieving this conservation:

  (i) variable volume tanks, e.g. floating roof tanks,
  (ii) variable pressure vessels,
  (iii) constant temperature tanks, e.g. underground storage.

In a floating roof tank, the vertical walls are of standard construction, but the roof floats directly on the liquid surface, thus blanketing the surface to reduce evaporation loss, corrosion and fire hazard. The most important part of the structure is the flexible seal between the floating roof and the tank shell, and if this is well designed evaporation losses will be negligible.

The surface area per unit volume of storage vessel reaches a minimum in the spherical tank. It is particularly suited to pressure applications, as the favourable stress distribution permits thinner walls than are acceptable in cylindrical tanks. The cost advantages in services, where expensive corrosion-resistant materials are to be employed, are apparent. Typical products normally stored in spherical vessels are the liquefied gases ammonia, liquefied petroleum gas (LPG), propane, butane and many of the volatile products of petroleum processing.

The attractions of the use of controlled temperature tanks for the storage of volatile liquids will be obvious; however, in practice more use is made of pressure vessels. The use of underground storage tanks has been mentioned and these are of two types. The most familiar example is the underground storage of petroleum at filling stations, and this type of storage may be suitable in many cases and provides constant temperatures at a reasonable cost. For very large-scale operations, use is sometimes made of caverns excavated from the rock strata below the plant location. Excavation may be undertaken by blasting, mining, or by dissolving salt deposits underground. An example of the last type has been described[65] and a cost quoted as 5 per cent of the cost of above ground storage of LPG. It is impossible to attempt any correlation of costs for this type of storage, but for other types of vessel costs are tabulated in Table 9–17 at the end of this section.

9.6.2 Gas Storage

Gases may be stored under four conditions:

(i) In constant pressure, variable volume gasholders.
(ii) In variable volume and slightly variable pressure gasholders.
(iii) In variable pressure gasholders (40–200 p.s.i.g.) [380–1500 kN/m²].
(iv) As liquids.

Gas holders of the first two types are most familiar for the storage of coke oven or towns gas and need little description. Use of the pressure vessel in either a spherical or cylindrical form has been discussed earlier and relevant costing data is presented later. The liquefaction of gases may be achieved by increasing the pressure, by refrigeration, or by a combination of the two. In recent years, bulk supply tanks of propane and butane in particular have become a common sight, and their storage as liquids under pressure is a familiar example of the first type of liquid gas storage. Storage examples of liquid gases under refrigeration are ethylene and methane. In the bulk storage and transportation of liquefied gases in large quantities it is now normal to liquefy the gas and to store it in a well insulated tank at, or slightly above, atmospheric pressure. Due to heat in leakage, continuous evaporation takes place at the surface and these vapours are withdrawn, compressed, condensed and then returned to the tank as liquid. This method of storage is employed in ocean-going tankers where single tanks capable of holding 15 000 ton [30 000 m³] of liquid gas are used and built to withstand a maximum pressure of 1.5 lb/in.² g [112 kN/m²]. The importance of insulation, materials of construction and aspects of mechanical design of this and other types of vessel have been discussed by Blenkin and Wardale.[66] The design codes which cover the range of pressure vessels included in this section are detailed in Table 9–15, and where a detailed design is to be undertaken the relevant code should be followed.

TABLE 9–15. DESIGN CODES FOR PRESSURE VESSELS

	British	American
Nominal pressure	B.S. 2654	A.P.I.–650
Low pressure Pressure vessels	— B.S. 1500 (1958) A.O.T.C. Rules Lloyds Register of Shipping	A.P.I.–620 A.S.M.E. Section VIII 'Unfired Pressure Vessels'

Table 9–16[67] has been included as a convenient guide for the selection of storage vessels for gases with a classification according to volume and pressure. Details of cost data for many of the types of storage tanks are presented in Table 9–17 and are corrected to values at June 1970 by means of the index in Chapter 2.

TABLE 9–16. RECOMMENDED VESSELS FOR GAS STORAGE[67]

Volume		Operating pressures (lb/in.² g) [kN/m²]					
(ft³)	[m³]	0–1 [101–108]	1–30 [108–310]	30–100 [310–790]	100–150 [790–1130]	150–300 [1130–2070]	over 300 [2070]
0–1 000	0–28	Tank	Tank or drum	Drum	Drum	Drum	Cylinder[1]
1 000–5 000	28–140	Tank	Drum	Drum	Drum	Cylinder	Cylinder
5 000–17 000	140–480	Tank or gas holder[3]	Spheroid[2]	Sphere[2]	Sphere	Not economical	Not economical
17 000–500 000	480–14 000	Gas holder[3]	Spheroid	Sphere	Sphere	Not economical	Not economical

[1] Gas cylinder. A welded or forged vessel with a ratio of length to diameter of between 5:1 and 50:1. Normally with hemispherical or specially designed heads.

[2] Spheroids and sphere usually in standard sizes in the range 32–120 ft diameter [10–38 m].

[3] Gas holders are usually in standard sizes from 10 000 to 10 000 000 ft³ capacity [280–280 000 m³].

TABLE 9–17. STORAGE TANK COST DATA

Description	Pressure (lb/in.² g)	[kN/m²]	X (A.A.)	[SI]	Range of X (A.A.)	[SI]	Y	n	C(A.A.)	C[SI]	Reference
Large storage tanks											
mild steel	Atmospheric		10 tons	kg	5–500	50 000–5 000 000	$	0.58	1930	9.25	68
stainless steel			tons	kg	50–500	50 000–500 000		0.57	1930	37.6	68
Vertical cylindrical field erected											
cone roof	Atmospheric		10³ U.S. gal.	m³	40–5000	150–19 000	$	0.63	894	386	64
floating roof								0.63	950	410	64
lifter roof								0.63	1120	484	64
Horizontal storage tanks											
mild steel	Atmospheric		Imp gal	m³	50–1000	0.25–5	£	0.63	3.75(X + 100)	425(X + 0.046)	69
mild steel			10³ gal	m³	1–15	5–70	£	0.63	290(X + 0.1)	478(X + 0.046)	69
stainless steel			Imp gal	m³	50–1000	0.25–5	£	0.66	15.2	2360	69
stainless steel			10³ gal	m³	1–15	5–70	£	0.66	1450	2240	69
Spherical mild steel pressure vessels	30	310	10³ U.S. gal	m³	10–400	40–1500	$	0.70	2800	1100	64
	100	790						0.70	3500	1380	64
	200	1500						0.70	4450	1750	64
Field erected mild steel pressure vessels	10	170	10³ft³	m³	60–280	1700–8000	£	0.77	1060	81	70
	20	240						0.77	1170	89	70
	30	310						0.77	1250	95	70
	50	450						0.80	1460	101	70
Horizontal mild steel cylindrical tanks	150	1050	10³ U.S. gal	m³	1–100	4–400	$	0.65	1680	706	64
	200	1500						0.65	1940	815	64
	250	1850						0.65	2220	932	64

9.6.3 Solids Storage

The handling and storage of solid materials is much more costly than for liquids,[67] and for this reason the volume of bulk solid storage at any stage in a process should be kept to a minimum.

The simplest means of storing solids is in a pile of material on the ground with handling performed by mechanical shovels. Obviously this scheme has limitations, and field erected silos are commonly used. These may be lined with suitable materials such as rubber, glass or plastic to suit the particular material, and loading and unloading may be carried out with belt conveyors.

Where a controlled feed rate of solids is required for a continuous process, an overhead storage bin is conveniently used. Bridging and packing of granular materials is a common problem with these bins and House[67] has listed some of the considerations which should be considered in their design. A cylindrical tank is to be preferred to a rectangular bin to minimize both equipment cost and problems of solids packing. The normal maximum size for such a vessel is 11.5 ft [3.5 m] diameter by approximately 30 ft [10 m] high. The outlet from the bin is best constructed as an eccentric cone with one straight vertical side and the total included angle of the cone not exceeding 45°. This angle should be reduced to 30° for materials which do not flow readily.

9.7 Nomenclature

C	cost	(£)	[$]
F	feed material size	(μ)	
H	total head loss	(ft)	[m]
M	molecular weight	(—)	[—]
n	exponent in cost equations	(—)	[—]
P	pressure	(lb/in.^{2})	[kN/m^2]
P	product material size	(μ)	
Q	volumetric liquid flow	(gal/min)	[m^3/s]
Q	capacity—equation (2–2)	(—)	[—]
S	specific gravity	(—)	[—]
T	absolute temperature	(°F)	[K]
W_b/W_a	entrainment ratio	(lb air/lb)	[kg/kg]
W	work input	(kWh/ton)	
W_i	work index	(kWh/ton)	
X	parameter in cost equations	(—)	[—]

Subscripts

a	component a
b	component b

REFERENCES

1. Perry, J. H. Ed. *Chemical Engineer's Handbook*, 4th ed. New York: McGraw-Hill Book Co., 1963.
2. Coulson, J. M. and Richardson, J. F. *Chemical Engineering*, Vol. 1, 2nd ed. Pergamon Press, London, 1967.
3. MacDonald, J. O. S. *Brit. Chem. Eng.* Oct./Nov. 1956, **3**, 298.
4. Pulsometer Engineering Co. Ltd. *Data for Pump Users.*
5. Institution of Chemical Engineers, London. *Capital Cost Estimation*, 1969.
6. Thurlow, C. III, *Chem. Eng.* 1965, **72**, 11, 117.

7. Thurlow, C. III, *Chem. Eng.* 1965, **72**, 12, 213.
8. Thurlow, C. III, *Chem. Eng.* 1965, **72**, 13, 133.
9. Rost, M. and Visich, E. T., *Chem. Eng.* 14 Apr., 1969, **76** 45.
9a. Holland, F. A. and Chapman, F. S. *Pumping of Liquids.* New York: Reinhold Publishing Corporation, 1966.
10. Beck, N. *Chem. Proc. Eng.* Apr. 1966, **47**, 172.
11. Begg, G. A. J. *Chem. Proc. Eng.* Apr. 1966, **47**, 153.
12. Wholley, R. F. *Chem. Proc. Eng.* Apr. 1968, **47**, 161.
13. Bauer, H. *Chem. Proc. Eng.* Apr. 1966, **47**, 164.
14. McCulloch, D. M. Paper to N. W. Branch Inst. Chem. Eng., 12 Dec. 1965.
15. Clarke, N. M. *Chem. Proc. Eng.* Apr. 1966, **47**, 167.
16. Bresler, S. A. and Smith, J. H. *Chem. Eng.* 1 June 1970, **77**, 161.
17. Stokes Microvac Pumps, Bulletin 526, Pennwalt-Stokes, Stokes Vacuum Components Dept. Philadelphia.
18. DeFrate and Hoerl. *Chem. Eng. Prog. Symp. Ser.* 1959, **55**, 21, 46.
19. Garrett, D. E. and Rosenbaum, G. P. *Chem. Eng.* 11 Aug. 1958, **65**, 125.
20. Van Hook, A. *Crystallisation—Theory and Practice.* New York: Reinhold Publishing Corp., 1961.
21. Mullin, J. W. *Crystallisation.* London: Butterworths, 1961.
22. Drew, T. B., Hoopes, J. W. Jnr. and Vermeulen, T. Eds. *Advances in Chemical Engineering*, 3. *Crystallisation from Solution.* New York and London: Academic Press, 1962.
23. Ostwald, Z. *Physik Chem.* 1897, **22**, 302.
24. Buckley, H. E. *Crystal Growth.* London: John Wiley and Sons, 1951.
25. Preckshot, G. W. and Brown, G. G. *Ind. Eng. Chem.* 1952, **44**, 1314.
26. Young, J. *Am. Chem. Soc.* 1911, **33**, 148.
27. *Ibid.*, 1913, **35**, 1067.
28. Ting, H. H. and McCabe, W. L. *Ind. Eng. Chem.* 1934, **263**, 1201.
29. Bamforth, A. W. *Chem. Proc.* 1970, **16**, 1.
30. Bamforth, A. W. *Industrial Crystallisation.* London: Leonard Hill, 1965.
31. Armstrong, A. J. *Chem. Proc. Eng.* Nov. 1970, **51**, 59.
32. Nelson, W. L. *Petroleum Refinery Engineering*, 4th Ed. New York: McGraw-Hill, 1958.
33. Kobe, K. A. and McKetta, J. J. *Advances in Petroleum Chemistry and Refining*, Vol. 10, Interscience, 1965.
34. Cole, J. W. 'Crystallisation'. Supplement to *Chemical Processing*, Oct. 1966.
35. Jeremiassen, F. British Patents. 240164 8 July, 1926, 392829 25 May, 1933.
36. Anon. *Chem. Proc. Eng.* Nov. 1970, **51**.
37. Newman, H. H. and Bennett, R. C. *Chem. Eng. Prog.* 1959, **55**, 1, 65.
38. Randolph, A. D. *Chem. Eng.* 4 May 1970, **77**, 80.
39. Miles, J. E. P. and Schofield, C. *Chem. Proc.* 1970, **16**, 8, 3.
40. Bourne, J. R. *The Chem. Eng. London.* 1964, No. 4/1 80, 202.
41. Valentin, F. H. H. *Chem. Proc. Eng.* 1967, **48**, 10, 69.
42. Griffin, M. J. *Chem. Proc. Eng.* 1969, **50**, 3, 69.
43. Lloyd, P. J. and Young, P. C. M. *Chem. Proc. Eng.* 1967, **48**, 10, 57.
44. Anon. *Chem. Proc. Eng.* 1969, **50**, 3.
45. Jones, J. E. M. *Chem. Proc. Eng.* 1967, **48**, 10, 66.
46. Frenkel, M. S. Brit. Patent No. 842692 and U.S. Rubber Co., U.S. Patent No. 2744207.
47. Sterbacek, Z. and Tausk, P. *Mixing in the Chemical Industry.* New York: Pergamon Press, 1965.
48. Chilton, T. H. and Genereaux, R. P. *A.I.Ch.E.J.* 1930, **25**, 102.
49. Clayton, C. G. *et al.* U.K. Atomic Energy Authority Research Report A.E.R.E.—R5569, 1968.
50. Penney, W. R. *Chem. Eng.* 1 June 1970, **77**, 171.
51. Holland, F. A. and Chapman, F. S. *Liquid Mixing and Processing in Stirred Tanks.* New York: Reinhold Publishing Corporation, 1966.
52. Jones, R. L. *Chem. Proc. Eng.* 1969, **50**, 3, 65.
53. Drew, J. W. and Ginder, A. F. *Chem. Eng.* 1970, **77**, 3, 100.
54. Harley, M. A. E. *Chem. Proc.* 1970, **16**, 3, 10.
55. Farrant, J. C. and North, R. *Chemical Engineering Practice*, Ed. Cremer, H. W. and Davies, T., Vol. 3, 48–96. London: Butterworths Scientific Publications, 1957.

56. Bond, F. C. *Am.I.M.E. Trans.* 1952, **193**, 484.
57. Bond, F. C. *Brit. Chem. Eng.* 1961, **6**, 6, 378.
58. Von Rittinger, P. R. *Lehrbuch der Aufbereitingskunde*. Berlin, 1867.
59. Kick, F. *Das Gesetz der proportionalem Widerstand und Siene Anwendung*. Leipzig, 1885.
60. Bond, F. C. *Brit. Chem. Eng.* 1961, **6**, 8, 543.
61. Harley, M. A. E. *Chem. Proc. Supplement*, 1969, **15**, 11, 529.
62. Hiorns, F. J. *Brit. Chem. Eng.* 1970, **15**, 12, 1565.
63. Mills, R. E. *Chem. Eng.* 16 March 1964, **71**, 133.
64. Guthrie, K. M. *Chem. Eng.* 1969, **76**, 6, 114.
65. Anon. *Canadian Petrol.* Feb. 1970, **11**, 71.
66. Blenkin, B. and Wardale, J. K. S. *Chem. Proc. Eng.* 1963, **44**, 7, 344.
67. House, F. F. *Chem. Eng.* 28 July 1969, **76**, 120.
68. Bauman, H. C. *Ind. Eng. Chem.* 1960, **52**, 12, 51A.
69. Anon. *Ind. Chem.* 1963, **39**, 2, 90.
70. Anon. *Ind Chem.* 1963, **39**, 5, 258.

Chapter 10: Scale-up of Process Equipment

10.1 Introduction

One of the prime functions of a chemical engineer is the development and design of equipment for the manufacture of materials in bulk on production-scale plant. This is based on data obtained in the laboratory initially at the bench scale. Such data are in many ways of limited value and serve to indicate parameters on an overall basis only. Essentially what is missing is the rate at which operations can be carried out—information which is absolutely vital to the chemical engineering of a production unit. For example, the fact that a mixture of reactants in a beaker can be brought to the required temperature with a bunsen burner in, say, 200 s is largely irrelevant in the design of a heat exchanger required for the production unit. The bench-scale test may provide data on the total heat involved, but the rate at which this heat is transferred is unknown. For this reason, pilot-scale plants are employed to provide the chemical engineering data so important in the design of production units. Such plant is essentially a copy of the proposed unit in which items of equipment are scaled down to say 1/100–1/10 full size. This not only provides design data for the full-scale plant, but is also used to produce small quantities of the end product with a view to development and assessment of properties and the saleable value of materials by way of market survey techniques. Even when the full-scale plant is in production, the pilot plant is used for studying proposed modifications to the process and product which may be impracticable or un-economic on the production sized unit.

The value of pilot-plant evaluation is almost entirely dependent on the ability to scale-up accurately. For example, if two liquids can be mixed satis-factorily in a pilot-scale vessel with a given motor and stirrer, it must be known how to achieve the same degree of mixing on the full-scale unit. Will geometric similarity alone suffice or will a proportionately larger motor be required? Are the turbulence characteristics the same in both vessels—is this an important factor? It is hoped in this chapter to provide an indication of the principles of scale-up of chemical plant which will go some way towards answering this sort of question. It must be pointed out at the outset that the whole value of pilot-plant work depends on accurate scale-up and the topic deserves a much fuller treatment than is possible in the present text. There are many comprehensive books on the subject and special mention is made of *Pilot Plants, Models and Scale-up Methods in Chemical Engineering* by Johnstone and Thring,[1] as well as an excellent series of articles by Holland published in *Chemical Engineering*.[2] The material in this latter publication forms the major part of the discussion in the present chapter.

Construction and development of pilot plant can be a significant proportion of the cost of a production unit, and recent trends have led to the replacing of both bench- and pilot-scale work by technical scale units, which in simple terms are dimensionally intermediate between the other scales of operation. Scale-up from the technical to the full-size plant thus requires an even greater degree of extrapolation and makes greater demands on the accuracy of scale-up techniques. Although the science and, in some respects, the art of scale-up has been developed over the years to a relatively high degree of perfection, there still remains a wide area of operation in which data for design can still only be obtained by building a full-scale unit. Such practice is hazardous to say the least, and large and expensive safety factors must be incorporated into the design. This state of affairs is particularly prevalent in furnace design, especially in water-tube boilers used in power stations, where the large increases in capacity dictate that the previous full-sized unit virtually becomes the pilot plant for the next unit, which may have double the output.

Certain operations require little other than obvious scale-up data. For example, if on the pilot unit a vessel is used as a buffer tank and contains 10 ft^3 [0.28 m^3] fluid, then with a scale-up factor of, say, 3, the full-size unit will have a design capacity of 30 ft^3 [0.85 m^3]. Similarly, where plant throughputs are increased by a factor of, say, 10, operating conditions, temperatures, pressures and so on remain the same and volumes and heat loads are increased by the same factor, 10. Where scale-up techniques are important is in the rate of carrying out operations, especially in reactions, mixing, heat and mass transfer, and a discussion of the equipment in which such operations are carried out forms the major part of this chapter.

10.2 Basic Principles of Scale-Up

The most important tool in scale-up is the principle of similarity—a concept originally formulated by Newton and expressed by equations of the form

$$\underline{A} = \phi(\underline{B}, \underline{C} \ldots) \tag{10-1}$$

where a dimensionless group $\underline{A}$ is a function of other dimensionless groups $\underline{B}$, $\underline{C}$, etc. Each dimensionless group represents a rule for scale-up and the actual groups involved depend upon the character of the regime. A dimensionless group is a very convenient way of correlating scientific and engineering data, and perhaps the best known example is specific gravity—the ratio of the weight of a given volume of a liquid to the weight of the same volume of water at 25°C [298 K]. Many dimensionless groups represent the ratio of applied to opposing forces in a system. For example, the Reynolds number is the ratio of the applied force to the viscous drag.

Most of the systems studied in pilot-plant investigations involve fluids, and in fluid dynamic regimes three types of similarity must be considered: geometric, kinematic and dynamic. Geometric similarity exists between two systems when the ratios of corresponding dimensions in one system are equal to those of the other—in simple terms both pilot and production units have the same shape. When two systems are kinetically similar, not only have they the same shape but the ratios of velocities between corresponding points are also equal. Finally, dynamic similarity exists when, in addition to being geometrically and kinetically similar, the ratios of forces between corresponding points are also equal in both systems. In many cases it is not possible to achieve dynamic similarity between the pilot and the production unit, and, in this case, data obtained on the former must be extrapolated to dynamically dissimilar conditions on the larger scale. In order to achieve this, use is made of the extended principle of similarity, which employs equations of the form

$$\underline{A} = K(\underline{B})^m(\underline{C})^n \tag{10–2}$$

where K is a constant depending on the geometry of the system and which is usually assigned a value determined experimentally.

In applying these scale-up principles there are two important conditions required for reliable scale-up:

(i) The regime must be relatively pure and scale-up on dynamic similarity should depend mainly on one dimensionless group. For example, in a fluid dynamic regime opposing forces should be due to viscosity or surface tension or gravity but not a combination of all three.

(ii) The regime must not change in going from the small to the large scale.

It is therefore important that in the design of pilot-scale equipment, the effects of certain dimensionless groups are minimized in favour of one particular group.

10.3 Scale-up of Heat Exchange Systems

10.3.1 General Principles

The basic rate equations for heat, mass and momentum transfer are all of the form described in equation (10–2). For example, the general equation for convective heat transfer is

$$Nu = C'Re^mPr^n \tag{10–3}$$

and, as a typical example, for the case where turbulent fluids are heated in clean, round pipes, this becomes

$$Nu = 0.0225 \, Re^{0.8}Pr^{0.4} \tag{10–4}$$

As explained previously, the constant is affected much more than the indices by the geometry of the system, and hence the equation is limited to a particular configuration. However, by comparing two systems, 1 and 2, of similar geometries, the constant is eliminated. Furthermore, if one makes the not unreasonable assumption that the temperatures at various points in the two systems, and hence the physical properties, are the same, then equation (10–3) reduces to

$$h_1d_1/h_2d_2 = (u_1d_1)^m/(u_2d_2)^m$$

or
$$h_1/h_2 = (u_1/u_2)^m(d_1/d_2)^{m-1} \tag{10–5}$$

Thus from measurements of the film coefficient h_1 at a velocity of u_1 in the pilot-scale unit, the equivalent coefficient in the full-scale unit can be predicted as a function of the velocity, knowing the dimensions of the geometrically and thermally similar systems. It is important to appreciate that in using such techniques the character of the flow pattern must be the same in both plants. The easiest way to achieve this is to arrange that $Re_1 = Re_2$, though problems may arise as such dynamic similarity may involve supersonic velocities on the small scale. In order to illustrate the application of the techniques two simple examples are now considered.

10.3.2 Scale-up of a Jacketed Vessel

Taking the case of a pilot-plant investigation involving a simple stirred reactor which is heated by steam in an external jacket, it may be assumed that the inside coefficient of heat transfer is controlling and that this is known from experiments together with the stirrer speed, power input and the physical properties of the fluid. It is desired to predict the performance of a full-size reactor which is geometrically similar, with the same type of agitator and baffles and with a linear dimension three times greater than the pilot-plant unit, i.e. $D_{t_2} = 3D_{t_1}$. In addition, the depth of liquid is geometrically the same in both cases; say $H_L = D_t$, the tank diameter. Neglecting chemical engineering considerations for the moment, it is fairly obvious that the area for heat transfer is nine times greater in the full-scale than in the pilot unit, and the factor relating the volumes of material is twenty-seven, i.e.

$$A_2 = 9A_1, \qquad V_2 = 27V_1$$

Now the heat load is directly proportional to the volume of material and therefore $Q_2 = 27Q_1$, but the area for heat transfer is only nine times greater. It is therefore three times harder to heat the material in the full-scale vessel, assuming the same overall coefficient and mean temperature difference.

From the chemical engineering viewpoint, the important factor is the inside film coefficient, h_i. This is given by[3]

$$h_i D_t/k = 0.74(\rho N D_i^2/\mu)^{0.67}(C_p\mu/k)^{0.33}(\mu/\mu_w)^{0.14} \qquad (10\text{-}6)$$

where D_t = tank diameter (ft) [m],
$\quad D_i$ = impeller diameter $(D_i = D_t/3)$ (ft) [m],
$\quad N$ = impeller speed (rpm) [Hz].

If the temperatures and hence the physical properties are the same in both systems, then equation (10–6) reduces to

$$(h_i D_t)_2/(h_i D_t)_1 = (N D_i^2)_2^{0.67}/(N D_i^2)_1^{0.67}$$

or $\qquad h_{i2}/h_{i1} = (D_{t1}/D_{t2})(N_2/N_1)^{0.67}(D_{i_2}/D_{i_1})^{1.33}$

If $D_i = D_t/3$ in both cases, then

$$h_{i_2}/h_{i_1} = (N_2/N_1)^{0.67}(D_{i_2}/D_{i_1})^{0.33} \qquad (10\text{-}7)$$

In this particular example, $D_{i_2} = 3D_{i_1}$ and hence

$$h_{i_2}/h_{i_1} = 1.44(N_2/N_1)^{0.67} \qquad (10\text{-}8)$$

The impeller tip speed, t_s is given by

$$t_s = \pi D_1 N$$

or
$$N = t_s / \pi D_1$$

Substituting for N in (10–8),

$$(h_{1_2}/h_{1_1}) = 1.44(t_{s_2}/t_{s_1})^{0.67}(D_{1_1}/D_{1_2})^{0.67}$$
$$= 0.685(t_{s_2}/t_{s_1})^{0.67} \tag{10–9}$$

Considering the basic equation for heat transfer, $Q = UA\Delta t_m$, and assuming that the inside film coefficient is controlling and $U \approx h_1$, then

$$(Q_2/Q_1) = (h_{1_2}/h_{1_1})(A_2/A_1)(\Delta t_{m_2}/\Delta t_{m_1})$$

As discussed previously, $Q_2/Q_1 = 27$, $A_2/A_1 = 9$ and for equal temperature conditions, $\Delta t_{m_2} = \Delta t_{m_1}$. Thus, in order for the full-sized unit to perform the scaled-up duty, h_{1_2}/h_{1_1}, *must* equal 3.

Considering equation (10–9), if the impeller tip speeds are equal in both the full-size and the pilot-scale units, then $h_{1_2}/h_{1_1} = 0.685$, which is $0.685 \times 100/3 = 23$ per cent of the value required. The only way of increasing h_{1_2} is to increase the impeller tip speed, though here a twofold increase is the maximum which can be tolerated, as shown in Table 10–1. With $t_s = 2t_{s_1}$, then in equation 10–9,

$$h_{1_2}/h_{1_1} = 0.685(2)^{0.67} = 1.09$$

or
$$(1.09 \times 100/3) = 36\% \text{ of required value}$$

TABLE 10–1. VALUES OF IMPELLER TIP SPEEDS[4]

Degree of agitation	Impeller tip speed t_s	
	(ft/min)	[m/s]
Low	500–650	2.5–3.3
Medium	650–800	3.3–4.0
High	800–1000	4.0–5.5

It must be concluded, therefore, that even with twice the impeller speed, the full-size unit is incapable of coping with the scaled-up duty, and in order to effect the same transfer of heat per unit mass of material in the system an external heat exchanger is required in the full-size unit. Without this additional item, either the temperatures in the two systems would not be equal or the batch time or size of batch handled would be vastly different.

10.3.3 Scale-up of heat exchangers

The very wide range of types and designs of heat exchangers renders the topic too complex for comprehensive treatment here, and the discussion will be confined to shell and tube units in which the inside film coefficient is controlling, i.e. $U \approx h_1$. In most cases this assumption introduces an error which is within the limits of accuracy of the calculations. Starting with the basic heat transfer equation

$$Q = UA \, \Delta t_m$$

for equal temperature conditions in the case cited,

$$Q_2/Q_1 = (h_iA_i)_2/(h_iA_i)_1 \tag{10-10}$$

As discussed fully in Holland's book, *Heat Transfer*,[5] the value of the inside film coefficient, h_i, may be obtained from a plot of the j_H factor against Reynolds number through the tubes as shown in Figure 10–1 where

$$j_H = (h_i/C_p\rho u_i)(C_p\mu/k)^{0.67}(\mu_w/\mu)^{0.14}$$

in which ρ = density (lb/cu ft) [kg/m³]
$\quad u_i$ = velocity (ft/hr) [m/s],
$\quad D$ = tube diameter (ft) [m],
$\quad L$ = tube length (ft) [m],
$\quad \mu$ = bulk viscosity (lb/ft hr) [N s/m²],
$\quad \mu_w$ = wall viscosity (lb/ft hr) [N s/m²],
$\quad C_p$ = specific heat capacity (BTU/lb °F) [J/kg K],
$\quad k$ = thermal conductivity (BTU/hr ft °F) [W/m K],
$\quad h_i$ = inside film coefficient (BTU/hr ft² °F) [W/m² K]

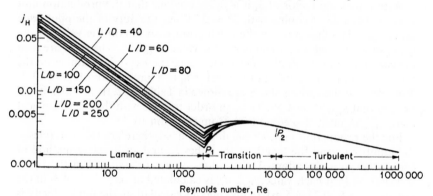

FIGURE 10–1. The j_H factor for heat exchangers[6]

Writing this equation for both the full-scale, 2, and the pilot-scale units, 1, and assuming constant temperature conditions,

$$h_{i_2}/h_{i_1} = (j_{H_2}u_{i_2})/(j_{H_1}u_{i_1}) \tag{10-11}$$

Now inside heat transfer area, $A_i = \pi d_i ln'$ for n' tubes of length l, and mass flow rate,

$$M = (\pi d_i^2/4)n'\rho u_i$$

Writing these two equations for both sized units and dividing the area term by the mass flow rate term,

$$(A_{i_2}/A_{i_1})/(M_2/M_1) = (j_{H_2}h_{i_1}/j_{H_1}h_{i_2})(l_2/d_{i_2})/(l_1/d_{i_1})$$

or $\quad (h_{i_2}A_{i_2}/h_{i_1}A_{i_1})/(M_2/M_1) = (j_{H_2}/j_{H_1})(l_2/d_{i_2})(l_1/d_{i_1}) \tag{10-12}$

M_2/M_1 is directly proportional to the ratio of the heat loads Q_2/Q_1 which from equation (10–10) is equal to $(h_{i_2}A_{i_2}/h_{i_1}A_{i_1})$; therefore the left-hand side of equation (10–12) reduces to unity and

$$j_{H_2}(l_2d_{i_2}) = j_{H_1}(l_1/d_{i_1}) \tag{10-13}$$

This simple relationship can be used in several ways. For example, from a complete analysis of the performance of the pilot-scale unit, the performance of the full-scale unit of dimensions l_2 and d_2 can be predicted or the dimensions calculated for a desired performance. A simple example may serve to illustrate the use of equation (10–13).

Example 10–1

33 gal/min [0.002 52 m³/s] of a fluid of viscosity 30 cP [0.03 N s/m²] and density 62.4 lb/ft³ [1000 kg/m³] is heated in a shell and tube exchanger with a total heat transfer area of 20 ft² [1.87 m²] consisting of 80 tubes, $\frac{3}{8}$ in. outer diameter × 22 BWG [9.5 mm OD × 0.7 mm wall] each 3 ft [0.915 m] long arranged in a single pass. The fluid is on the tube side of this pilot-scale unit. It is desired to design a full-size exchanger which is related by a scale-up factor of 3 on the linear dimension. What is the recommended length and number of tubes on the production sized unit if $\frac{3}{4}$ in. OD × 18 BWG [19.0 mm OD × 1.2 mm] tubes are to be specified?

With a scale up factor of 3, it is fairly obvious that the production unit must be capable of coping with $3^3 = 27$ times the duty of the pilot-scale exchanger. The first stage in the calculation is to obtain the value of the j_H factor for the pilot unit and hence the value of $j_H l_1/d_1$. The steps are then the calculation of the fluid velocity and hence the Reynolds number on the tube side and then the value of the factor j_H is obtained from Figure 10–1. The results at the various stages are shown in Table 10–2, from which it will be seen that $j_{H_1} l_1/d_1 = 1.39$. Thus, in order to cope with the scaled up heat load, $j_{H_2} l_2/d_2$ must also equal 1.39, from equation 10–13.

For the production unit, although d_{i_2} is fixed, there are two variables—the number n_2' and the length of the tubes l_2, and an iterative method of calculation must be adopted. Six values of n_2' are assumed in the range 200–700, and for each value, $j_{H_2} l_2/d_{i_2}$ is calculated at six values of l_2 in the range 2.5–15 ft [0.75–4.5 m]. In making this calculation, the fluid velocity is given by

$$u_{i_2} = 1037/n_2' \text{ ft/s}$$

$$[= 316/n_2' \text{ m/s}]$$

and the Reynolds number by

$$Re = 174\ 200/n_2'$$

taking care that the flow is within the same regime as the pilot unit, i.e. laminar. The values of j_H are then obtained from Figure 10–1 as before. The calculated values of $j_{H_2} l_2/d_{i_2}$ are plotted as a function of tube length l_2 for the various assumed tube numbers n_2' in Figure 10–2. The value of $j_{H_1} l_1/d_{i_1}$ is shown on the same figure and it will be seen that equation (10–13) is satisfied at several values of n_2'. For example, $n_2' = 400$, though in this case l_2 is nearly 15 ft [4.57 m] and two tube side passes would be required in the production unit. With 700 tubes, the length is about 8.5 ft [2.59 m] to satisfy $j_{H_1} l_1/d_{i_1} = 1.39$ and only one pass would be required. Thus several specifications are possible and the final choice would probably be made on the basis of an optimization procedure.

TABLE 10–2. SCALE-UP OF HEAT EXCHANGERS[7]

	Pilot exchanger		Production unit	
	(A.A.)	[SI]	(A.A.)	[SI]
Heat load scale-up factor	1	1	27	27
Tube outer diameter	0.375 in.	9.5 mm	0.750 in.	19.0 mm
Tube inner diameter	0.319 in. (d_{i_1} = 0.0266 ft)	8.1 mm (d_{i_1} = 0.0081 m)	0.652 in. (d_{i_2} = 0.054 ft)	16.6 mm (d_{i_2} = 0.0166 m)
Cross sectional area for flow/tube	0.0799 in.^{2}	0.000 057 6 m^2	0.3339 in.^{2}	0.000 216 m^2
Total cross sectional area for flow	0.0444 ft^2	0.004 13 m^2	0.002 32n_2 ft^2	0.000 216n_2 m^2
Length of tubes	l_1 = 3 ft	l_1 = 0.915 m	l_2	l_2
l/d_t	113	113	18.4l_2	60.2l_2
Heat transfer area/unit length of tube	0.0835 ft^2/ft	0.0255 m^2/m	0.1707 ft^2/ft	0.0521 e_2
Total heat transfer area	20.05 ft^2	1.87 m^2	0.1707$l_2 n_2'$ ft^2	0.0521 l_2 n_2'm^2
Volumetric flow rate	0.0892 ft^3/s	0.002 52 m^3/s	2.405 ft^3/s	0.0681 m^3/s
Fluid velocity	u_{i_1} = 2.01 ft/s	u_{i_1} = 0.613 m/s	u_{i_2} = 1037/n_2' ft/s	u_{i_2} = 316/n_2' m/s
Re	166	166	174 200/n_2	174 200/n_2
j_H	0.0123	0.0123		
j_H/d_t	1.39	1.39		

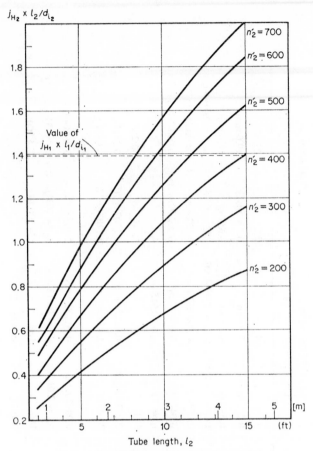

$j_{H_2} \times l_2/d_{i_2}$

Value of $j_{H_1} \times l_1/d_{i_1}$

$n'_2 = 700$

$n'_2 = 600$

$n'_2 = 500$

$n'_2 = 400$

$n'_2 = 300$

$n'_2 = 200$

Tube length, l_2

FIGURE 10–2. Scale-up of heat exchangers[7]

This example serves to illustrate that in the case of an almost completely controlling heat transfer coefficient on the tube side, the heat transfer area cannot, even to a first approximation, be scaled up directly on the heat load scale-up ratio for the same temperature conditions. For example, with 500 tubes, each 11.75 ft [3.58 m] long, the inside heat transfer area is 1002 ft² [93.1 m²] which is 50 times greater than the pilot exchanger, although the heat load on the production unit is only 27 times greater.

It is convenient at this stage to consider the case where kinematic similarity prevails as well as geometric. In this case, $Re_1 = Re_2$, $j_{H_1} = j_{H_2}$ and hence in equation (10–13),

$$l_1/d_{i_1} = l_2/d_{i_2}$$

Similarly in equation (10–11),

$$h_{i_2}/h_{i_1} = u_{i_2}/u_{i_1} = d_{i_1}/d_{i_2} \qquad \text{(from } Re_1 = Re_2\text{)}$$

Proceeding as before, it can be shown that

$$Q_2/Q_1 = (h_{i_2}A_2)/(h_{i_1}A_1) = (l_2n'_2)/(l_1n'_1) = (d_{i_2}n'_2)/(d_{i_1}n'_1)$$

from which

$$n'_2 = (Q_2/Q_1)(d_{i_1}/d_{i_2})n'_1 \qquad (10\text{–}14)$$

Example 10–2

Assuming the conditions described in Example 10–1, what number and length of tubes should be specified for the production unit so that the Reynolds number is the same on the tube side of both units?

In this case, $n'_1 = 80$, $Q_2/Q_1 = 27$ and

$$d_{i_1}/d_{i_2} = 0.319/0.652$$
$$[= 8.1/16.6] = 0.488$$

Thus in equation (10–14),

$$n'_2 = 27 \times 0.488 \times 80 = 1057$$

Tube length
$$l_2 = l_1 \times d_{i_2}/d_{i_1}$$
$$= 3/0.488 = 6.14 \text{ ft}$$
$$[= 0.915/0.488 = 1.88 \text{ m}]$$

Hence 1057 tubes each 6.14 ft [1.88 m] long would be specified.

10.4 Reactor Design

10.4.1 Fundamentals of Chemical Reaction

In specifying the size of a chemical reactor involving a homogeneous liquid phase, it is necessary to know the rate and character of the transfer process as well as the type of equipment and the quantities of materials involved. These depend largely on the chemical reaction being carried out, the fundamentals of which are only partially understood. In most cases, the design engineer must be satisfied with an empirical rate equation rather than precise data on the order and rate of reaction. The main types of reaction likely to be encountered in reactor design are now considered.

(i) First Order Reactions

In this case, the rate of reaction is proportional to the concentration of a single reactant. Thus for the reaction $A \rightarrow R + S$ (or $A + B \rightarrow R + S$, if B is in excess and has an essentially constant concentration), the reaction rate r for a constant volume system is given by

$$r = -dc_A/dt = k_1 c_A \qquad (10\text{–}15)$$

where c_A is the molar concentration of A and k_1 is the reaction coefficient. Thus, plotting $\log r$ against $\log c_A$ produces a straight line with a slope 1.0. If the initial concentration of A is c_{A_0}, then the concentration of A at any time is given by

$$c_A = c_{A_0} - c_R$$

and the rate of formation of R is

$$dc_R/dt = k_1 c_A$$

Combining these two equations and integrating on the basis of $c_R = 0$ at $t = 0$, the following equation which gives the concentration of R at time t is obtained:

$$\log (c_{A_0}/(c_{A_0} - c_R)) = k_1 t \qquad (10\text{--}16)$$

(ii) Second Order Reactions

In a second order reaction, the rate of reaction is proportional to the product of the concentration of two reactants. Thus, for the reaction $A + B \rightarrow R + S$, at constant volume, $r = -dc_A/dt = k_2 c_A c_B$. Proceeding as in the previous section, the concentration of product R as a function of time is given by

$$\{1/(c_{A_0} - c_0)\} \log \{c_{B_0}(c_{A_0} - c_R)/c_{A_0}(c_{B_0} - c_R)\} = k_2 t \qquad (10\text{--}17)$$

Where the initial reaction mixture contains identical concentrations of A and B (i.e. $c_{A_0} = c_{B_0}$) then equation (10–17) becomes:

$$c_R/(c_{A_0} - c_R)c_{A_0} = k_2 t \qquad (10\text{--}18)$$

Similarly the reaction rate can be written as

$$\log r = \log k_2 + 2 \log c_A$$

Thus if the reaction is second order, a plot of $\log r$ and $\log c_A$ should give a straight line of slope 2.0.

(iii) Third Order Equations

In this case $r = -dc_A/dt = k_3 c_A c_B c_C$, again for a constant volume system.

This discussion greatly simplifies the analysis of the rate of chemical reaction, although it has been shown that the order of a reaction is not necessarily synonymous with its molecularity.[8] In many cases, an experimental determination of the reaction coefficient might result in a reaction of a fractional order such as

$$r = -dc_A/dt = k(c_A)^{1.7} \qquad (10\text{--}19)$$

Another simplification which has been made is that the reaction is irreversible. This is not the general case and usually equilibrium is reached when the rates of the forward and reverse reactions become equal. For example, in the reaction $A + B \rightleftharpoons R + S$, with the reactions in both directions being second order,

$$r = k_2(c_A c_B - c_R c_S/K') \qquad (10\text{--}20)$$

where K', the equilibrium constant, is defined as $K' = [c_R][c_S]/[c_A][c_B]$; the square brackets indicating equilibrium concentrations. For high conversions, i.e. high values of K', then

$$r \approx k_2 c_A c_B \qquad (10\text{--}21)$$

which also applies at the start of the reaction when c_R and c_S are very small.

10.4.2 Reactor Types

Reactors, whether they consist of a simple stirred tank or a complex tubular device, are best classified according to the mode of operation, that is batch or continuous. The choice between these two types of processing unit depends largely on the quantity of material being handled, the value of the product, and the difficulties encountered in determining the degree of conversion and in generally controlling the reaction. Economics tend to favour continuous operation, especially in the case of bulk production, though the advantages may be offset by the need for expensive control equipment. As the latter becomes more sophisticated and reliable, many traditional batch processes will be converted to automated continuous operation—brewing and steel-making are two obvious examples. Nevertheless, there will always be a number of products which may best be produced on a batchwise system—small quantities of high cost materials in the pharmaceutical industries, for example, where one reactor might be used for the production of a wide variety of materials in turn.

The main types of reactor are now considered on this basis, without a detailed description of specific design for a particular duty. A good source of data on this topic is *Chemical Reactor Theory* by K. Denbigh.[9] The following discussion refers basically to homogeneous liquid phase reactors.

(i) *The Batch Stirred Tank Reactor*

In its simplest form, the reaction is carried out in a vessel fitted with an agitator, and provision may be made for heating or cooling the material by means of either a coil or an external jacket. The reactants are usually added to the empty vessel and the products are withdrawn on completion of the reaction. Although the temperature and composition will therefore vary with time, it may be assumed that in a well agitated system the composition and temperature will be uniform throughout the tank at any particular instant. It can also be assumed that the reaction takes place at constant volume, if density changes are negligible.

The general equation for this type of reactor is

$$t = c_{A_0} \int_0^{x_A} (v_0/v)(dx_A/r) \tag{10–22}$$

where t is the time from the start of the reaction (hr) [s],

c_{A_0} is the initial mole concentration of reactant A (lb mol/ft^3) [kmol/m^3],

v_0 is the initial volume of reaction mixture (ft^3) [m^3],

v is the volume of reaction mixture at time t (ft^3) [m^3],

x_A is the number of moles of A reacted at time t as a fraction of moles of A initially present,

r is the rate of reaction expressed as moles of A converted per unit time (lb mol/ft^3 hr) [kmol/m^3 s],

For a constant volume system this becomes:

$$t = \int_{c_A}^{c_{A_0}} dc_A/r \tag{10–23}$$

If, for example, the experimentally determined reaction rate is $r = k(c_A)^{1.7}$ as in equation (10–19), then the time for reactant A to be reduced to a concentration c_A is given by

$$t = (1/0.7k)(1/c_A^{0.7} - 1/c_{A_0}^{0.7}) \qquad (10\text{--}24)$$

The characteristics of operating this type of reactor are shown in Table 10–3 taken from Holland.[10]

(ii) *The Continuous Stirred Tank Reactor*

This type of reactor also consists of an agitated tank and hence the composition is again uniform throughout the reaction mix. In addition, however, the composition is uniform with time once steady state conditions have been established, and reactants flow continuously into the reactor whilst the product is continuously withdrawn.

For the steady state condition, the basic design equation is

$$v_t/a' = (x_{A_0} - x_{A_1})/r \qquad (10\text{--}25)$$

where v_t is the volume of reaction mixture (ft^3) [m^3],

a' is the feed rate of reactant A (lb mol/hr) [kmol/s],

x_{A_0}, x_{A_1} is the number of moles of A converted in the outlet and inlet streams respectively as a fraction of a'.

Where thorough mixing prevails, this equation can be rewritten as

$$(v_t/a'v_a) = (c_{A_0} - c_{A_1})/t \qquad (10\text{--}26)$$

in terms of the corresponding molar concentrations of reactant A. In this equation, v_a is the volume of feed liquid per mole of A. The term $(v_t/a'v_A)$ is the mean residence time or holding time of the reactor and for the reaction rate cited in equation (10–19), for example, this becomes

$$(v_t/a'v_A) = (c_{A_0} - c_{A_1})/\{k(c_{A_0})^{1.7}\} \qquad (10\text{--}27)$$

Equations of this type are extremely useful in determining values of the reaction coefficient, k at the bench or pilot scale of operation.[11],[12]

(iii) *The Continuous Tubular Flow Reactor*

As shown in Table 10–3, reactants are fed continually into the reactor whilst products are withdrawn, and although composition varies with position in the reactor, it may be assumed that at steady state the properties at any point in the system do not vary with time. The simplest form of this type of reactor is a tube with an L/D ratio such that axial mixing is negligible.[13]

The basic equation for the continuous tubular flow reactor is:

$$(v_r/a') = \int_0^{x_A} dx_A/r \qquad (10\text{--}28)$$

where v_r is the reactor volume corresponding to the conversion x_A (ft^3) [m^3],

x_A is the moles of reactant converted as a fraction of a' (lb mol/lb mol/hr) [kmol/kmol/s]

Table 10-3. Operating Characteristics of Various Reactor Types

Type	System	Variable composition with time	Variable composition within reactor	Uniform temperature throughout process
Batch		✓	✗	✗
Tank flow		✗	✗	✓
Tubular flow		✗	✓	✗
Semi-batch		✓	✗	✓

The term (v/a') is the hypothetical residence time of reactant A in the reactor if the feed is expressed as a volume rather than in moles. The inverse of this term is the space velocity, which is the maximum permissible feed rate per unit reactor volume.

(iv) *The Semi-Batch Reactor*

In the semi-batch reactor the basic equipment is again a stirred vessel, though in this case one of the reactants is fed in batchwise whilst the other is introduced continuously. The rate at which this is carried out can be used to control not only the reaction rate, but also the temperature of the mix in the case of an exothermic reaction.

10.5 Scale-up of Chemical Reactors

10.5.1 General

The basic principles of scale-up were considered in section 10.2 and it was noted that in a fluid dynamic system, geometric, kinematic and dynamic similarity were important. In scaling up chemical reactors, two further types of similarity must be considered:

   (a) thermal similarity, in which geometric and kinematic similarity exist and, in addition, temperature differences between corresponding points in the systems have a constant ratio with one another;
   (b) chemical similarity where, in addition to being geometrically, kinematically, and thermally similar, concentration differences between corresponding points in the two systems have a constant ratio to one another.

In practice the aim is to arrange for temperatures and concentrations to be equal at corresponding points. Also the following two ratios should be constant on both the pilot and full scale reactors:

   (i) rate of chemical conversion/rate of bulk flow,
   (ii) rate of chemical conversion/rate of molecular diffusion, though this is usually neglected in favour of (i).

As an example of equation (10–1), Damköhler[2] has derived the following relationship for a continuous reactor operated at a temperature where radiation effects may be neglected:

$$(rl/c_A u) = \phi\{(\rho u d_r/\mu), (C_p \mu/k), (\mu/\rho D), (q c_A M_w/\rho C_p T)\} \quad (10\text{--}29)$$

where r is the rate of reaction (lb mol A reacted/ft^3 hr) [kmol/m^3 s],
   c_A is the concentration of A at distance l ft [m] along the reactor (lb mol/ft^3) [kmol/m^3],
   u is the mean fluid velocity along reactor (ft/hr) [m/s],
   q is the heat of reaction (BTU/lb product) [kJ/kg product],
   D is the diffusion coefficient of A (ft^2/hr) [m^2/s],
   d_r is the inside diameter of reactor (ft) [m],
   T is the mean temperature of reactants (°F) [K],
   M_w, ρ, μ, k and C_p are the molecular weight, density, viscosity, thermal conductivity, and specific heat capacity of the reaction mass.

The significance of each of the groups in Equation 10–29 is shown in Table

10–4. As explained in section 10.2, each of these dimensionless groups, represents a rule for scale-up which is reliable only if a particular relationship is dominant. To determine this particular relationship, the character of the chemical and physical processes must be known. Frequently a number of controlling factors exist and the relevant dimensionless groups require individual scale-up rules which may be incompatible. The relative sizes of the resistances involved have an important bearing in the design method

TABLE 10–4. DIMENSIONLESS GROUPS RELEVANT TO REACTOR DESIGN

Number	Group	Significance
—	$rl/c_A u$	converted material/reactant concentration at a given point in the reactor
Reynolds	$\rho u d_1/\mu$	applied forces/viscous drag forces
Prandtl	$C\rho\mu/k$	heat transferred by bulk transport/heat transferred by conduction
Schmidt	$\mu/\rho D$	material transferred by bulk transport/material transferred by molecular diffusion
—	$qc_A M_w/\rho\phi T$	potential chemical heat/sensible heat content per unit volume

used. For example, in very slow reactions chemical resistance is dominant and the system is said to be subject to a chemical regime. Similarly, for very rapid reactions diffusional resistance may be the rate-controlling process and the system is subject to a fluid dynamic regime. Scale-up is most difficult where the resistances are of comparable magnitude, i.e. where the system is subject to a mixed regime. Sometimes it is possible to change the character of the regime in order to facilitate scale-up as, for example, in a stirred reactor, where decreased agitation can render the dynamic regime dominant.

In scaling up reactors, it is important that the factors which affect the reaction rate should theoretically be equal in both sizes of unit, though in practice this is difficult. Perhaps the most important of these factors is that of heat addition or removal from the system; the problem having already been considered in section 10.3.2. Unless the rate of heat transfer per unit mass is the same in both cases, the corresponding temperatures will differ on the two scales, resulting in either different reaction rates, and thus degree of conversion, or, in the case of a reversible reaction, the end point and hence the yields of product. Another important factor which should be the same on both scales is fluid velocity. The reaction rate will not be affected by velocity in a chemical regime, but in a fluid dynamic regime the reaction rate depends on the rates of heat or mass transfer, which depend on the fluid velocity. For example, where fluids are heated in turbulent flow in tubes, the Dittus–Boulter equation applies,

$$\mathrm{Nu} = 0.0225\mathrm{Re}^{0.8}\mathrm{Pr}^{0.4}$$

When the same material is subjected to the same temperatures on two sizes of unit, then

$$h_{i_2}/h_{i_1} = (u_{i_2}/u_{i_1})^{0.8}(d_{i_1}/d_{i_2})^{0.2}$$

and if the same sized tube is used in both units, then

$$h_{i_2}/h_{i_1} = (u_{i_2}/u_{i_1})^{0.8} \tag{10–30}$$

The index of the ratio of velocities in calculating the heat (and mass) transfer coefficient ratio is known as the 'velocity index'. Values of this index for a tubular flow chemical reactor, subject to a fluid dynamic regime are:

homogeneous—turbulent	0.8
homogeneous—laminar	0.325
heterogeneous (liquid/liquid or gas/liquid)	3–5

10.5.2 Scale-up of Tubular Flow Reactors

The design equation for a tubular flow reactor has been considered as equation (10–28) and it will be noted that this is the same as the equations for a batch reactor, (10–22) and (10–23). In theory it should therefore be possible to design a large-scale tubular-flow reactor from data obtained in a pilot-scale batch reactor, though in practice it is difficult to equate the factors affecting the reaction rate in the two cases. In order to determine which factors affect r, the following procedure may be adopted:

(i) a series of continuous flow experiments is carried out on a small tube and the conversion x_A is plotted against the average residence time of the reactants for various temperatures, thus enabling the optimum temperature conditions to be evaluated.

(ii) The same experiments are then carried out on a second tube twice the length of the first, but of equal diameter, and the plots compared with those for the first tube at corresponding temperatures.

It is important that the flow pattern is the same in both tubes, i.e. fully streamline or fully tubulent. From these experiments it may be concluded that if the conversions are the same in both cases for equal average residence time (thus $Re_2 = 2Re_1$), then strict kinematic similarity is not necessary for scale-up. Although this is frequently the case, it is rare for equal temperatures and residence times not to be a vital feature of scale-up of reactors. The effects of differing temperatures on the two scales have been considered in section 10.5.1, and in addition the possibility of competing side reactions not observable in the pilot unit occurring on the production unit should be considered.

To keep the temperature of the reaction mass equal on the two units, equal rates of heat transfer per unit reaction mass are required. This has already been considered in some detail in section 10.3.3, where the following equation was developed:

$$j_{H_2}(l_2/d_{i_2}) = j_{H_1}(l_1/d_{i_1}) \qquad (10–13)$$

This equation is true for either turbulent or laminar flow and its use for the latter regime is illustrated by the following example.[14]

Example 10–3

A tubular reactor consisting of a single tube 1 in. outer diameter [25.4 mm] and 2 ft [0.610 m] long is capable of producing 30 000 000 lb/yr [0.49 kg/s] of a liquid product of density 62.4 lb/ft³ [1000 kg/m³] in a standard

8000 hr year. The liquid linear velocity is 5.4 ft/s and the flow is laminar if the viscosity is greater than 17.5 cP [0.0175 N s/m²]. Will a tube of the same diameter but eight times the length be able to cope with eight times the throughput?

For laminar flow inside a tube,

$$j_{\mathrm{H}} = C/R_e{}^{0.675} \propto C/(u_i d_i)^{0.675} \tag{10-31}$$

The value of the constant C depends on the l/d ratio of the tube as shown in Figure 10–1. Thus, combining equations (10–13) and (10–31),

$$(l_2/d_{i_2})C_2/(u_{i_2}d_{i_2})^{0.675} = (l_1/d_{i_1})C_1/(u_{i_1}d_{i_1})^{0.675} \tag{10-32}$$

where the physical properties of the fluid are the same on both scales. For equal residence times, $l_2/u_{i_2} = l_1/u_{i_1}$ and hence equation (10–32) becomes

$$C_2(u_{i_2}/u_{i_1})^{0.325} = C_1(d_{i_2}/d_{i_1})^{1.675} \tag{10-33}$$

Thus to ensure equal temperatures of the reaction mass on both the pilot and production units, equation (10–33) must be satisfied. This equation applies for streamline flow only and for the case where the average residence

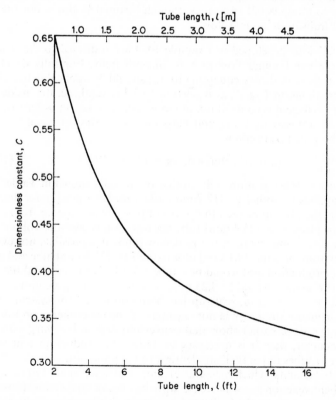

FIGURE 10–3. Values of C for tubular reactors, 1 in. outer diameter, 16 B.W.G. [25.4 mm]

time is the same for both scales of reactor. Where the tube diameter is the same in both cases, equation (10–33) becomes

$$C_2(u_{i_2}/u_{i_1})^{0.325} = C_1 \qquad (10\text{–}34)$$

Values of C for a 1 in. [25.4 mm] outer diameter × 16 B.W.G. tube are shown in Figure 10–3—these have been obtained from Figure 10–1 using the relevant value of d_1.

It is required in this problem to scale up from a 2 ft [0.610 m] tube to a 16 ft [4.88 m] long tube—a factor of 8, and hence for an increase of throughput also by a factor of 8, the residence times will be equal in both units. Thus, from Figure 10–3, for a 1 in. [25.4 mm] outer diameter tube, 16 ft [4.88 m] long, $C_2 = 0.332$, and hence in (10–34)

$$C_1 = 0.332(8u_{i_1}/u_{i_1})^{0.325}$$
$$= 0.332 \times 1.967 = 0.653$$

From Figure 10–3, for a 1 in. [25.4 mm] outer diameter tube, 2 ft [0.610 m] long, $C_1 = 0.65$; hence equation (10–33) is satisfied and the temperature of the reaction mass will be the same on both units. The pressure drops will differ though this is not usually significant for liquid phase systems.

For non-Newtonian liquids, the ratio u_1/d_1 must be the same in both units in order to attain equal viscosities, though fortunately this is not the case for heat transfer in laminar flow.

The relations developed in Example 10–3 are restricted to the scale-up of units where laminar flow prevails in both units. For fully developed turbulent flow, in theory equation (10–13) should be applicable but only if the approximation $U \approx h_1$ as in section 10.3.3 is valid, i.e. the inside heat transfer coefficient is controlling. In more realistic cases, in order to ensure equal heat-transfer rates per unit mass on both scales for equal temperatures, the scale up criteria is

$$\text{ratio of volume flows} = (U_1 A_1)_2/(U_1 A_1)_1 \qquad (10\text{–}35)$$

The use of this equation is illustrated by the data shown in Table 10–5, which applies to both $\frac{1}{2}$ in. [12.7 mm] and $\frac{3}{4}$ in. [19.1 mm] outer diameter pilot reactors. It will be seen that it is not possible to scale up from a $\frac{1}{2}$ in. [12.7 mm] to a 1 in. [25.4 mm] tube because in this case the volume/flow ratio is always greater than the heat/flow ratio. It is possible, however, to scale up from the $\frac{3}{4}$ in. [19.1 mm] tube to the 1 in. [25.4 mm] tube and in this case the production unit would be a tube 4.3 ft [1.31 m] long and the volumetric throughput would be 4.25 times that of the pilot-scale unit.

So far in this section, scale-up has been considered for systems where strict kinematic similarity is not essential. If the experiments on the pilot reactor described earlier show that conversion does in fact vary with Reynolds number, then it is necessary to design the production unit with a number of tubes in parallel, all identical to the pilot unit.

In tubular units, which have the advantage of small space requirements, and where reaction rates decrease with increasing conversion (the usual case), they benefit by the high reaction rate in the first part of the reactor, where conversions are low.[1]

TABLE 10–5. SCALE-UP DATA FOR STAINLESS STEEL TUBULAR REACTORS[14]

(Basis: 0.5 s average residence time in both pilot and production units.
    Average linear velocity in pilot unit = 4 ft/s [1.22 m/s]
    Liquid specific gravity = 1, viscosity = 1 cP [0.001 N s/m²]
    Subscript 2 refers to a 1 in. [25.4 mm] outer diameter, 16 B.W.G. reactor of
    length l_2)

linear velocity of liquid (ft/s)		6	7	8	10
	[m/s]	1.83	2.13	2.44	3.05
length of production unit (ft)		3	3.5	4	5
l_2	[m]	0.914	1.068	1.219	1.524
Re_2		40.400	47.100	53.900	67.300
j_{H2}		0.003 13	0.003 05	0.002 97	0.002 85
inside coefficient, h_{12} (BTU/hr ft² °F)		424	483	537	644
	[W/m² K]	2407	2742	3049	3657
$1/(U_iA_i)_2$	(hr ft² °F/BTU)	0.005 45	0.004 32	0.003 55	0.002 57
	[m² K/W]	0.000 96	0.000 76	0.000 63	0.000 45

Scale-up from 2 ft [0.610 m] long, 0.5 in. [12.7 mm] outer diameter × 16 B.W.G. tube

$(U_iA_i)_2/(U_iA_i)_1$	3.83	4.84	5.88	8.19
Ratio of volume flows	8.29	9.66	11.05	13.82

Scale-up from 2 ft [0.610 m] long, 0.75 in. [19.1 mm] outer diameter × 16 B.W.G. tube

$(U_iA_i)_2/(U_iA_i)_1$	2.47	3.12	3.79	5.28
Ratio of volume flows	2.95	3.44	3.93	4.92

10.5.3 Scale-up of Batch Reactors

As described in section 10.4.2, equation (10–22) may be used for the design
of batch reactors providing the volume of the reaction mass remains constant. Providing all the factors affecting the reaction rate, r, are equal, the
integral may be evaluated in a pilot plant and used in designing the production unit. It has been shown in section 10.3.2, that a simple agitated batch
reactor with a controlling inside heat transfer coefficient can only be scaled
up within very narrow limits to give the same heat transfer rate per unit mass
for the same temperatures. This can be obtained, however, by the installation of an external heat exchanger and a recirculation loop—a system
which also permits dynamic similarity. This is not important for processes
with a chemical regime, but highly desirable in the scale-up of diffusion
rate controlled processes.

10.5.4 Scale-up of other reactor types

Where the inside heat transfer coefficient is controlling, both continuous
stirred tank and semi-batch reactors may be scaled-up on the basis of
dynamic similarity and equal heat transfer rates per unit mass, though a
recirculation loop with an external heat exchanger is usually required. For
slow reactions continuous stirred-tank reactors tend to be large, as the
reaction takes place at the slow rate of the required exit conversion. However,
the reaction can take place under optimum temperature and concentration
conditions and the ease of control in reducing undesirable side-reactions

is an important advantage. Several reactors in series may be used to reduce the volume per reactor.

10.6 Scale-up of Liquid Mixing Systems

10.6.1 Introduction

The general equation for fluid mixing contains some thirteen terms; however, many of these have little effect as far as scale-up is concerned, and for geometrically similar systems, the following equation for the power number, written in the general form of equation (10–2), is applicable:[15]

$$N_P = \text{const. } Re^m Fr^n \qquad (10\text{–}36)$$

Rewriting the Power, Reynolds, and Froude numbers in full,

$$P/(\rho N^3 D_i^5) = C(\rho N D_i^2/\mu)^m (N^2 D_i/g)^n \qquad (10\text{–}37)$$

where N = impeller speed (rev/hr) [Hz],
D_i = impeller diameter (ft) [m],
P = power to impeller (ft lb/hr) [W].

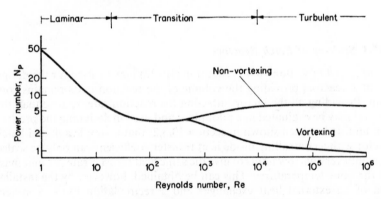

FIGURE 10–4. Correlation of mixing data

The Power number is plotted against Re in Figure 10–4, from which it will be noted that in the laminar region $Fr^n = 1$ and

$$N_P = \text{const. } Re^m$$

As an example of the use of this general correlation for geometrically similar systems the following problem based on Holland[16] will be worked.

Example 10–4

A tank is available with the dimensions shown in Figure 10–5, the total capacity being 5000 gal [18.93 m³]. The impeller rotates at 84 rpm [1.4 Hz] with a tip speed of 1056 ft/min [5.37 m/s]. Will a 15 hp [11.2 kW] motor be adequate in mixing a fluid of mass viscosity 400 cP [0.4 N s/m²]?

This problem cannot be solved numerically without further data, and this is best obtained on a smaller pilot-scale tank of (say) 1/1000 the volume.

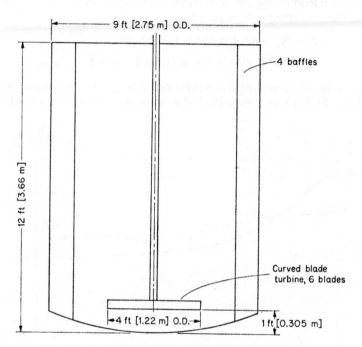

FIGURE 10–5. Tank dimensions

The data is applicable to the larger tank as a plot of N_P, and Re is independent of the tank volume for geometrically similar systems. By using several liquids with different viscosities, Re may be varied in the pilot tank and a plot obtained as shown in Figure 10–6. From this curve it is possible to obtain a curve of power consumed and fluid viscosity as follows:

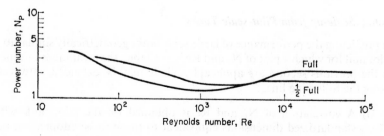

FIGURE 10–6. Power relations for pilot tank

(i) for a given viscosity, μ,

$$Re = \{62.4 \times 84 \times 60 \times (4)^2\}/\mu$$

$$[= \{1000 \times 1.4 \times (1.22)^2\}/\mu]$$

(ii) using this value of Re, a value for N_P which is applicable to the production scale unit is obtained from Figure 10–6.

(iii) this gives the power required as

$$P = N_P \times 62.4 \times (84 \times 60)^3 \times (4)^5 \quad \text{(ft lb/hr)}$$

$$[= N_P \times 1000 \times 1.4 \times (1.22)^5] \quad [W]$$

These values are then plotted as a function of the chosen viscosity as shown in Figure 10–7. From Figure 10–7, it is fairly obvious that for a fluid with

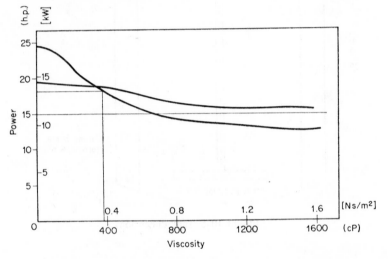

FIGURE 10–7. Power requirement as a function of fluid viscosity[16]

$\mu = 400$ cP [0.4 N s/m²], the power required is greater than 15 hp [11.2 kW] and hence the motor would burn out. However, the tank could be used with fluids where the viscosity is in excess of 800 cP [0.8 N s/m²].

10.6.2 Scale-up from Pilot-scale Tanks

In predicting the performance of large-scale tanks geometrically similar to a pilot unit for which a plot of N_P and Re is available, the techniques outlined in the previous section are applicable. These may be extended, however, along the following lines:

(i) A correlation of N_P and Re is obtained on the pilot tank with standardized dimensions equivalent to those to be adopted on the production unit. Typical dimensions are shown in Table 10–6.

TABLE 10–6. TANK WITH STANDARDIZED CONFIGURATIONS

Tank diameter	D_t
Depth of liquid	D_t
Impeller diameter	$D_i = D_t/3$
Width of impeller blade	$D_i/5$
Thickness of baffles	$D_t/10$
Height of impeller above tank base	D_i

(ii) For the liquid to be mixed on the production unit, the physical properties are known together with the dimensions of the larger tank. Thus from the N_P and Re correlation, the impeller speed and the power required to achieve the same degree of mixing may be calculated. From (10–37),

$$N = \text{Re } \mu/\rho D_i^2 \quad \text{(revs/hr) [Hz]}$$

$$P = N_P \rho N^3 D_i^5 \quad \text{(ft lb/hr) [W]}$$

It is now possible to plot the impeller speed and the required power for a series of production tanks of various volumes, V_t, as shown in Figure 10–8.

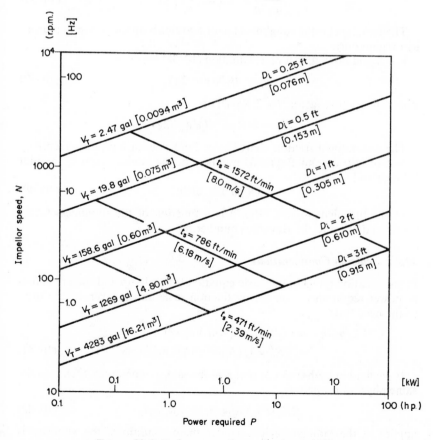

FIGURE 10–8. Performance of a production unit[16]

For a given tank, the impeller diameter is known from Table 10–6 and hence the impeller tip speed, t_s, may be calculated from

$$t_s = \pi D_1 N$$

and plotted on the same diagram. In a similar way lines of constant Re and Fr may also be added to the figure. Such a figure permits the motor and the impeller speed to be selected for a given tank volume in order to achieve the same degree of mixing as in the pilot-scale unit. It must be stressed that Figure 10–8 applies to one fluid only and must be redrawn in order to evaluate the performance of production units with other fluids.[16]

10.7 Scale-up of Fluid Flow Systems

10.7.1 Introduction

The basic equation for fluid flow is given by the ratio of head loss to velocity head and this is a function of both the Reynolds number and the length to diameter ratio of the pipe. That is,

$$\frac{h_f}{u^2/2g} = \phi \left(\mathrm{Re}_1 \frac{l}{d} \right) \left(\frac{\varepsilon}{d} \right) \tag{10–38}$$

The term (ε/d) is the roughness factor and this is applicable in the turbulent region only.

For liquid flowing in a pipe, equation (10–38) becomes

$$h_f = 4f(l/d)(u^2/2g) \tag{10–39}$$

where the friction factor, f is $2(R/\rho u^2)$ or

$$h_f = 8(R/\rho u^2)(l/d)(u^2/2g)$$

The well-known friction chart[17] for fluid flow in a pipe is essentially a plot of $(R/\rho u^2)$ and Reynolds number. In the laminar region, the plot is linear and

$$(R/\rho u^2) = 8/\mathrm{Re} \tag{10–40}$$

In the turbulent region, $(R/\rho u^2)$ is a function of the roughness factor (ε/d) in addition to the Reynolds number.

10.7.2 Scale-up Considerations

From equation (10–39), the basic equation for scale-up and the prediction of power requirements on the production scale unit is given by the ratio of the head lost:

$$h_{f2}/h_{f1} = (f_2/f_1)(l_2/l_1)(d_1/d_2)(u_2^2/u_1^2)$$
$$= \{(R/\rho u^2)_2/(R/\rho u^2)_1\}(l_2/l_1)(d_1/d_2)(u_2^2/u_1^2) \tag{10–41}$$

If the fluid in both scales of unit is at the same temperature, then with the same physical properties,

$$(R/\rho u^2)_2/(R/\rho u^2)_1 = u_1 d_1/u_2 d_2 \tag{10–42}$$

applies in the laminar region. Under these conditions, the viscosity is proportional to the shear rate which in turn is proportional to the ratio

u/d. Hence, if the viscosity is the same in both systems, $u_2/u_1 = d_2/d_1$ and equation (10–42) becomes

$$(R/\rho u^2)_2/(R/\rho u^2)_1 = u_1^2/u_2^2$$

which on substitution in (10–41) gives

$$h_{f2}/h_{f1} = (l_2/l_1)(d_1/d_2)$$

or, writing $l/d = N'$, the number pipe diameters,

$$h_{f2}/h_{f1} = N_2'/N_1' \tag{10–43}$$

This equation applies to non-Newtonian fluids in the laminar region only as indicated in the following example.[18]

Example 10–5

A liquid of specific gravity 1.0 and viscosity 140 cP [0.14 N s/m²] is flowing through a 2 in. [50.8 mm] nom. pipe at a mean velocity of 2 ft/s [0.610 m/s]. It is required to scale up the pipe to 6 in. nom. bore [152 mm], the viscosity being the same in both systems. Is equation (10–43) applicable?

The internal diameter of 2 in. [50.8 mm] nominal pipe is 2.067 in. [52.50 mm] and for 6 in. [152 mm] pipe, 6.065 in. [154.1 mm]. If the viscosity is to be the same in both systems, then $u_2/u_1 = d_2/d_1$. Therefore

$$u_2 = 2 \times (6.065/2.067) = 5.870 \text{ ft/s}$$
$$[= 0.610 \times (154.1/52.5) = 1.790 \text{ m/s}]$$

From this stage it is now possible to calculate the volumetric throughput on both units and hence the Reynolds number in both pipes. The results are shown in Table 10–7. The Reynolds number in the production unit is

TABLE 10–7. SCALE-UP OF FLOW IN A PIPE

	Pilot unit		Production unit	
	(A.A.)	[SI]	(A.A.)	[SI]
Nominal pipe diameter	2 in.	50.8 mm	6 in.	152 mm
Pipe inner diameter	2.067 in.	52.5 mm	6.065 in.	154.1 mm
Linear velocity	2 ft/s	0.610 m/s	5.87 ft/s	1.79 m/s
Velocity ratio	1.0		2.935	
Volumetric flow	17.4 gal/min	1318 cm³/s	440 gal/min	33 330 cm³/s
Flow ratio	1.0		25.3	
Reynolds number	228	228	1975	1975

1975, which is just on the verge of the transition region in which equation (10–43) does not apply. It is fairly obvious that with a viscosity of less than 140 cP [0.14 Ns/m²] on the larger scale, the flow would be turbulent and equation (10–43) would not be applicable. In practice with calculations of this type, one would start with the desired flow rate scale-up ratio and compute the pipe diameter, which would give the same u/d ratio on both the large and small scales. To facilitate the use of equation (10–43), the friction losses through pipe fittings and valves are given in Table 10–8 of this text and further information is available in the literature.

TABLE 10–8. FRICTION LOSSES THROUGH PIPE FITTINGS

Fitting	Friction loss (number of pipe diameters)
90° elbows	30–40
90° square elbows	60
T-piece	60–90
globe valves: fully open	60–300
gate valves: fully open	7
¾ open	40
½ open	200
¼ open	800
unions and couplings	negligible

So far the discussion has been limited to laminar flow of non-Newtonian fluids. There is unfortunately no simple relationship such as equation (10–43) for turbulent flow, although equation (10–41) may be used for Newtonian fluids in this regime. Fortunately, most non-Newtonian fluids are fairly viscous, and in general low viscosity fluids may be treated as Newtonian. In cases where rheological studies may be made conveniently, scale-up methods proposed by Bowen[19] and Metzner[20] may be used for both laminar and turbulent systems. These do not depend on the constancy of the u/d ratio discussed here.

10.8 Nomenclature

A	surface available for heat transfer	(ft^2)	[m^2]
$\underline{A}$	dimensionless group	(—)	[—]
a'	feedrate of reactant A	(lb mol/hr)	[kmol/s]
$\underline{B}$	dimensionless group	(—)	[—]
C	constant in equation (10–31)	(—)	[—]
C'	constant in equation (10–3)	(—)	[—]
$\underline{C}$	dimensionless group	(—)	[—]
C_p	specific heat capacity of fluid	(BTU/lb °F)	[kJ/kg K]
c_A	molar concentration of reactant A	(lb mol/ft^3)	[k mol/m^3]
D	diffusion coefficient	(ft^2/hr)	[m^2/s]
D_i	impeller diameter	(ft)	[m]
D_t	diameter of tank	(ft)	[m]
d	tube diameter	(ft)	[m]
d_r	diameter of reactor	(ft)	[m]
Fr	Froude number ($N^2 D_i/g$)	(—)	[—]
f	friction factor defined in equation (10–39)	(—)	[—]
g	acceleration due to gravity	(ft/hr^2)	[m/s^2]
h	film coefficient of heat transfer	(BTU/hr ft^2 °F)	[W/m^2 K]
h_f	frictional head loss	(ft)	[m]
j_H	heat transfer factor defined in Figure 10–1	(—)	[—]
K	constant in equation (10–2)	(—)	[—]
K'	equilibrium constant defined in equation (10–20)	(—)	[—]
k	thermal conductivity of fluid	(BTU/hr ft^2 °F/ft)	[W/m K]
$k_1 k_2$	reaction coefficients	(1/hr)	[1/s]

l	tube length	(ft)	[m]
M	mass flow rate of fluid	(lb/hr)	[kg/s]
M_w	molecular weight	(lb/lb mol)	[kg/k mol]
m	index	(—)	[—]
N	impeller speed	(rev/min)	[Hz]
N'	number of pipe diameters (l/d)	(—)	[—]
N_P	Power number $(P/\rho N^3 D_1^5)$	(—)	[—]
Nu	Nusselt number (hd/k)	(—)	[—]
n	index	(—)	[—]
n'	number of tubes	(—)	[—]
P	power to impeller	(ft lb/hr)	[W]
Pr	Prandtl number $(C_p\mu/k)$	(—)	[—]
Q	heat load	(BTU/hr)	[W]
q	heat of reaction	(BTU/lb)	[kJ/kg]
R	tangential shearing force	(pdl/ft²)	[N/m²]
Re	Reynolds number $(du\rho/\mu)$ or $(\rho N D_1^2/\mu)$	(—)	[—]
r	reaction rate	(lb mol/ft³ hr)	[kmol/m³ s]
T	mean temperature	(°F)	[K]
t	time	(hr)	[s]
t_s	impeller tip speed	(ft/min)	[m/s]
Δt_m	log mean temperature difference	(°F)	[K]
U	overall coefficient of heat transfer	(BTU hr/ft² °F)	[W/m² K]
u	fluid velocity in pipe	(ft/hr)	[m/s]
V	volume of vessel, or volumetric flow	(gal)	[m³]
V_t	volume of tank	(ft³)	[m³]
v	volume of reaction mixture	(ft³)	[m³]
v_r	reactor volume corresponding to conversion x	(ft³)	[m³]
x	fraction reacted after time t, defined in equation (10–22)	(—)	[—]
ρ	fluid density	(lb/ft³)	[kg/m³]
ε	roughness factor defined in equation (10–38)	(ft)	[m]
μ	fluid viscosity	(lb/ft hr)	[N s/m²]

Subscripts

A, B, C	reactants A, B and C
i	inside film conditions
t	conditions at time t
w	conditions at the tube wall
o	initial state
1	pilot-scale unit
2	production-scale unit

REFERENCES

1. Johnstone, E. and Thring, M. W. *Pilot Plants, Models and Scale-up Methods*. New York: McGraw-Hill Book Co., 1957.
2. Holland, F. A. *Chem. Eng.* 17 September 1962, **69**, 179; *Chem. Eng.* 26 November 1962, **69**, 119; *Chem. Eng.* 24 December 1962, **69**, 71; *Chem. Eng.* 7 January 1963, **70**, 73.
3. Brooks, G. and Su, G. J. *Chem. Eng. Prog.* 1959, **55**, 10, 54; *Chem. Eng. Prog.* 1960, **56**, 30, 237.
4. Holland, F. A. *Chem. Eng.* 26 Nov. 1962, **69**, 121.
5. Holland, F. A., Moores, R. M., Watson, F. A. and Wilkinson, J. K. *Heat Transfer*. London: Heinemann Educational Books Ltd., 1970.
6. Brown, G. G. *Unit Operations*. London: Wiley and Sons, 1950.
7. Holland, F. A. *Chem. Eng.* 26 Nov. 1962, **69**, 122.
8. Mickley, H. S., Sherwood, T. K. and Reed, C. E. *Applied Mathematics in Chemical Engineering*, 2nd ed. New York: McGraw-Hill Book Co., 1957.

9. Denbigh, K. G. *Chemical Reactor Theory*. Cambridge: Cambridge University Press, 1965.
10. Holland, F. A. *Chem. Eng.* 7 Jan. 1963, **70**, 75.
11. Hammelt, F. and Young, J. *J. Amer. Chem. Soc.* 1950, **72**, 280.
12. Denbigh, K. G. and Page, F. M. *Farad Soc. Dis.* 1954, **17**, 145.
13. Smith, J. M. *Chemical Engineering Kinetics*. New York: McGraw-Hill Book Co., 1956.
14. Holland, F. A. *Chem. Eng.* 7 Jan. 1963, **70**, 78.
15. Rushton, J. H. *Chem. Eng. Prog.* 1957, **53**, 9, 485.
16. Holland, F. A. *Chem. Eng.* 17 Sept. 1962, **69**, 181.
17. Coulson, J. M. and Richardson, J. F. *Chemical Engineering*, Vol. 1, 2nd ed. London: Pergamon Press, 1967.
18. Holland, F. A. *Chem. Eng.* 24 Dec. 1962, **69**, 72.
19. Bowen, R. L. *Chem. Eng.* 12 June 1961, **68**, 243; 24 July 1961, **68**, 143.
20. Metzner, A. B. *Chem. Eng. Prog.* 1954, **50**, 243.

Part 4: Overall Considerations

Part 4: Overall Considerations

Chapter 11: The Complete Plant

11.1 Introduction

So far in this book, the discussion has been limited to individual plant items with little mention of the complete plant unit. There are, however many aspects of a commercial venture for the production of materials in bulk which merit separate consideration, and indeed, they are in many cases of vital significance to the whole project. Such aspects are best dealt with on the basis of the completed plant, whether it be an addition to an existing complex or a new production unit. One good example of this overall concept is costing. As was shown in Chapter 2, although plant items may be costed individually it is the total cost of the complete plant which really matters and all the experience factors are expressed as a function of this parameter. Having evaluated the cost of individual items, the cost of the complete plant is then determined in order to obtain the manufacturing cost on which depends the success of the entire venture. Having reached the stage at which the individual plant items are designed, one must consider how these are best linked together to form the complete unit, and here considerations of plant layout are of prime importance. Plant layout is often a neglected topic and yet it has a vital bearing on the success of the complete project. There is little point for example, in optimizing a heat exchanger to the nth degree, if it is ultimately sited with insufficient space to allow removal of the tube bundle. Similarly, no matter how efficient a column is, it is of little value if provision is not made in laying out the plant for erecting and maintaining the column. Sensible layout of plant items does not only add to the aesthetic qualities of the plant and operational convenience, but can have an important bearing on the profitability of the venture. For example, costs can be reduced considerably by making use of gravity flow of process materials, especially where solids are involved—the cost of providing bucket elevators may indeed be a major cost item, possibly far more significant than smaller plant items which have been designed with great care and forethought. Although of a mundane nature, process pipework can be a very large percentage of plant costs and sensible layout can again effect substantial savings. One only has to consider the area taken up by storage facilities and pipework at a modern refinery compared with that allocated to processing units, to appreciate the substantial gains to be made in careful pipework design and layout.

With the completed unit, a further factor of some importance is the proposed location, not only within an existing complex but also geographically. This is usually a decision for senior management, but it can have great significance for the designer, especially in relation to the supplies of services, cooling water and so on, and also the disposal of effluent streams, where local legislation can have an important bearing on the treatment operations required and hence the cost of the plant and, ultimately, the product. The prevailing climate may also affect the structure of the plant, especially such factors as materials of construction and whether or not the plant is to be enclosed or open to the elements.

This type of factor relating to the completed plant has been included in this section largely for convenience and also to serve as an introduction to considerations, which may indeed have more effect on the cost of a project than many of the topics considered in other sections. The choice of parameter relating to the completed plant has been personal and this chapter can only indicate one or two of the more important aspects of the topic.

11.2 Location of Chemical Plant

If a plant is to be installed within an existing complex, the general pattern is to arrange for process materials to pass from one plant to the next as in a form of production line, with storage facilities for raw materials and finished products on either side and allowance for buffer storage between production units. Access by road and rail is important and adequate proximity to maintenance and administration facilities must be allowed for. Considerations of this nature are best dealt with under plant layout (section 11.3) and the present section is largely concerned with the geographical location of plant, the comments being in the main particularly relevant to the U.K.

The location of a chemical plant can have a very marked effect on the success or otherwise of a commercial venture. The guiding principle is that the plant should be located where the cost of production and distribution of products is a minimum, bearing in mind factors such as space for future expansion and the general amenities of the district. It is convenient to list the more important factors influencing plant location as follows:

(i) Source of Raw Materials

The supply of raw materials is one of the most important factors in deciding the location of plant, especially where large quantities are involved. Consider, for example, the proximity of steelworks in the U.K. to the major coalfields, and the large complexes of petrochemical plant which are associated and situated around large oil refineries. Looking at the world situation, it is current practice to build plants in countries with large resources of raw materials and to transport the finished products. Fertilizer materials provide one example in that it is obviously more economic to transport phosphoric acid than phosphate rock, which may contain 50 per cent of worthless material. This trend has resulted in a major growth in the numbers of vessels for the marine transport of materials in recent years. Location of plant near the source of raw materials permits not only a reduction in transport charges, but also in the capital invested in storage facilities.

(ii) Consumer Markets

Equally as important as proximity to sources of raw material is the ability to get the finished product to the consumer as quickly and as cheaply as possible. Indeed, a compromise is often necessary between proximity to the source of feedstocks and ease of access to the consumer, and the choice usually depends on the quantities involved and the costs of transporting the raw material and the product. Fertilizer plants again provide a good example in that they are often located in agricultural areas, as the raw materials may be transported relatively cheaply in bulk whereas the finished product must be distributed widely in smaller quantities—obviously the more expensive operation. The consumer will naturally buy from a local source where possible as this is usually cheaper and, again, his investment in storage capacity will be less as stocks may be replenished fairly readily. In considering the marketing of the finished product, the sale and distribution of by-products, often a crucial factor in the profitability of a process, must also be taken into account.

(iii) *Sources of Power*

In most chemical processes the demands for fuel and power are high, and where large quantities of coal and oil are required location near a source of supply is essential for economic operation. Where a large amount of electrical power is involved this is an even more important consideration, and a decision must be made as to whether this will be self-generated or purchased. This largely depends on the price which can be negotiated with the electricity authority, bearing in mind the benefits to be obtained by purchasing power at constant load. It is not unusual for a chemical manufacturer to sell power back to the grid. Obviously, where very large power requirements are involved location in close proximity to hydro-electric generators is most advantageous.

(iv) *Climate*

The climate can have an important bearing on the economic operation of a process. Where severe climatic conditions are likely to be encountered, the cost of protective buildings and facilities for personnel must be included in the plant costing. Similarly, in hot conditions the installation of cooling towers, air conditioning and possibly refrigeration plants are important considerations.

(v) *Transport Facilities*

The choice of means of transport of both raw materials and finished products is really dependent on the type and quantity of material involved. If possible, three modes of transport, i.e. road, rail and sea, should be available, and certainly close proximity to either a railhead or a docks complex is important. The choice is complicated by factors such as freight charges, geographical layout of rail routes, handling and storage facilities, motorway networks and so on. A minor consideration is the availability of air transport as a rapid means of communication for personnel with other plants and central offices. Only in exceptional circumstances is air transport considered as a means of transporting process materials.

(vi) *Water Resources*

Most of the process industries require large quantities of water, not only for process use but also in cooling, washing and steam generating operations. For economic operation location of plant near a river or lake is essential, and even here unlimited amounts of water usage are rarely permitted and the cost of water cooling is an important factor. As an alternative wells are used as a means of water supply, though these may be subject to seasonal fluctuations, in which case a reservoir must be installed. It is important to consider, whatever the source, the temperature and quality of the water as additional purification costs may be incurred.

(vii) *Effluent Disposal*

Closely allied with water supplies is the problem of waste disposal, whether this be solid, liquid or gaseous. There are quite severe restrictions on waste disposal and these differ with a particular location. It is necessary to consider the permissible tolerance levels for water, land and air dispersal at a particular site and, as in so many cases, the resulting choice is something of a compromise, the added cost of the effluent treatment being balanced against the other convenience factors of the location.

(viii) *Labour*

Although the general trend is for increased automation in chemical plant, in addition to staff and engineering personnel, many processes still require a reasonably large labour force, especially where a shift system is in operation. It is therefore important in locating plant, to consider the prevailing pay rates in a particular region, the competing industries, the turnover rate, and the skill and intelligence of workers. In remoter areas, the possibility of providing housing and social amenities must be taken into account.

(ix) *Financial Considerations*

In an attempt to promote the industrial development of a particular region, financial incentives by way of development grants may be available for the capitalization of new projects. Such a grant can have a vital bearing on choice of location as, indeed, is the intention. Building regulations may be an additional factor of significance and, on an international basis, local taxation practices must be considered. In selecting a location for a 'grass roots' plant, future expansion must be allowed for, especially in congested industrial areas.

(x) *Fire and Flood Protection*

As discussed previously, in order to allow for adequate cooling water supplies chemical plants are usually situated close to rivers and estuaries; a fact which carries with it the inherent risk of flooding. In selecting a site within a chosen location, a check on the regional history of such events may be worthwhile. It is important that adequate fire services should be available in the district and also that the risk of fire on adjacent sites is well known.

The terrain and topography, together with the price of land and buildings, must also be considered.

(xi) Proximity to Industrial Centres

Location of plant near an important industrial centre has the advantage of a large supply of labour and a variety of transportation facilities. It may be that other nearby industries may offer materials and services at reduced prices and there are many instances of process materials being piped directly from, say, an oil refinery to adjacent companies. In rural areas land is usually cheaper and the rate of turnover of labour and local rates lower. In times of war and civil strife a rural location has certain fairly obvious benefits.

11.3 Layout of Chemical Plant

11.3.1 Introduction

As with so many aspects of design, the layout of chemical plant is not an exact science but rather an art, as it embraces a high degree of experience coupled with the need to anticipate the human element in both operation and maintenance. It is, however, perhaps the most important single factor in a process plant project, in that a carefully planned, functional arrangement of equipment, buildings and pipework is the key to economical construction and efficient operation. In addition, sensible and aesthetically pleasing layouts can make major contributions to safety and sound community relations, bearing in mind that on a modern plant, which may be operated for twenty years, any errors in the initial layout of equipment will be costly to rectify later. With the wide variety of processes and plant operated today there is no one ideal layout, and the best one can do at this stage is to indicate one or two important considerations, with the aim of producing the most economical design with operation at the required output, adequate safety to life and property and reasonable maintenance costs. Many of the decisions on plant layout are taken by senior personnel and yet a knowledge of the basic concepts is important to all who are involved, because of their operating or design functions.

11.3.2 Basic Principles of Plant Layout

Two basic types of plant must be considered—the 'grass roots' plant and, more usually, additions or replacements to existing plants. With the latter, the 'ideal' layout must often be adapted to fit a restricted site, and even with the former the space available is rarely unlimited. At the outset, it is necessary to know the scale and scope of the project, details of the labour available, services, including maintenance, and to allow for possible future extensions. There are two general ways in which plant equipment is located—firstly the 'grouped' pattern where vessels, exchangers, columns, pumps, etc., are positioned together in separate areas for ease of operation and maintenance, and secondly the 'flow line' pattern, where equipment is laid out as arranged on the flow sheet. The former is more often used on larger plants where large numbers of similar units are employed, though in practice a compromise between the two methods is usually the case.

Bearing in mind the aims outlined in section 11.3.1, the following guidelines may be employed:

(i) *Minimum Labour Demands*

In most process operations, the cost of labour can be a very significant, if not a dominant, proportion of the manufacturing costs, and any step at reducing the payroll costs to a minimum, consistent with plant efficiency and safety, is highly desirable. Effective plant layout is an important factor in this aim. Automatic control and metering of process variables have resulted in substantial reductions in labour demands, especially where, as in most cases, the controls are installed in a central unit. Even so, manpower is required outside the control room, for example in process operation, on sampling and routine analysis. Where solids handling is involved, manpower is required for unblocking chutes, elevators and so on, and also in routine maintenance such as greasing and repacking valves. On most batch processes and on start-up and shut-down of continuous processes, the labour demands are high. Considerable savings can be made in laying out the plant in order that as many process operations as possible are integrated, resulting in a minimum operating staff. One example is the grouping of all pumps in a single bank together with the flowmeters, switchgear and a common drainage channel. In this way, maintenance time is kept to a minimum and the number of standby pumps required can be reduced. In general, the maintenance schedules for every item on the plant should be well understood in order that plant requiring frequent attention can be easily accessible.

(ii) *Elevation of Equipment*

It is important to decide which equipment should be elevated at an early stage. This is usually dictated by pump-suction and other process requirements, though elevation of plant is always expensive and in general should be limited to whatever is absolutely necessary for satisfactory operation of the process. There are cases, however, where elevation of plant has important advantages, especially where solids handling is involved and where gravity flow may be employed. Figure 11–1, which shows the layout of a compound fertilizer plant, provides such an example. The solid feeds fall from hoppers on to constant weigh-belt feeders and are then conveyed to a single (the secondary) elevator by which they are transferred to the blunger or mixer. Here the solids meet the ammonium phosphate slurry, which is also gravity fed, and together the material passes, again by gravity, into the dryer. The final product is then passed by (the primary) elevator to the top of the screening area, from which it falls through the plant and is transported to storage. In practice, a plant of this type is laid out with the reactor section on the second floor, the storage hoppers on the first floor together with the blunger, and the dryer on the ground floor—this being a very heavy plant item. In this way, considerable use is made of gravity flow and only two main elevators are necessary. Such bucket elevators are expensive to

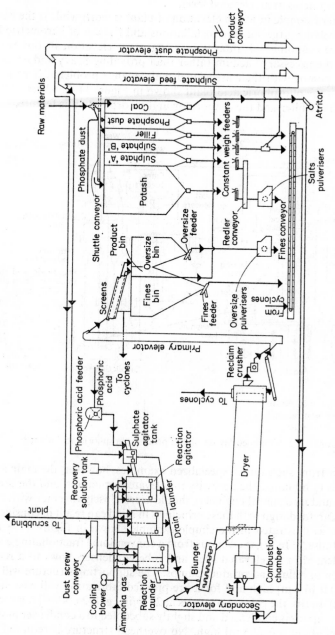

FIGURE 11–1. Simplified layout of compound fertilizer plant[1]

install and maintain, and the saving more than offsets the cost of installing
other plant items at an elevated level.

Another example in which elevation of plant is worthwhile is the case of
plant involving large vacuum installations and the use of barometric legs.
There are economic advantages in using plant elevation in this way to
provide the necessary vacuum rather than providing heavy and expensive
vacuum pumps with their associated maintenance problems. This is illustrated in the phosphoric acid plant shown in Figure 11–2, where both the
large Prayon filter and the evaporator unit incorporate this feature.

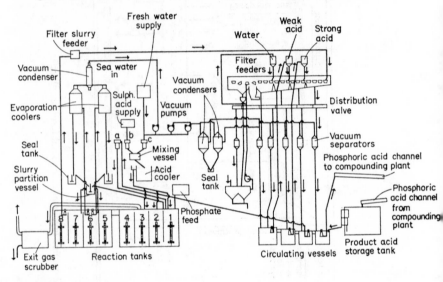

a = Phosphoric acid feeder
b = Sulphuric acid feeder
c = Fresh water feeder

FIGURE 11–2. Simplified layout of wet process phosphoric acid plant[1]

Apart from such rather specialized examples, however, the main aim in
layout is to minimize the necessity for elevated structures, and the process,
project and mechanical engineers should work closely together with this in
mind. Careful design with these principles in mind can result in considerable
savings, as illustrated in the simple example shown in Figure 11–3 taken
from House. [2] In (a), the vapour from a conventional fractionating column
is condensed in an elevated shell and tube condenser, flows to a receiver
and then to reflux and product pumps. The supporting structure will be at
least 20 ft [6.1 m] high and for a fire-proofed fabrication will cost around
£10 000. In (b), the column operates at 25 lb/in.$^2 g$ [172 kN/m^2] top pressure
and the vapour is condensed and slightly subcooled in a condenser mounted
on piers about 8 ft [2.5 m] high. No overhead structure is required, and,
although the column and condenser must be heavier to cope with the increase
in pressure, the column diameter will be less due to the smaller vapour
volume and the condenser will be smaller because of the greater mean temperature difference. In addition, the entire condenser is more easily removed
with a mobile crnae.

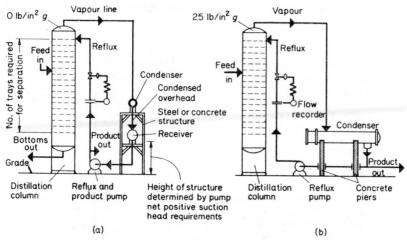

FIGURE 11–3. Layout of distillation equipment[2]

(iii) *Operating Convenience*

As outlined previously, considerable savings may be made if plant requiring frequent attention, whether of an operating or a maintenance aspect, is grouped together for ease of access. Similarly, all operating hazards must be well understood so that the safest arrangement of equipment can be devised and potentially dangerous plant items can be reached (or evacuated) quickly in an emergency. In this respect, the most economical layout for a unit area is usually the most compact area, consistent with adequate clearances between pieces of equipment. A rectangular layout with a central overhead pipe rack permits equipment to be installed along both sides of the pipeway with ease of access. It is important to lay out items so that equipment with removable parts can be dismantled without the need to move long lengths of pipe or other equipment. Free access by hoisting equipment is essential and suggested preliminary clearances between plant items are given in Table 11–1.[2]

(iv) *Layout of Specific Plant Items*

It is convenient to locate pumps in line along each side of an access way with the motors aligned outwards for easy access. Similarly, large fixed equipment, such as towers and reactors, should be positioned in a line away from the pumps, and it is most important that the problems encountered during erection of such equipment are carefully studied beforehand. Where it is necessary to service equipment with large mobile cranes, for example in the case of large heat exchangers, these are usually located around the perimeter of the plant, adjacent to a main roadway which has adequate width and overhead clearance for such equipment.

Compressors, which are expensive items, should be installed so as to permit rapid dismantling and reassembly, thereby alleviating the need to provide standby equipment. This may be achieved by using compressors

TABLE 11–1. CLEARANCES FOR PRELIMINARY LAYOUT[2]

	Horizontal		Vertical	
	(ft)	[m]	(ft)	[m]
Main road to plant limits	30	9.2	18	5.5
Access ways to plant limits	25	7.6	16	4.9
Railways to plant limits	50	15.3	23	7.0
Overhead pipe racks:				
Main	15	4.6	16	4.9
Secondary	10	3.1	12	3.7
Others	—	—	7	2.1
Clearance between:				
Pumps < 25 hp [19 kW]	2.5	0.76	12	3.7
Pumps > 25 hp [19 kW]	3	0.92	14	4.3
Compressors and other items	10	3.1	*	*
Adjacent vertical vessels	10	3.1	—	—
Adjacent horizontal vessels				
< 10 ft [3.1 m] dia.	4	1.2	4	1.2
> 10 ft [3.1 m] dia.	8	2.4	—	—
Horizontal heat exchangers	4	1.2	3	0.92
Fired heater and other items	50	15.3	—	—
Control room and main plant	30	9.2	—	—

* As required by maintenance.

with bottom suction and discharge connections and supporting the machine on a concrete platform say 8 ft [2.5 m] above ground level. This is obviously more costly, but the cost must be offset against the saving in rapid overhauls.

It is important to note that in many cases the foundations often exceed the dimensions of the equipment they support and due allowance must be made for this. As discussed briefly previously, the control panel for the plant should be located for optimum accessibility and in such a position that the piping and wiring are as simple as possible.

(v) *Layout of Process Units*

Within a large chemical plant complex, individual process units should be separated not only for more efficient operation and maintenance, but also with a view to avoiding possible spread of explosion and fire. At the outset a master plan must be made for immediate and future process areas, utilities, storage, transport facilities, offices, laboratories and workshops. This type of arrangement is basic in most oil refineries, where the crude distillation units are grouped in one area, cracking and reforming in another and further processing units in a third distinct section of the complex. Location of similar units in one section of the plant not only simplifies operation, maintenance and utility distribution, but also reduces piping and pumping costs where the product from one process unit is fed straight to another. Plants for the production of chemicals and petrochemicals are much more complex, in general terms, than oil refineries and tend to grow with the development of new products and markets and with the advance of technology. The layout of process units must provide for the addition of future units, whether they be in active development or merely a future project.

The division of a plant into process units is usually by roadways varying from 50 to 100 ft [15–30 m] in width, and it is important that any part of the plant should be accessible from at least two different directions. Many serious fires have been caused by vehicles striking tanks containing inflammable materials, and proper arrangement of roads and clearances between process units is important. Some suggestions for minimum spacings between hazardous process units, taken from House,[2] are given in Table 11–2.

TABLE 11–2. MINIMUM SPACING BETWEEN HAZARDOUS PROCESS UNITS[2]

Hazard	Recommended Spacing	
	(ft)	[m]
Medium flammability/medium pressure	50–100	15–31
High flammability/high pressure	100–150	31–46
Direct fired boilers and furnaces	50–100	15–31
Flare stacks	100–120	31–37
Loading facilities	50–100	15–31
Public roads/railways	100	31
Cooling towers	100	31
Storage tanks	75–100	23–31

11.3.3 Drainage and Waste Disposal

In the early stages of planning the layout of a plant, consideration must be given to both the installation and operation of adequate drainage and disposal systems. The ideal site is one which is almost level, with a gentle slope running towards a stream or river which can cope with peak flows of storm water. As may be expected, the site chosen very rarely fulfils these requirements, and in any event it is almost impossible to guarantee that surface drainage will never be contaminated with process materials. The most economical solution to the problem is usually the separation of the drainage from different areas of the plant into several separate collecting systems.

(i) In areas which are undeveloped or occupied by offices and buildings involving no process materials, the drainage may be isolated in a separate system and disposed of without treatment.

(ii) In areas where contaminated waste water is not normally produced, but where, in the event of an accident or fire, some process material may pass into the drains, the waste should be collected in a holding system where routine analysis of the effluent is carried out before discharge either to the public sewer or, if required, a treatment plant.

(iii) Where an area is subject to continuous or intermittent spillage of process materials, this should be paved, curbed and provided with a separate chemical sewer leading to the treatment plant. The latter, which may vary from a simple gravity oil–water separator to complex biological processing or even complete incineration, should be located fairly centrally in the plant.

11.3.4 Services

In general, buildings of an ancillary nature as far as the process is concerned, such as offices, workshops, canteen and power supply, should be located so as to afford maximum convenience with minimum interference with operation of the plant. There is no real need for offices to be close to the processing unit, and since uncontrollable flames and sparks are common to both laboratories and workshops, these should be well clear of any hazardous area. Where a relief device can vent inflammable or noxious fumes in an emergency, this should be located down wind of the administrative facility. Storage areas should be positioned for ease of access from public roads and railways and, again, remote from hazardous areas.

The generation and distribution of services must be maintained in any emergency, especially continued operation of the power supply, steam and water supplies, and hence these facilities must also be located in a completely safe area. The quality of the terrain is an important factor in plant layout, in that the best soil-bearing characteristics must be selected for the processing areas and buildings. A sloping site may be used to advantage in locating storage tanks, where full use should be made for the gravity loading and conveyance of fluids.

Where roads are to be used by all types of vehicles at all times, these must be surfaced and main two-way roads should be at least 20 ft [6.1 m] wide with a minimum centre-line radius of 30 ft [9.3 m] to permit the turning of 3–4 axle vehicles. Unsurfaced roads, where soft spots are built up with gravel, are usually muddy in rains and rutted at other times and are only acceptable for occasional maintenance or infrequent light loads such as fire patrols. Railways are mainly the concern of the local operating company, who will specify rail size, gradients, radii, clearances and the dimensions of wagons, etc. Allowance must be made for storage of the company's vehicles and also for future extensions.

11.4 Piping Layout

11.4.1 General

Closely allied to the layout of items of processing equipment is piping layout, and together these can be the biggest single cost saver in refineries and chemical plant, assuming the equipment design possibilities have been exhausted. Savings may result not only from economies in use of piping, but also in the costs of pumping, compression and utilities. In essence, there are three types of lines to be considered:

(i) Process Flow Lines

These carry the main flow of process materials and pass through the various plant items such as furnaces, reactors and columns, often with exchangers and pumps between them. Such lines are shortest where towers are arranged as close to each other as possible, consistent with the necessary clearances, in the process flow sequence. With smaller interconnecting lines towers can be spaced further apart, as other economies might be realized, such as the shortening of cooling-water lines, where condensers are grouped together. This grouping also permits a common supporting structure. Figure 11–4, taken from Kern,[3] shows different layouts of towers whereby main processes flow lines can be shortened. There are many cases where the process flow is split into two or three parallel streams, and in certain plants subsidiary flows such as refrigeration circuits must be taken into account.

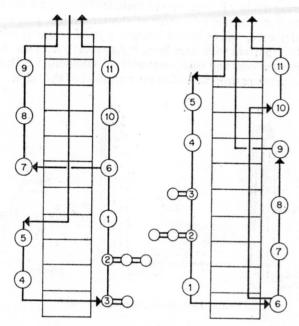

FIGURE 11–4. Alternative layout of columns in saving process pipe lengths[3]

(ii) Interconnecting Lines

This group of pipes are those connecting closely related equipment; for example towers with reboilers and condensers. In general, these are of large diameter and savings are more significant than with group (i).

(iii) Feed and Product Lines

These are usually of small diameter and are of minimum length where the equipment at which they terminate is close to the storage facilities.

11.4.2 Economy of Piping Layout

The calculation of economic pipe diameter is usually based on the cost of bare pipe, and although pipe sizes provide a readily available basis for comparison, accurate costs depend on the weight, type of material, insulation and construction. For this reason an installed cost per unit length is a more realistic basis for calculation. In very simple terms, installed costs are about double that of mild steel pipe.[4] Details of the calculations have been given elsewhere in this book, and in this section a few general rules which are useful in estimating line sizes and fittings will be considered.

Valve sizes generally correspond to the pipe diameter, though in many cases control valves are one size smaller, provided that sufficient head is available. The choice of valve size is one of optimization between the saving in valve costs with smaller sizes and the increased cost of larger diameter piping required to offset the increase in pressure drop. Similarly, when metering fluids it is necessary to optimize between a more expensive orifice plate, but with the advantage of shorter straight pipe runs and a cheaper pitot tube, where longer straight pipe runs are required. This is especially important with large lines. Safety relief valves should be connected to process lines and equipment with the minimum length of piping, as excessive pressure drop will affect the operation of the valve.

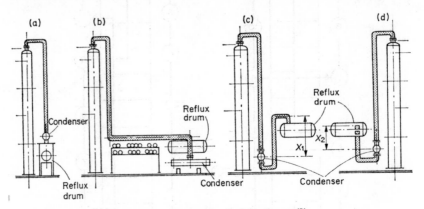

FIGURE 11-5. Overhead piping layout[3]

With overhead reflux circuits, there are many variations in layout; typical examples are given in Figure 11–5. The simplest arrangement is shown in (a), where the condenser is situated above the reflux drum. If, as an alternative, it is desired to dispense with the supporting structure and to place the condenser at ground level, as in (b), this involves more elbows and longer pipe runs. As the pressure available is usually limited, a larger pipe diameter is often required for this layout, offsetting to some extent the saving in omitting the supporting structure. In the layout shown in (d) a smaller pipe size may be possible because the static head back pressure (X) is reduced by increasing the height of the condenser. In this layout, the reflux drum is entered from below by way of a standpipe. Even with the same pipe diameters, (c) is a better layout than (d).

Compressor piping deserves an entire section for comprehensive treatment, though one or two points are worthy of note here. Wherever possible, pipes should be connected directly and bends used rather than elbows. These give less friction loss and less vibration, as does the omission of T-junctions. Close integration of intercoolers minimizes piping, and in the same way knockout drums should be adjacent to the machine. It is preferable to stack after-coolers to permit a direct gas flow and all equipment should be in the process flow sequence. Where compressor lines are supported, such supports should be quite independent of any other structure or foundation because of the vibration problem.

Pumps should generally be placed adjacent to their suction vessel and it is advantageous to have common access to all the pumps in the plant for convenience in operation and maintenance. Suction piping should be free of loops and pockets and is generally one or two sizes larger than the pump suction nozzle. Table 11–3, taken from Kern,[3] shows the pressure drops at

TABLE 11–3. ECONOMICAL PRESSURE DROPS FOR PUMP DISCHARGE LINES[3]

Flowrate		Pressure drop per unit length			
		Steel pipe		Alloy pipe	
(gal/min)	[cm³/s]	(lb/in.²/ 100 ft)	[kN/m²/ 100 m]	(lb/in.²/ 100 ft)	[kN/m²/ 100 m]
0–208	0–1575	2.5–10	57–226	6–15	136–340
208–583	1575–4417	1.5–7	34–158	4–11	90–249
583+	4417+	0.9–4	20–90	2–7	45–158

various flow rates, which give economical pump discharge header sizes. Leads to the header may be one size smaller, though they must be larger than the pump nozzle.[5]

11.4.3 Overhead Piping

The main artery of any plant is the overhead pipe network, or yard piping as it is termed in the United States, which is normally carried on overhead gantries or racks, though where long runs between processing units are involved, the pipework may be sited at, or below, ground level. Such racks carry not only long process lines, but also the main utilities—water, steam, gas and so on—together with the relief and blow-down headers. In many cases the electrical power supply and instrument lines are carried on the same structure. Careful consideration of plant layout can result in considerable savings in this enormous amount of pipework, with the consequent effect on operating costs. Kern[3] has defined the critical dimensions which influence piping cost from a rack or yard piping layout standpoint, and these are shown in Figure 11–6.

Dimension A is the total length of the plant area (or yard) and this depends on the amount and size of equipment, as well as the buildings and other structures. Considerable savings may be accrued with sensible plant layout aimed at reducing A to a minimum. For example, stacking heat exchangers, locating them under drums or supporting them on towers, fitting closely located towers with common platforms and locating equipment under the piperack are just a few ways in which A, and hence process and utility lines, is reduced, as well the passage of pipes through the area. Any equipment, for example a control room, which is not directly associated with the pipe rack, should be positioned at one end of the plant, thereby avoiding unnecessary bypass lengths of piping.

Pipe lengths between the rack and adjacent process equipment can be minimized by careful selection of dimensions B and C in Figure 11–6, and similarly the minimum values of D and E should be chosen for the most economic arrangement. In general, where a pipe changes direction it should also change elevation, though some large-diameter pipes may make a flat turn when entering at the edge of the pipe rack.

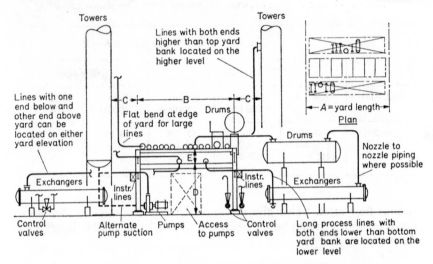

FIGURE 11–6. Layout of overhead pipe-racks[3]

Piping calculations and layout are an extremely complex subject, so much so that large numbers of engineers are employed by design and contracting organizations whose sole activity is the design and layout of pipe networks. With large, expensive piping, even the smallest details can run into thousands of pounds, and the more penetrating the analysis of the problem is, the more likely it will lead to the most economical piping design solution. This section is intended to merely indicate the scope of the topic and the way in which it impinges on the work of the plant designer. A great deal of information is available in the literature,[6]–[12] especially in journals relating to the petroleum industry.

11.5 General Considerations

11.5.1 Maintenance of Equipment

Many of the problems, which arise during both routine maintenance of equipment and emergency and planned shutdowns, are a direct result of errors in the original design of the layout and siting of plant items. As discussed in section 11.3, the design engineer should allow sufficient space for routine maintenance and also specify the necessary facilities such as power plants and overhead lifting equipment in his design. Experience has shown that the cheapest equipment to install is not always the cheapest to maintain. It may be that dismantling operations are extremely time-consuming and special knowledge, skills and tools are required. These factors, as well as the availability and cost of spares, should be taken into account when preparing the design specification, as should the design of standby equipment and the location of greasing points. Such factors are of prime importance where plant is operated under extremes of temperature and pressure and where a high degree of routine maintenance is involved.

11.5.2 Structural Design

Although the provision of foundations, buildings and structural steel-work are mainly within the province of specialist engineers, the chemical engineer should be aware of the limitations involved when drawing up his design. The important factor is correct design of foundations, the main aim being to avoid settling by careful distribution of the loads carried. This settling tendency depends not only on the magnitude of the load, but also on the degree of vibration and the bearing characteristics of the soil. Typical values for the latter parameter are:

> soft clay: 1 ton/ft² [23 930 MN/m²]
>
> sand, hard clay: 4–10 ton/ft² [95 720–239 300 MN/m²]
>
> rock: 30 ton/ft² [718 000 MN/m²]

Reinforced concrete is commonly used for foundations and it is important that this should extend below the frost line; in all construction work, specialized cements and coatings which are resistant to heat and chemical attack should be specified where appropriate. The roofing of plants and buildings is usually most economic when a flat design is used. A felt material coated with coal tar pitch is usually acceptable, though asphalt should be used where the slope is greater than 0.5 in./ft [4.2 cm/m].

There are many other factors to be considered, for example the corrosion effects of effluent and process materials, the optimized cost of construction, provision for future extensions, likely climatic conditions including extremes of wind and temperature, and the installation of adequate lighting, draining and fume extraction. Even though a functional design is more important than some elaborate architectural creation, it must be remembered that the overall quality of any structure is apparent long after the initial cost has been forgotten.

11.5.3 Manning

The costing of labour requirements has already been considered in Chapter 2 where, for a first approximation, it was taken as 10 per cent of the manu-facturing cost per year. As with the majority of the topics covered in the present chapter, detailed estimation of the labour required on a given plant is a task for specialists, although one or two relevant factors may be appro-priate at this point.

On chemical plant, the process operators are usually paid a flat hourly rate to which is added various increments depending on their level of respon-sibility. In addition, certain payments may be made for working in dirty or very hot conditions (colloquially known as 'dust money' or 'danger money'), these being negotiable usually with the trade union involved. Seniority also carries bonuses and top operators are usually awarded staff status with associated benefits. Weekend working carries extra payments, and where plants are operated continuously a three-shift system is the usual practice. In this, three shifts per day are operated with one group of operators off and the normal paid hours will vary from 40- 52 per week. It is useful in assessing the labour demand to bear in mind the need for four sets of operators when operating continuously on this type of system.

In considering the process operators required, there are essentially three types of duty involved:

(i) routine—this includes the logging of data, manual control of operating conditions, sampling and testing and cleaning.

(ii) occasional—duties on plant start-up and shut-down, major alterations in process variables such as change in product composition and flowrates.

(iii) emergency—cleaning blockages and dealing with unscheduled stoppages or, for example, fires.

It is important that the manning of the plant is adequate to cope with situations under (ii) and (iii), even if men are underemployed under normal operating conditions. In addition to process operators, fitters are required to cope with minor breakdowns and routine maintenance, although they may be able to cope with more than one unit during the normal operation of a plant.

During the normal working day, the shift labour will be supplemented by a labour squad and also by workshop personnel during planned shutdowns. In general, the labour requirements for a given process depend on the degree of automatic control, the complexity of the process, the sensitivity of the plant to change in process variables, and the type of material handled. For example, the plant shown in Figure 11–1, where a considerable amount of solids handling is involved, requires eight men on shift. In contrast, a single man may be able to control a complete crude oil distillation unit, mainly as a result of sophisticated automatic control equipment and the constancy of process variables.

11.5.4 Health and Safety

Even in today's enlightened climate, safety aspects of operation are thought to be the concern of one or two specialists and quite outside the realm of the plant designer. It is rather valueless, however, to design and optimize a plant, with the expenditure of considerable time and effort, if the plant is unsafe or hazardous to operate, and it should be the prime aim of the designer to incorporate all possible safety devices into his design. In chemical plants, the main hazards are toxic and corrosive chemicals, explosions, fires and accidents common to all industrial activities, such as those involving falls and mechanical equipment. The design engineer must be aware of these dangers and his ultimate plant design must provide the maximum protection for personnel with the minimum chance of accident. In general, chemicals are hazardous materials in that strong hydrating agents, acids and bases and oxidizing agents can all destroy living tissue, and the eyes, nose and throat are particularly sensitive to dusts, mists and many gases and vapours. It is vital to specify the installation of first-aid points, breathing apparatus, goggles, eye wash and easily operated and functional safety sprays. In laying out the plant, ease of access from the control room should be incorporated, even if this results in an increase in piping costs.

Apart from chemicals, fire and explosion are possibly the major hazards in chemical plant and these should be avoided at all cost. Sensible location of plant items can go a long way to reducing the spread of fire, and hazardous operations should be housed in separate buildings constructed with fire-proof walls and 'collapsible' roofing. In such processes all sources of ignition should be avoided, and many of these can be eliminated in the original design specification—for example, by the use of flame-proof electrical equipment and by avoiding the build-up of static electricity and spontaneous combustion. Fire and temperature alarms together with fire-fighting equipment should be specified, and the design should allow for protected walk-ways and work areas—always with ease of access.

All vessels should be tested at 1.5–2.0 times the design pressure, and where high pressure processes are involved, spring loaded valves and rupture discs must be incorporated in the specification. The efficiency of personnel as a function of quality of working conditions is well known and adequate facilities, particularly satisfactory ventilation, must be included where appropriate.

11.6 Conclusion

This chapter has attempted to consider some aspects of design which are common to a process plant taken as a single unit. The choice of topics discussed is quite arbitrary and is intended to serve as an introduction to the concepts involved. Wide and complex considerations such as scheduling, erection of plant, materials of construction and contracting in general have been totally ignored, as in many ways these are the concern of a particular organization and certainly beyond the present text. This book has not been concerned with detailed aspects of current industrial practice, but is an attempt to translate the academic principles of chemical engineering into pieces of hardware which are linked into the complete plant unit. The plant designer must be aware of costs in all his activities, as without financial profit the whole exercise is quite worthless. However, it is far more important to appreciate that unless the design contributes to the betterment of man, society and the environment, the venture is less than worthless and further technological advance is in vain.

REFERENCES

1. Anon. *Chem. and Ind.* 5 April 1958, 406.
2. House, F. F. *Chem. Eng.* 28 July 1969, **76**, 120.
3. Kern, R. *Hydrocarb. Proc.* 1966, **45**, 10, 119.
4. Mendel, O. *Chem. Eng.* 1961, **68**, 1, 190.
5. Braca, R. M. and Hoppel, J. *Chem. Eng.* 1953, **60**, 180.
6. Surdi, J. L. and Romain, D. *Hydrocarb. Proc. and Pet. Ref.* 1964, **43**, 116.

7. Thomas, J. W. *Hydrocarb. Proc. and Pet. Ref.* 1965, **44,** 2, 153.
8. Driskell, L. R. *Pet. Ref.* 1960, **39,** 7, 127.
9. Kern, R. *Pet. Ref.* 1958, **37,** 3, 136.
10. Kern, R. *Pet. Ref.* 1960, **39,** 2, 137.
11. Kern, R. *Pet. Ref.* 1960, **39,** 12, 139.
12. Kern, R. *Hydrocarb. Proc. and Pet. Ref.* 1961, **40,** 5, 195.

Appendix: Common Conversion Factors to SI Units

Length

1 in.: 25.4 mm
1 ft: 0.305 m
1 yd: 0.914 m

Time

1 hr: 3.6 ks
1 day: 86.4 ks
1 year: 31.5 Ms

Mass

1 lb: 0.4536 kg
1 ton: 1016 kg

Volume

1 in.^{3}: 16.39 cm^3
1 ft^3: 0.0283 m^3
1 U.K. gal: 4546 cm^3
1 U.S. gal: 3785 cm^3

Area

1 in.^{2}: 645 mm^2
1 ft^2: 0.0929 m^2

Temperature difference

1°F (°R): 0.556 K(°C)

Energy, heat

1 ft lb: 1.356 J
1 cal: 4.187 J
1 BTU: 1.055 J
1 kWh: 3.6 MJ
1 therm: 105.5 MJ

Calorific value, latent heat

1 BTU/lb: 2.326 kJ/kg
1 BTU/ft^3: 37.26 kJ/m^3

Velocity

1 ft/s: 0.305 m/s
1 mile/hr: 0.447 m/s

Mass and volume flow

1 ft^3/s: 0.0283 m^3/s
1 ft^3/h: 7.866 cm^3/s
1 U.K. gal/min: 75.76 cm^3/s
1 U.S. gal/hr: 1.263 cm^3/s
1 lb/hr: 0.126 g/s
1 ton/hr: 0.282 kg/s
1 lb/ft^{2}hr: 1.356 g/s m^2

Pressure

1 lbf/in.^{2}: 6.895 kN/m^2
1 atmos: 101.33 kN/m^2
1 ft water: 2.99 kN/m^2
1 in. water: 249 N/m^2
1 mm Hg: 133.3 N/m^2

Power, heat flow

1 hp: 745.7 W
1 BTU/hr: 0.293 W

Physical properties

1 poise: 0.1 N s/m^2
1 lb/ft hr: 0.413 mN s/m^2
1 stoke: 10^{-4} m^2/s
1 dyne/cm^2: 10^{-3} J/m^2
1 lb/ft^3: 16.02 kg/m^3

Thermal properties

1 BTU/hr ft^2: 3.155 W/m^2
1 BTU/hr ft^2 °F: 5.678 W/m^2 K
1 BTU/hr ft^2 °F/ft: 1.731 W/m K
1 BTU/lb °F: 4.187 kJ/kg K

Index